CRITICAL POINT THEORY
IN GLOBAL ANALYSIS
AND DIFFERENTIAL TOPOLOGY

This is Volume 33 in
PURE AND APPLIED MATHEMATICS
A Series of Monographs and Textbooks

Editors: SAMUEL EILENBERG AND HYMAN BASS

A list of recent titles in this series appears at the end of this volume.

Critical Point Theory in Global Analysis and Differential Topology

AN INTRODUCTION

MARSTON MORSE

INSTITUTE FOR ADVANCED STUDY
PRINCETON, NEW JERSEY

STEWART S. CAIRNS

DEPARTMENT OF MATHEMATICS
UNIVERSITY OF ILLINOIS
URBANA, ILLINOIS

ACADEMIC PRESS
New York San Francisco London 1969
A Subsidiary of Harcourt Brace Jovanovich, Publishers

ACADEMIC PRESS, INC.
111 Fifth Avenue, New York, New York 10003

United Kingdom Edition published by
ACADEMIC PRESS, INC. (LONDON) LTD.
24/28 Oval Road, London NW1

LIBRARY OF CONGRESS CATALOG CARD NUMBER: 69-12275

Second Printing, 1978

PRINTED IN THE UNITED STATES OF AMERICA

In Grateful Acknowledgment
of
Inspiration
from the Works of
HENRI POINCARÉ
JACQUES HADAMARD
GEORGE DAVID BIRKHOFF

PREFACE

From its beginning critical point theory has been concerned with mutual relations between topology and geometric analysis, including differential geometry. Although it may have seemed to many to have been directed in its initial years toward applications of topology to analysis, one now sees that the road from topology to geometric analysis is a two-way street. Today the methods of critical point theory enter into the foundations of almost all studies of analysis or geometry "in the large."

Mathematicians are finding that the study of global analysis or differential topology requires a knowledge not only of the separate techniques of analysis, differential geometry, topology, and algebra, but also a deeper understanding of how these fields can join forces.

It is the object of this book to add to this understanding in a new way, a way that lays rigorous and revealing foundations.

The reader may be interested in diverse problems: in the Poincaré problem when $n = 3$ or 4, in the existence of equilibria in conservative fields of forces, in the existence of periodic orbits, in global aspects of Lie theory, or even in the possibility of new approaches to homology or homotopy theory by way of critical point theory.

In the Introduction we refer to notable recent discoveries by masters of global analysis. This book aims to reformulate and establish some of the first theorems underlying these advances.

Among those who appear to have a major interest in the methods here presented are mathematical physicists. Our studies of focal points and of equilibrium points of Newtonian potentials contribute to geometric optics and dynamics. The possibility of new global topological attacks on quantum mechanics has been sensed by many who are familiar with critical point theory.

This book should be understandable to a mature first-year graduate student who has taken introductory courses in modern algebra, analysis, and general topology. The course in algebra should have familiarized the student with the elements of group theory and with fields and rings. In analysis a knowledge is needed of classical implicit function theorems and of existence theorems for ordinary differential equations.

In an application of the theory to critical chords of compact differentiable manifolds, given late in the book, a knowledge of how a short minimizing geodesic arc varies with its end points is briefly outlined and used. There are many places in the literature where the student can find the geometric analysis needed to clarify this use of geodesic arcs.

The reader will find the book a source of problems and fields of study. This is true both in analysis and topology. The student whose preference is for analysis will find, for example, several problems at the end of §32 on "Equilibrium Points of an Electrostatic Potential." One whose major interest is topology will be challenged by our treatment of the homology of differentiable manifolds without any use of global triangulations of the manifolds. Our treatment must be supplemented in many ways.

This book could be used for individual study or as a basis for a graduate course. There are four parts:

Part I. Analysis of nondegenerate functions.
Part II. Abstract differentiable manifolds.
Part III. Singular homology theory.
Part IV. Other applications of critical point theory.

Part I is concerned with the existence of nondegenerate functions on regular manifolds in euclidean spaces. See Morse [2].*

In Part II abstract differentiable manifolds M_n are defined. Studies of focal points of regular manifolds and of dynamical systems on M_n are followed by the fundamental homotopy theorems of §23.

* References will be indicated by giving the author's name followed by a number, usually [1], giving the number of the paper as listed in our bibliography.

Part III starts with a development of singular homology theory over an arbitrary field, based on the original definitions of Eilenberg [1]. It continues with a determination of homology groups by an inductive process which avoids any use of a global triangulation of M_n.

Part IV applies the theory to the "critical chords" of an arbitrary compact regular manifold, to projective spaces both complex (Milnor [2]) and real, to Stein manifolds (Andreotti and Frankel [1]), and to electrostatic potentials.

The first of three appendices contains preliminary definitions. The second supplements the analysis in Part I. The third appendix is described in the Introduction.

The student who approaches critical point theory for the first time may find the Introduction a little formidable. We recommend that such a student make a first reading of the Introduction as a historical document and a later reading for a more complete understanding. For the less advanced student another recommendation will be helpful. Theorem 10.1 is basic among the "equilibrium" theorems and also leads to Theorem 12.4, a first theorem on "cobordism." Theorem 10.1 is not otherwise needed. It is accordingly recommended that in a first reading the student omit §11, "Proof of Theorem 10.1 under Boundary Conditions B."

This book is concerned with nondegenerate functions. The study of degenerate functions or integrals has a large history. One of the greatest contributions to this theory was made in 1929 by the Russian mathematicians Snirel'mann and Ljusternik [1]. References to these mathematicians and to other distinguished Russian contributors will be found in our bibliography. In particular, see Ljusternik [1].

We are indebted to those mathematicians who have labored to clarify the field. The book by Seifert and Threlfall [1] has had and still has a large influence. One must also pay one's respect to Munkres for his book [1] on *Elementary Differential Topology*.

The authors are indebted to Rev. John Blanton, S.J., and Dr. Gudrun Kalmbach for their painstaking analysis, criticism and correction of the manuscript of this book.

The text of this book had its origin in lectures by Morse at the University of Rennes in the spring of 1965 and in lectures at the City University of New York in the academic year 1965-66. The authors, Cairns and Morse, began their collaboration in September of 1967.

Professor Morse's research was supported in part by U.S. Army Research Office–Durham grant DA-31-124-ARO-D-455.

Professor Cairns' research was supported in part by National Science Foundation grant GP8640.

M. M.

February, 1969 S. C.

CONTENTS

INTRODUCTION

This introduction to critical point theory covers finite-dimensional differentiable manifolds, and will be followed by an exposition of global variational theory. The distinction between global analysis and differential topology corresponds to the contrast between a theory of *equilibria* in analysis and the role of *nondegenerate* (ND) *functions* as a structural basis for homotopy or homology theory on differentiable manifolds.

Equilibria. The theory of equilibria is concerned with the existence and classification with regard to stability of points of equilibrium of a "function of forces" such as a "Newtonian potential," the existence and density of focal points of "regular" r-manifolds in $(n + r)$-Euclidean spaces, including the real manifolds defined by the complex-structured Stein manifolds (Andreotti and Frankel [1]), and the existence of extremal chords, arcs, orbits, and minimal surfaces under diverse boundary conditions.

ND **Functions on** M_n. The existence of infinitely many such functions was affirmed in the first theorem by Morse [2]. It has been made clear by the notable applications and extensions of critical point theory by Bott, Milnor, Smale, and Thom that the existence of ND functions on a differentiable manifold M_n gives an *initial* topological structure on M_n whose *modification* leads to the most fundamental results. We shall strengthen this conclusion by stating a major theorem.

The singular homology groups (Eilenberg [1]) *on M_n over an arbitrary field can be determined up to an isomorphism by giving a suitable* ND *function f on M_n without making use of any global triangulation of M_n* .

A ND function f on M_n is termed *suitable* if each of its critical values is assumed at just one critical point of f. The homology groups of M_n over a prescribed field are determined by the indices k of the respective critical points of f and the homology characteristics of the $(k - 1)$-spheres bounding the *universal* k-caps, which we have associated with p_a (see Appendix III for a definition of universal k-caps).

A triangulation of a differentiable manifold M_n exists, but is not needed or used. The motivation in dispensing with triangulations was simplicity and not abstract generality.

A paper by the authors on "*Singular homology theory over* $\mathbf{Z}$ *on an untriangulated manifold*" will follow.

Some Recent Advances. It is clearly impossible in a book of this size, intended as an introduction, to go deeply into the many fields of application of critical point theory. We shall nevertheless indicate how the works of some of the principal contributors are introduced.

Bott [1] and Bott and Samelson [1] have shown how the critical point theory can be effectively applied in the homology theory of Lie groups. An introduction to these contributions is given in Bott's lectures [3] at Bonn. In Theorem A by Bott [3] one finds a "homotopy equivalence" of basic importance. Inspired by Bott's formulation of Theorem A, we have obtained a similar homotopy equivalence in Corollary 23.3. Our formulation makes use of "bowls" and of the theorems on retracting deformations of § 23. We shall turn to the contributions of Bott and Samelson on "Symmetric Spaces" [1] in a later exposition of variational analysis.

The discovery by Smale [2] of how to solve the Poincaré problem when $n > 4$ was accompanied by major advances in the structural analysis of differentiable manifolds (Smale [3]). Smale makes use of handlebodies in his approach to the problem. At the end of § 23 we recall how such handlebodies were introduced (but not named) by Morse [1].

One of the major auxiliary theorems used in solving the Poincaré problem for $n > 4$ is Theorem B of Smale [1]. Smale here affirms that there exist ND functions on the given compact differentiable manifold whose critical values have the numerical order of the indices of the corresponding critical points.

Somewhat later Morse [12, 16] made a detailed analysis of bowls, f-fiber-bundles, and the alteration of critical values. This analysis shows how the alteration of critical values of f and the domain of altered values can be made to depend on the nature of the "*bowls*" ascending and descending from the critical points of f. The critical points remain invariant with their indices. Theorem 39.4 gives one of the theorems of Morse [12]. Corollary 39.1 of Theorem 39.4 recovers Theorem B of Smale [1].

Milnor's contributions and lectures are most illuminating. Our Theorem 16.8 comes from Milnor's paper [1]. In § 35 we follow Milnor in determining the homology groups of the complex projective n-space.

In his 1965 lectures Milnor [3] presents his proof and extensions of Smale's "h-cobordism theorem." Milnor departs considerably from Smale's mode of proof. The applications of the "h-cobordism theorem" by Smale and Milnor are of the greatest importance from the point of view of the critical point theory.

The earlier studies of cobordism by Thom [1] are of continued interest both historically and mathematically.

Theorem 12.4. In approaching the problem of the cobordism of two differentiable manifolds (see § 12) critical point theory has entered naturally in the form of Theorem 12.4, stated as a first necessary condition for the cobordism of Σ' and Σ'' (see Wallace [1] and Milnor [3]).

A proof of Theorem 12.4 can be modeled after our proof of Theorem 12.1. Such a proof of Theorem 12.4 depends in part on an existence theorem for ND functions with ND boundary value functions. Theorem 9.1 of Morse [15] is such a theorem. However, our proof of Theorem 12.4, as abstracted in § 12, depends to a greater degree on an extension of Theorem 10.1, in which the underlying manifold E_{n+1} is replaced by a general differentiable manifold M_{n+1}.

For the original proof of Theorem 10.1 see Morse and Van Schaack [1]. Our proof of the extension of Theorem 10.1 when E_{n+1} is replaced by M_{n+1} is modeled almost exactly on the proof of Theorem 10.1 given in § 11.

Topologically ND Functions. A word must be added concerning the mathematical possibilities which may attend the development of singular homology theory without the use of any triangulation or block subdivision of CW type. This is both realistic and desirable, especially when one recalls that the spaces of analysis to which the critical point theory is naturally applied are often not even locally compact, and that nonisotopic deformations and retractions are natural instruments, particularly in variational theory.

Morse's [8] definition and use of topologically ND functions, together with the methods of Morse's lectures [4] in 1947, make the following fairly evident: If a compact topological manifold M_n is provided with a topologically ND function f, the determination, up to an isomorphism, of the singular homology groups of M_n can be carried through and results in essentially the same homological relations as in the differentiable case.

According to Kervaire [1] there are compact topological manifolds that are triangulable, but which admit no differentiable structure. To this we add the results of Eells and Kuiper [1] that a compact combinatorial manifold is the domain of a topologically ND function. A crucial question is then: Do there exist topological manifolds Γ_n which neither admit a differentiable structure nor are *known* to be triangulable, but which are the domains of a topologically ND function f? The homology groups of any such manifolds are then determinable with an explicitness not previously known to be possible.

Other Developments. An extension of part of the theory to Hilbert space has been made by Palais [1]. Other mathematicians, including Smale, have obtained results of a similar abstract nature.

The paper of Eells and Kuiper [1] is a model of research on an inverse problem: On what manifolds are there ND functions with just three critical points? Many more such problems should be studied.

PART I

ANALYSIS OF NONDEGENERATE FUNCTIONS

§ 1

DIFFERENTIABLE MAPPINGS

Let E_n and U_m be Euclidean spaces of points x and u, respectively, with coordinates $x_1, \ldots, x_n$ and $u_1, \ldots, u_m$. A mapping ψ of an open subset X of E_n into a "subspace" Y of U_m is said to be of class C^r, $r \geqslant 0$, if the mapping of X into U_m with the same values as ψ is of class C^r, in the sense of Appendix I. A mapping ψ of class C^r, is of class C^q for $0 \leqslant q < r$. For brevity, we term a mapping of class C^r a *C^r-mapping*. A mapping ψ of an arbitrary nonempty subspace X of E_n into a subspace Y of U_m is said to be of class C^r if ψ admits an extension over an open neighborhood of X in E_n which is a mapping of class C^r into a subspace of U_m.

A 1–1 mapping is termed *biunique*; it may be discontinuous. We shall define a diffeomorphism (abbreviated "diff") of a nonempty open subset X of E_n onto a subspace Y of U_n in two ways and prove these two definitions equivalent.

Definition 1.1. *Diffs.* A biunique mapping $\psi : X \to Y$ of a nonempty open subset X of E_n onto a subspace Y of U_n will be called a *C^r-diff* if both ψ and its inverse $\varphi : Y \to X$ are C^r-mappings, $r > 0$.

Definition 1.2. *Diffs.* A biunique mapping $\psi : X \to Y$ of a nonempty open subset X of E_n onto a subspace Y of U_n will be called a *C^r-diff*, $r > 0$, if ψ is of class C^r, and if its Jacobian vanishes at no point of X.

There is no assumption that Y is open in either definition. However, the openness of Y follows from the second definition and the classical implicit function theory locally applied. We shall prove the following lemma:

Lemma 1.1. *The above two definitions of a diff of X onto Y are equivalent.*

It is trivial that a mapping which is a diff in the sense of Definition 1.2 is a diff in the sense of Definition 1.1. It remains to show that a mapping ψ which satisfies the conditions of Definition 1.1 has a nonvanishing Jacobian at each point of X.

If φ is the inverse of ψ, then if x is an arbitrary point of X,

$$(\varphi \circ \psi)(x) = x, \qquad x \in X, \tag{1.1}$$

(see Appendix I). By hypothesis, φ admits an extension φ^e as a mapping of class C^1 of an open neighborhood Y^e of Y in U_n into E_n. If one sets

$$\varphi^e(u) = (\varphi_1{}^e(u),..., \varphi_n{}^e(u)), \qquad u \in Y^e,$$

$$\psi(x) = (\psi_1(x),..., \psi_n(x)), \qquad x \in X,$$

it follows from the identity (1.1) and the chain rule that at each point $x \in X$, for k and μ on the range 1,2,..., n

$$\frac{\partial \varphi_k{}^e}{\partial u_j} \frac{\partial \psi_j}{\partial x_\mu} = \delta_k{}^\mu, \tag{1.2}$$

where $\delta_k{}^\mu$ is the Kronecker delta.

We are following a convention of tensor algebra whereby the term inscribed on the left of (1.2), with index j repeated in both factors, is summed for j on its range 1, 2,..., n.

If J_φ and J_ψ are the Jacobians of the mappings φ^e and ψ, it follows from (1.2) that at each point $x \in X$

$$J_\varphi(u)\, J_\psi(x) = 1, \qquad u = \psi(x).$$

Hence $J_\psi(x) \neq 0$ at each point $x \in X$.

Lemma 1.1 follows.

A particular consequence of the equivalence of the two definitions is that under the conditions of either definition $\psi(X)$ is an open subset of U_n and that ψ^{-1} is a diff of $\psi(X)$ onto X.

Let E_n, E_n', E_n'' be Euclidean n-spaces. By means of either definition one readily proves the following: If ψ is a diff of an open subset X of E_n onto an open subspace Y of E_n' and φ is a diff of Y onto an open subspace Z of E_n'', then $\varphi \circ \psi$ is a diff of X onto Z.

Definition 1.3. *Critical Points.* Given a real-valued function $x \rightarrow f(x)$ of class C^1 defined in an open subset X of E_n, a *critical* point of f is a point $x \in X$ at which each of the first-order partial derivatives of f vanishes. A point of X which is not critical is termed *ordinary*.

§ 2

LAWS OF THE MEAN

We shall have need of integral laws of the mean in various forms.

Lemma 2.1. *Let $x \to f(x)$ be a real-valued function of class C^1 defined on an open convex subset X of E_n. If u and x are in X,*

$$f(x) - f(u) = (x_i - u_i) A_i(x, u), \tag{2.1}$$

summing with respect to i, where i has the range 1, 2,..., n, and

$$A_i(x, u) = \int_0^1 f_{x_i}(u + t(x - u))\, dt, \tag{2.2}$$

so that

$$A_i(u, u) = f_{x_i}(u). \tag{2.3}$$

For an elementary proof of this lemma see Jordan [1], p. 251.
One can represent the right side of (2.1) as the "product"

$$(x - u) \cdot A(x, u) \tag{2.4}$$

of the vector

$$(x - u) = (x_1 - u_1, x_2 - u_2, \ldots, x_n - u_n)$$

and the vector

$$A(x, u) = (A_1(x, u), A_2(x, u), \ldots, A_n(x, u)). \tag{2.5}$$

The *norm* of a vector z in E_n is defined as its length and denoted by $\| z \|$.

It follows from (2.1) that

$$|f(x) - f(u)| \leqslant \| x - u \| \, \| A(x, u) \|. \tag{2.6}$$

Corresponding to any compact subset K of X there accordingly exists a positive constant H_K such that

$$|f(x) - f(u)| \leqslant \| x - u \| H_K, \qquad x \in K, u \in K. \tag{2.7}$$

The following is an important consequence of Lemma 2.1:

Corollary 2.1. *Corresponding to any compact subset K of X and to the closed subset K_0 (supposed not empty) of critical points of f in K, there exists a monotone continuous mapping*

$$t \to \varphi(t) : R_+ \to R_+, \qquad R_+ = \{t \in R \mid t \geqslant 0\}, \tag{2.8}$$

such that $\varphi(0) = 0$, $\varphi(t) > 0$ for $t > 0$, and

$$\| A(x, u) \| \leqslant \varphi(\| x - u \|), \qquad u \in K_0, x \in K. \tag{2.9}$$

Definition of φ. Let d be the diameter of K. It will be sufficient to define $\varphi(t)$ for $t \in [0, d]$ and then set $\varphi(t) = \varphi(d)$ for $t > d$. The relation (2.9) involves no values of φ other than those for which $t \in [0, d]$.

For $u \in K_0$ and $t \in [0, d]$ set

$$M(u, t) = \max\{\| A(y, u) \| \mid y \in \text{the convex hull of } K, \ \| y - u \| \leqslant t\}. \tag{2.10}$$

So defined, M is uniformly continuous, and for fixed $u \in K_0$ monotone increasing with respect to t. For each $t \in [0, d]$ set

$$\varphi(t) = \max\{M(u, t) \mid u \in K_0\} + t, \tag{2.11}$$

and for $t > d$ set $\varphi(t) = \varphi(d)$.

So defined, φ is continuous and monotone increasing. For each $u \in K_0$, $A(u, u) = \mathbf{0}$, in accord with (2.3), so that $M(u, 0) = \varphi(0) = 0$. It follows from (2.10) that

$$\| A(x, u) \| \leqslant M(u, \| x - u \|), \qquad x \in K, u \in K_0, \tag{2.12}$$

and from (2.11) that $\varphi(t) > 0$ for $t > 0$ and

$$M(u, t) \leqslant \varphi(t), \qquad u \in K_0\,;\, t \in [0, d]. \tag{2.13}$$

Relation (2.9) follows from (2.12) and (2.13).

This establishes Corollary 2.1.

The preceding lemma and corollary are concerned with a mapping of the set X of Lemma 2.1 into R, the axis of reals. We extend this lemma and corollary to a mapping into E_n as follows.

Let S be the $(n - 1)$-sphere of *directions* α, that is, of points $\alpha = (\alpha_1, \alpha_2, \ldots, \alpha_n)$ such that $\| \alpha \| = 1$.

Lemma 2.2. *Let there be given a mapping*

$$x \to F(x) = (F_1(x), \ldots, F_n(x)) : X \to E_n \tag{2.14}$$

of class C^1 of an open convex subset X of E_n into E_n. If α is an arbitrary direction, x and u arbitrary points in X, then

$$\alpha \cdot F(x) - \alpha \cdot F(u) = (x - u) \cdot A(x, u, \alpha), \tag{2.15}$$

where the vector-valued mapping

$$(x, u, \alpha) \to A(x, u, \alpha) : X \times X \times S \to E_n \tag{2.16}$$

is continuous and so chosen that

$$A(u, u, \alpha) = \left(\alpha \cdot \frac{\partial F}{\partial x_1}(u), \ldots, \alpha \cdot \frac{\partial F}{\partial x_n}(u)\right). \tag{2.17}$$

Lemma 2.2 results from setting $f = \alpha \cdot F$ in Lemma 2.1. Relation (2.15) follows from (2.1), where

$$A_i(x, u, \alpha) = \int_0^1 \alpha \cdot \frac{\partial F}{\partial x_i}(u + t(x - u))\, dt, \tag{2.18}$$

in accord with (2.2), understanding that

$$\frac{\partial F}{\partial x_i} = \left(\frac{\partial F_1}{\partial x_i}, \ldots, \frac{\partial F_n}{\partial x_i}\right), \qquad i = 1, \ldots, n. \tag{2.19}$$

Relation (2.17) follows from (2.18).

A consequence of (2.15) is that

$$| \alpha \cdot F(x) - \alpha \cdot F(u)| \leqslant \| x - u \| \| A(x, u, \alpha)\| \tag{2.20}$$

for each direction α and for x and u in X. It follows from (2.20) that corresponding to any compact subset K of X there exists a positive constant H_K such that

$$| \alpha \cdot F(x) - \alpha \cdot F(u)| \leqslant \| x - u \| H_K, \qquad x \in K; u \in K; \alpha \in S. \tag{2.21}$$

An Extension of Corollary 2.1. Given $u \in X$, the n linear conditions

$$\alpha \cdot \frac{\partial F(u)}{\partial x_i} = 0, \qquad i = 1,..., n, \tag{2.22}$$

on $(\alpha_1 ,..., \alpha_n)$ are satisfied for some direction α and point $u \in X$ if and only if the Jacobian

$$J(u) = \frac{D(F_1 ,..., F_n)}{D(x_1 ,..., x_n)} (u) = 0 \tag{2.23}$$

(Goursat [1], p. 93).

The set $(K \times S)_0$. Given a compact subset K of X, we introduce the subset

$$\sigma = \{u \in K \mid J(u) = 0\} \tag{2.24}$$

of K, assuming that $\sigma \neq \varnothing$. For each point $u \in \sigma$ there is at least one direction α such that conditions (2.22) are satisfied. Let $[K \times S]_0$ denote the ensemble of such pairs (u, α). For $(u, \alpha) \in [K \times S]_0$, u is a critical point of $\alpha \cdot F$.

We state the following extension of Corollary 2.1:

Corollary 2.2. *Corresponding to the compact subset K of the convex set X of Lemma* 2.2 *and to* $[K \times S]_0$, *supposed nonempty, there exists a continuous monotone mapping* $t \to \varphi(t)$ *of* R_+ *into* R_+ *such that* $\varphi(0) = 0$, $\varphi(t) > 0$ *for* $t > 0$, *and*

$$\| A(x, u, \alpha)\| \leqslant \varphi(\| x - u \|), \qquad (u, \alpha) \in [K \times S]_0 ; x \in K. \tag{2.25}$$

The proof is similar to the proof of Corollary 2.1, $(u, \alpha) \in [K \times S]_0$ replacing $u \in K_0$.

The following lemma is essential in reducing f to its canonical form near a critical point (see Jordan [1], p. 251):

Lemma 2.3. *If f is a real-valued function of class C^{r+2}, $r \geqslant 0$, in an open convex subset X of E_n and if the origin $\mathbf{0}$ is a critical point of f, then*

$$f(x) - f(\mathbf{0}) = a_{ij}(x)\, x_i x_j\,, \qquad x \in X, \tag{2.26}$$

where

$$a_{ij}(x) = \int_0^1 (1-t)\, f_{x_i x_j}(tx)\, dt. \tag{2.27}$$

We see that each coefficient function a_{ij} is of class C^r on X. Moreover, for i, j on the range $1,\ldots, n$ we have $a_{ij}(x) = a_{ji}(x)$ and

$$2a_{ij}(\mathbf{0}) = f_{x_i x_j}(\mathbf{0}). \tag{2.28}$$

§ 3

REAL, SYMMETRIC, QUADRATIC FORMS

We shall be concerned in this section with a symmetric quadratic form

$$Q(x) = a_{ij}x_ix_j\,, \qquad a_{ij} = a_{ji}\,, \tag{3.1}$$

in n variables $x_1\,,\ldots,\,x_n$. The coefficients a_{ij} are assumed real. The matrix of coefficients will be denoted by $\mathbf{a}$ or by $\| a_{ij} \|$, and the determinant of this matrix by $| a_{ij} |$.

The *rank* r of $\mathbf{a}$ is called the rank of $Q(x)$. The integer $n - r$ is called the *nullity* of $\mathbf{a}$ and of $Q(x)$. The following definition is particularly important:

Definition 3.1. *The Index of* $Q(x)$. By the *index* of $Q(x)$ is meant the maximum integer k such that $Q(x)$ is negative-definite on some k-plane π_k meeting the origin.

We shall state without detailed proof a number of fundamental theorems concerning $Q(x)$, referring the reader to standard introductory books on algebra for the proofs of these theorems.

Theorem 3.1. *If in the quadratic form* (3.1) *the coordinates are subjected to a homogeneous linear transformation*

$$x_i = c_{ij}y_j\,, \qquad i = 1,\ldots, n,$$

with n-square matrix $\mathbf{c}$, *there results a new symmetric quadratic form* $P(y)$ *with matrix* $\mathbf{c}'\mathbf{a}\mathbf{c}$, *where* $\mathbf{c}'$ *is the transpose of* $\mathbf{c}$ (cf. Bôcher [1], p. 129).

Theorem 3.2. *Under a real, nonsingular, linear transformation of the variables x the index, rank, and nullity of a quadratic form are invariant.*

The principal theorems of this section are concerned with the index of $Q(x)$ and reductions of $Q(x)$ by nonsingular linear transformations to canonical forms. These reductions are of three types: (1) *The Lagrange method of reduction.* (2) *Reduction by orthogonal transformations.* (3) *The Kronecker method of reduction.*

The *Lagrange* method (1) will be extended in § 4 to real-valued functions f no longer quadratic, but restricted to a neighborhood of a critical point of f. *Orthogonal transformations* (2) are needed to reduce $Q(x)$ to a canonical form essential in defining *centers* of *principal normal curvature* of regular differentiable manifolds M_{n-1} in E_n. We shall need only one of *Kronecker's* theorems (3) and its corollary. It is essential in the study in § 11 of the critical points under boundary conditions B.

Theorem 3.3. *There exists a real, nonsingular, linear transformation of the variables x to variables y by virtue of which $Q(x)$ is reduced to a form*

$$-y_1^2 - \cdots - y_k^2 + y_{k+1}^2 + \cdots + y_r^2, \tag{3.2}$$

where k and r are, respectively, the index and rank of $Q(x)$ (see Bôcher [1], p. 148).

The Lagrange method, as presented in § 45 of Bôcher, suffices to prove Theorem 3.3. It is recommended that the reader familiarize himself with this method, inasmuch as this method will be extended in § 4 to obtain the first fundamental "Reduction Theorem" for f near a critical point.

Orthogonal Transformations. Definition 3.2. Two n-square matrices $\mathbf{a}$ and $\mathbf{b}$ with real elements a_{ij} and b_{ij} are termed *similar* if there is a real, nonsingular, n-square matrix $\mathbf{p}$ such that

$$\mathbf{b} = \mathbf{pap}^{-1}. \tag{3.3}$$

We shall refer to the n-square *diagonal* matrix I_n whose elements are the Kronecker deltas δ_{ij}.

Definition 3.3. By the *characteristic values* or *roots* of an n-square matrix $\mathbf{a}$, is meant the n roots $\lambda_1, \ldots, \lambda_n$ of the equation

$$| a_{ij} - \lambda \delta_{ij} | = (\lambda_1 - \lambda) \cdots (\lambda_n - \lambda) = 0. \tag{3.4}$$

Equation (3.4) is termed the *characteristic* equation of $\mathbf{a}$.

The following lemma is easily proved (see Birkhoff and Mac Lane [1], p. 249):

Lemma 3.1. *The characteristic values of similar n-square matrices are equal.*

Definition 3.4. *Orthogonal Transformations.* A linear transformation of form (3.2) is termed *orthogonal* if for each point $x \in E_n$ and its image y, $\| x \| = \| y \|$.

For a proof of the following lemma see Birkhoff and Mac Lane [1], p. 258:

Lemma 3.2. *A real $n \times n$ matrix $\mathbf{a}$ is the matrix of an orthogonal transformation T if and only if $\mathbf{aa}' = I_n$.*

From the relation $\mathbf{aa}' = I_n$ follows the relation $\mathbf{a}'\mathbf{a} = I_n$, so that $\mathbf{a}$ and $\mathbf{a}'$ are inverses and T is nonsingular.

The following theorem is essential (see Birkhoff and Mac Lane [1], p. 314):

Theorem 3.4. (i) *The characteristic values of a real symmetric matrix are real.*

(ii) *A quadratic form $Q(x)$ with real symmetric matrix $\mathbf{a}$ may be reduced by an orthogonal transformation to a form*

$$\lambda_1 y_1^2 + \lambda_2 y_2^2 + \cdots + \lambda_n y_n^2, \tag{3.5}$$

where the coefficients λ_i are the roots of the characteristic equation of $\mathbf{a}$.

As a consequence, the *index* k of $Q(x)$ is the number of roots λ_i which are negative, counting each root with its multiplicity as a root, since Theorem 3.2 implies that the index of $Q(x)$ is the index of the form (3.5), while Definition 3.1 of the index of a quadratic form implies that the index of the form (3.5) is the number of roots λ_i which are negative.

We add the following theorem:

Theorem 3.5. *A quadratic form $Q(x)$ and a quadratic form obtained from $Q(x)$ by subjecting the variables x to an orthogonal transformation have matrices with the same characteristic values.*

If $\mathbf{a}$ is the matrix of $Q(x)$ and $\mathbf{c}$ the matrix of the orthogonal transformation, the new form has the matrix

$$\mathbf{c}'\mathbf{a}\mathbf{c} = \mathbf{c}^{-1}\mathbf{a}\mathbf{c}.$$

The matrices of the two forms are thus similar and so have the same characteristic values.

The following theorem will be useful in our study of Riemannian forms:

Theorem 3.6. *If $\mathbf{a}$ is a symmetric, nonsingular, n-square matrix, the characteristic values of $\mathbf{a}$ and of $\mathbf{a}^{-1}$ are reciprocals.*

It follows from Theorem 3.4 that if $\mathbf{c}$ is a suitably chosen orthogonal matrix,

$$\mathbf{c}'\mathbf{a}\mathbf{c} = \mathbf{b}, \tag{3.6}$$

where $\mathbf{b}$ is a diagonal matrix whose diagonal elements are the characteristic values of $\mathbf{a}$. From (3.6) it follows that

$$\mathbf{c}^{-1}\mathbf{a}^{-1}\mathbf{c} = \mathbf{b}^{-1}. \tag{3.7}$$

The characteristic values of $\mathbf{b}^{-1}$ are the reciprocals of those of $\mathbf{b}$, that is, of those of $\mathbf{a}$. By Lemma 3.1 the characteristic values of $\mathbf{b}^{-1}$ and of $\mathbf{a}^{-1}$ are equal.

This establishes Theorem 3.6.

Corollary 3.1. *If $Q(x)$ is a positive-definite quadratic form with matrix $\mathbf{a}$, the quadratic form whose matrix is $\mathbf{a}^{-1}$ is also positive-definite.*

A Kronecker Theorem

Theorem 3.7. *Let $Q(x)$ be a quadratic form in variables $x_1, \ldots, x_n$ with coefficients a_{ij} and with A_{nn} the cofactor of a_{nn} in $| a_{ij} |$. If $A_{nn} \neq 0$, variables $y_1, \ldots, y_n$ can be introduced by a nonsingular linear transformation T under which $x_n = y_n$ and*

$$Q(x) = \sum_{1}^{n-1} a_{hk} y_h y_k + \frac{| a_{ij} |}{A_{nn}} y_n^2.$$

For proof see Bôcher [1], p. 141.

Corollary 3.2. *If $| a_{ij} | A_{nn} > 0$, the index of the form Q is the index of the form $\sum_1^{n-1} a_{hk} y_h y_k$.*

§ 4

THE REDUCTION THEOREM FOR f

Introduction. The reduction theorem, presented here as Theorem 4.1, was first stated as Lemma 10.1 by Morse [5], p. 44. The original proof assumed that f was of class C^3 in a neighborhood of a nondegenerate critical point at the origin. For an extension to Hilbert space see Palais [1].

The theorem, as reformulated here, is valid even when f is of class C^2. In this extreme case we show that a composition of f with an L^0-homeomorphism (see Definition 4.3) is a canonical quadratic form associated with f at the origin. Our method has the advantage that it "reduces" f in the sense of Theorem 4.2 even when the origin is a degenerate critical point, provided the rank of the quadratic form "underlying" f at the origin is not zero.

Kuiper [1] has given a proof of the reduction theorem in the nondegenerate case which makes the reduction by a diff even when f is of class C^2. We have not followed Kuiper's interesting method for two reasons. His method does not seem applicable in the degenerate case, and in the nondegenerate case the theorem presented here aids in a verification of Kuiper's proof which seems necessary for completeness.

Definition 4.1. *Nondegenerate Functions.* A critical point $x = a$ of a real-valued C^2-function on an open subset X of E_n will be termed ND (*nondegenerate*) if the Jacobian

$$\frac{D(f_{x_1}, \ldots, f_{x_n})}{D(x_1, \ldots, x_n)} = | f_{x_i x_j}(a) | \neq 0. \tag{4.1}$$

If each critical point of f is ND, f itself will be termed ND.

We shall verify Lemmas 4.1 and 4.2.

Lemma 4.1. *If f is of class C^2 and* ND, *its critical points are isolated.*

Proof. If $x = a$ is a critical point of f, (4.1) holds by hypothesis, so that by the classical implicit function theorem the equations

$$f_{x_1}(x) = f_{x_2}(x) = \cdots = f_{x_n}(x) = 0,$$

have no solution other than $x = a$ in a sufficiently small neighborhood of a.

Definition 4.2. Let f be a real-valued function of class C^2 defined in a neighborhood in E_n of a critical point x^0. The *index* and *nullity* of x^0 as a critical point of f are taken as the index and nullity of the quadratic form with matrix

$$\|f_{x_i x_j}(x^0)\|.$$

Lemma 4.2. *Let f and F be real-valued functions of class C^2 defined respectively, on open subsets U and V of E_n. If there exists a C^2-diffeomorphism*

$$x \to \varphi(x) = (\varphi_1(x), \ldots, \varphi_n(x))$$

of V onto U such that $F = f \circ \varphi$, then:

(i) *The critical points of f and F correspond under φ in a* 1–1 *manner.*

(ii) *The indices and nullities of corresponding critical points are equal.*

Proof of (i). By hypothesis $F(x) = f(\varphi(x))$ for $x \in V$. Hence by the chain rule, for $x \in V$

$$F_{x_i}(x) = f_{x_k}(\varphi(x)) \frac{\partial \varphi_k(x)}{\partial x_i}, \qquad i = 1, \ldots, n;\ x \in V. \tag{4.2}$$

Since the Jacobian of φ never vanishes, by hypothesis, Cramer's rule applied to (4.2) implies that grad f (cf. Appendix I) vanishes at $\varphi(x)$ if and only if grad F vanishes at x. Thus (i) is true.

Proof of (ii). Suppose that a critical point x^0 of F corresponds under φ to a critical point $y^0 = \varphi(x^0)$ of f. Set

$$\mathbf{a} = \|f_{x_i x_j}(y^0)\| \qquad \text{and} \qquad \mathbf{b} = \|F_{x_i x_j}(x^0)\|$$

and let $\mathbf{c}$ be the Jacobian matrix of φ at x^0. If one differentiates both sides of (4.2) with respect to x_j, one finds that at x^0

$$\mathbf{b} = \mathbf{c}'\mathbf{a}\mathbf{c}.$$

Statement (ii) now follows from Theorems 3.1 and 3.2.

Definition 4.3.a. L^r-**Homeomorphisms**,† $r \geqslant 0$. A homeomorphism

$$x \to \varphi(x) = (\varphi_1(x), \ldots, \varphi_n(x)) : N \to V \tag{4.3}$$

of an open neighborhood N in E_n of the origin $\mathbf{0}$ onto an open subspace V of E_n such that

$$\varphi_i(x) = c_{ij}(x)\, x_j\,; \qquad i = 1, \ldots, n;\ x \in N; \tag{4.4}$$

where each mapping c_{ij} is of class C^r on N and $|\,c_{ij}(x)| \neq 0$, will be called an *L^r-homeomorphism.*

Definition 4.3.b With an L^r-homeomorphism φ we associate the nonsingular transformation

$$\varphi^0 : y_i = c_{ij}(\mathbf{0})\, x_j\,, \qquad i = 1, 2, \ldots, n, \tag{4.5}$$

calling φ^0 the *linear homeomorphism underlying* φ.

If φ is an L^r-homeomorphism for which $r > 0$, the Jacobian matrix of φ at the origin is the matrix $\|\,c_{ij}(0)\|$ of (4.5).

Lemma 4.3. (i) *A C^{r+1}-diff,* $r \geqslant 0$,

$$x \to \varphi(x) : X \to \varphi(X) \subset E_n \tag{4.6}$$

of an open convex neighborhood X of the origin which leaves the origin fixed defines an L^r- homeomorphism of a sufficiently small N in which (4.4) *holds with*

$$c_{ij}(x) = \int_0^1 \frac{\partial \varphi_i}{\partial x_j}(tx_1, \ldots, tx_n)\, dt, \qquad x \in X. \tag{4.7}$$

(ii) *For $r \geqslant 0$ the inverse ψ of an L^r-homeomorphism φ with domain N is an L^r-homeomorphism with domain $\varphi(N)$.*

(iii) *If $r > 0$, an L^r-homeomorphism φ is a C^r-diff.*

Statement (i) follows with the aid of Lemma 2.1.

---

† The letter L is used here to connote a type of pseudolinearity.

To verify (ii), suppose that φ has the form (4.4) and set $\|c_{ij}(x)\| = \mathbf{c}(x)$ The matrix $\mathbf{c}^{-1}(x)$ then exists for $x \in N$. If one sets $y = \varphi(x)$ for $x \in N$, then by hypothesis $x = \psi(y)$ for $x \in \varphi(N)$. Set

$$\mathbf{c}^{-1}(x) = \mathbf{c}^{-1}(\psi(y)) = \| b_{ij}(y)\|, \qquad y \in \varphi(N).$$

By Cramer's rule the n equations $y_i = c_{ij}(x)x_j$, with $x \in N$, admit the solution

$$x_i = b_{ij}(y)\, y_j = \psi_i(y), \qquad y \in \varphi(N); i = 1, 2, \dots, n.$$

Statement (ii) follows.

Statement (iii) is true, since φ is of class C^r, r > 0, and has an inverse of class C^r.

This completes the proof of Lemma 4.3.

If φ and ψ are inverse L^r-homeomorphisms, then $\varphi \circ \psi$ is the identity map of a neighborhood of the origin. It follows that [for notation see (4.5)]

$$\varphi^0 \circ \psi^0 = I \tag{4.8}$$

is the identity map of E_n onto E_n.

L^r-homeomorphisms of Diagonal Type. An L^r-homeomorphism φ whose underlying linear homeomorphism φ^0 has a matrix of diagonal type will be said to be of *diagonal* type.

If φ and ψ are inverse L^r-homeomorphisms, it follows from (4.8) that φ is of diagonal type if and only if ψ is of diagonal type.

Quadratic forms are generalized by Q^r-functions (quadratic functions) defined as follows:

Definition 4.4a. **Q^r-Functions.** A real-valued function

$$x \to \Gamma(x) = a_{ij}(x)\, x_i x_j, \qquad a_{ij}(x) = a_{ji}(x); x \in V, \tag{4.9}$$

defined on an open neighborhood V of the origin and such that each function a_{ij} is of class C^r on V will be called a *Q^r-function*, $r \geqslant 0$.

Definition 4.4b. Given the Q^r-function Γ, the quadratic form

$$\Gamma^0(x) = a_{ij}(0)\, x_i x_j,$$

will be called the *quadratic form underlying Γ*.

Lemmas 4.4 and 4.5 will be used in proving Theorem 4.1.

Lemma 4.4. *Given an L^r-homeomorphism $\varphi : N \to V$ as in* (4.3) *and* (4.4) *and a Q^r-function Γ with domain V as in* (4.9), *the composite function $\Gamma \circ \varphi$ is well-defined on N and is a Q^r-function whose underlying quadratic form is $(\Gamma^0 \circ \varphi^0)(x)$.*

Proof. One finds that

$$(\Gamma \circ \varphi)(x) = b_{ij}(x)\, x_i x_j\,, \qquad x \in V,$$

where the coefficients $b_{ij}(x)$ have a matrix

$$\| b_{ij}(x)\| = \| c_{ij}(x)\|' \, \| a_{ij}(\varphi(x))\| \, \| c_{ij}(x)\|. \tag{4.10}$$

Setting $x = \mathbf{0}$ in (4.10), we see that the quadratic forms $(\Gamma \circ \varphi)^0(x)$ and $(\Gamma^0 \circ \varphi^0)(x)$ have the same matrix, thereby establishing the lemma.

Q^r-Functions of Diagonal Type. A quadratic form $Q(x)$ whose matrix is diagonal will be said to be of diagonal type. A Q^r-function Γ for which the underlying quadratic $\Gamma^0(x)$ is of diagonal type will be said to be of *diagonal type*.

With this understood, Lemma 4.4 implies the following:

Lemma 4.5. *A well-defined composition $\Gamma \circ \varphi$ of an L^r-homeomorphism φ and a Q^r-function Γ, both of diagonal type, is a Q^r-function of diagonal type.*

The following notion of d^r-equivalent Q-functions is central in the proof of Theorem 4.1:

Definition 4.5. **d^r-Equivalent Q^r-Functions.** Two Q^r-functions Γ' and Γ'' will be said to be *d^r-equivalent* if there exists an open neighborhood U of the origin and an L^r-homeomorphism $\varphi : U \to \varphi(U) \subset E_n$ such that

$$\Gamma'(\varphi(x)) = \Gamma''(x), \qquad x \in U.$$

The relation of d^r-equivalence among Q^r-functions is seen to be symmetric, reflexive, and transitive.

Theorem 4.1 is a form of the first Reduction Theorem, simplified by taking the critical point of f at the origin and the corresponding critical value as zero.

Theorem 4.1. *Let f be a real-valued function of class C^{r+2}, $r \geqslant 0$, on an open neighborhood N of the origin in E_n.*

If the origin is an ND critical point of f of index k with critical value zero, there exists an L^r-homeomorphism $\varphi : X \to \varphi(X) \subset E_n$ of an open neighborhood $X \subset N$ of the origin such that $\varphi(\mathbf{0}) = \mathbf{0}$ and

$$f(\varphi(x)) = -x_1^{\,2} - \cdots - x_k^{\,2} + x_{k+1}^2 + \cdots + x_n^{\,2}, \qquad x \in X. \quad (4.11)$$

We shall establish this theorem by verifying statements (i)–(iii):

(i) *If $U \subset N$ is an open convex neighborhood of the origin, then $f \mid U$ is a Q^r-function Γ.*

(ii) *Γ is d^r-equivalent to a Q^r-function*

$$x \to \Gamma_0(x) = a_{ij}(x)\, x_i x_j\,, \qquad x \in V, \quad (4.12)$$

with underlying quadratic form,

$$e_1 x_1^{\,2} + \cdots + e_n x_n^{\,2} = -x_1^{\,2} - \cdots - x_k^{\,2} + x_{k+1}^2 + \cdots + x_n^{\,2}, \quad (4.13)$$

where k is the index of the origin as a critical point of f.

(iii) *Γ_0 is d^r-equivalent to the quadratic form* (4.13).

Proof of (i). (i) follows at once from Lemma 2.3.

Proof of (ii). Let Λ be a linear homeomorphism which reduces the quadratic form $\Gamma^0(x)$ underlying Γ to the quadratic form (4.13). Λ exists by virtue of Theorem 3.3. Let λ be the restriction of Λ to so small an open neighborhood V of the origin that $\Gamma \circ \lambda = \Gamma_0$ is a well-defined Q^r-function on V. It follows from Lemma 4.4 that

$$\Gamma_0^{\,0}(x) = (\Gamma^0 \circ \lambda)(x), \qquad x \in V, \quad (4.14)$$

so that the Q^r-function Γ_0 has the form (4.13) as underlying quadratic form.

This establishes (ii).

Proof of (iii). By virtue of (ii) $a_{ii}(\mathbf{0}) = e_i$ for $i = 1,\ldots, n$ and $a_{ij}(\mathbf{0}) = 0$ if $i \neq j$. Following the Lagrange method of reduction of a quadratic form, we introduce the mapping

$$x \to \theta_1(x) = \frac{a_{11}(x)\, x_1 + a_{12}(x)\, x_2 + \cdots + a_{1n}(x)\, x_n}{|\, a_{11}(x)|^{1/2}}, \qquad x \in N_1 \quad (4.15)$$

where $N_1 \subset V$ is an open convex neighborhood of the origin so small that $a_{11}(x) \neq 0$ for $x \in N_1$. We note that

$$\frac{\partial \theta_1}{\partial x_1}(0) = e_1 \neq 0. \tag{4.16}$$

The L^r-homeomorphism Θ. Let

$$x \to (\theta_1(x), x_2, \ldots, x_n) = \Theta(x) \in E_n, \qquad x \in N_1, \tag{4.17}$$

be the L^r-homeomorphism of N_1 in which θ_1 is given by (4.15). This L^r-homeomorphism has an inverse which is an L^r-homeomorphism of the form

$$x \to (\psi_1(x), x_2, \ldots, x_n) = \Psi(x), \qquad x \in \Theta(N_1). \tag{4.18}$$

The Form of $\Gamma_0 \circ \Psi$. The first step in determining $\Gamma_0 \circ \Psi$ is similar to the first step in the Lagrange reduction of a quadratic form. For $x \in N_1$ we see that

$$\Gamma_0(x) = a_{ij}(x)\, x_i x_j = \frac{[a_{11}(x)\, x_1 + \cdots + a_{1n}(x)\, x_n]^2}{a_{11}(x)} + R_1(x), \tag{4.19}$$

where

$$R_1(x) = x_h x_k \left[a_{hk}(x) - \frac{a_{1h}(x)\, a_{1k}(x)}{a_{11}(x)}\right], \qquad x \in N_1, \tag{4.20}$$

summing terms in h, k over the range 2, 3,..., n. We can write

$$\Gamma_0(x) = e_1 \theta_1^{\,2}(x) + R_1(x). \tag{4.21}$$

Since Ψ is the inverse of Θ, (4.21) gives the identity

$$(\Gamma_0 \circ \Psi)(x) = e_1 x_1^{\,2} + \sum_2^n a_{hk}^{(2)}(x)\, x_h x_k, \qquad x \in \Theta(N_1), \tag{4.22}$$

where the mappings $a_{hk}^{(2)}$ introduced here are of class C^r on $\Theta(N_1)$. The identity (4.22) shows that Γ_0 is d^r-equivalent to the Q^r-function $\Gamma_1(x)$, given by the right side of (4.22).

Moreover, the quadratic form underlying $\Gamma_1(x)$ is again the form

(4.13). This follows from the fact that the quadratic form underlying the Q^r-function $R_1(x)$ of (4.20) is the form

$$e_2 x_2{}^2 + \cdots + e_n x_n{}^2, \tag{4.23}$$

while $\Psi^0(x) = (e_1 x_1, x_2, \ldots, x_n)$ [cf. (4.16) and (4.18)].

The Induction. Given an integer s such that $1 < s \leqslant n + 1$, we assume that Γ_0 is d^r-equivalent to a Q^r-function with values

$$\Gamma_{s-1}(x) = e_1 x_1{}^2 + \cdots + e_{s-1} x_{s-1}^2 + \sum_s^n a_{hk}^{(s)}(x)\, x_h x_k \tag{4.24}$$

on some open neighborhood of the origin and with underlying quadratic form (4.13), and prove this statement true if s is replaced by $s + 1$. The assumption is valid if $s = 2$, as (4.22) shows.

By hypothesis $a_{ss}^{(s)}(\mathbf{0}) = e_s$, since the form (4.13) is assumed to underlie Γ_{s-1}.

Hence for suitable choice of an open neighborhood N_s of the origin there exists an L^r-homeomorphism,

$$x \to (\theta_1(x), \ldots, \theta_n(x)) = \Theta(x),$$

in which $\theta_i(x) = x_i$ for $i \neq s$ and

$$\theta_s(x) = \frac{a_{ss}^{(s)}(x)\, x_s + a_{s,s+1}^{(s)}(x)\, x_{s+1} + \cdots + a_{sn}^{(s)}\, x_n}{|\, a_{ss}^{(s)}(x)|^{1/2}}, \qquad x \in N_s. \tag{4.25}$$

Following the method used for $s = 2$, we find that Γ_{s-1} is d^r-equivalent to a Q^r-function Γ_s with values $\Gamma_s(x)$ given by (4.24), with $s + 1$ replacing s, and with underlying quadratic form (4.13).

By induction Γ_0 is d^r-equivalent to Γ_n. But Γ_n, as given by (4.24) when $s = n + 1$, is identical to the form (4.13), so that Theorem 4.1 follows.

The Degenerate Case. The method of proof of Theorem 4.1 suffices, with trivial changes, to prove a theorem in which the origin is a critical point of f which may be degenerate.

Theorem 4.2. *Let f be a real-valued function of class C^{r+2}, $r \geqslant 0$, defined on an open neighborhood N of the origin in E_n, vanishing at the* *origin. If the origin is a critical point of index k and rank* ν *with* $0 < \nu \leqslant n$, *there exists an* L^r*-homeomorphism of form*

$$x \to \varphi(x) = (\varphi_1(x), \ldots, \varphi_\nu(x), x_{\nu+1}, \ldots, x_n)$$

onto $\varphi(X)$, *of an open neighborhood* $X \subset N$ *of the origin* $\mathbf{0}$, *leaving the origin fixed, and such that*

$$f(\varphi(x)) = -x_1^2 - \cdots - x_k^2 + x_{k+1}^2 + \cdots + x_\nu^2 + \sum_{\nu+1}^{n} A_{hk}(x)\, x_h x_k\,, \qquad x \in X,$$

where each mapping A_{ij} *is of class* C^r *on* X *and vanishes at the origin.*

Recall that φ is a C^r-diff if $r > 0$, but, as an L^r-homeomorphism, it is something more than a C^r-diff. When $\nu = n$ Theorem 4.2 reduces to Theorem 4.1.

§ 5

REGULAR MANIFOLDS IN EUCLIDEAN SPACES

In Part II abstract differentiable manifolds will be defined, each with a differentiable structure.

However, the differentiable manifolds most relevant to the developments of Part I are differentiable manifolds in E_r which are "regular" in E_r. We shall define these manifolds in a manner which will show that they are differentiable manifolds in the sense of the general definition of Part II. Our definition of regular manifolds will show that two of the conditions which must be imposed on the general manifolds of Part II are automatically satisfied by regular manifolds in E_r. We refer to the "compatibility" and the "countability" conditions.

Definition 5.0. *Presentations.* If V is an open subset of some Euclidean space E_n, a homeomorphism

$$v \to F(v) : V \to X$$

onto an open subspace† X of a topological n-manifold† will be called a *presentation* $(F : V, X)$ of X.

Definition 5.1a. *A C^m-Immersion in E_r.* Suppose that $0 < n \leqslant r$. A C^m-mapping, $m > 0$,

$$v \to F(v) = (F_1(v),..., F_r(v)) : V \to E_r \qquad (V \text{ open in } E_n), \qquad (5.1)$$

† For definition see Appendix I.

such that the $r \times n$ functional matrix

$$\left\| \frac{\partial F_i(v)}{\partial v_j} \right\| \tag{5.2}$$

has the rank n at each point $v = (v_1, ..., v_n) \in V$ is called a *C^m-immersion* of V in E_r.

Definition 5.1b. A *C^m-embedding.* A C^m-immersion (5.1) which is a homeomorphism onto a subspace X of E_r is called a *C^m-embedding* of V and a *regular C^m-presentation* of X in E_r.

It is important to note that the above definition of a C^m-embedding in E_r would be essentially changed in meaning if the phrase "a homeomorphism onto X" were replaced by the phrase "a continuous biunique mapping onto X." In fact, F might then fail to be a homeomorphism onto X, as simple examples when $n = 1$ and $r = 2$ would show.

Let Γ_n be a topological n-manifold.

Definition 5.2. *Families of Presentations Covering Γ_n.* A family

$$(F_k : V_k, X_k)_{k \in \alpha} \qquad (V_k \text{ open in } E_n) \tag{5.3}$$

of presentations of open subspaces X_k of Γ_n whose union is Γ_n is termed a *family of presentations covering Γ_n*.

Definition 5.3. *Regular C^m-Manifolds in E_r. Let Γ_n, $0 < n \leqslant r$, be a topological n-manifold which is a subspace of E_r and is coverable by a family* (5.3) *of regular C^m-presentations.*

The set of all regular C^m-presentations of open subspaces of Γ_n defines a "differentiable structure" $\mathcal{D}$ on Γ_n. Γ_n, provided with the differentiable structure $\mathcal{D}$, is called a regular C^m-manifold M_n in E_r with "carrier" $| M_n | = \Gamma_n$ and set of presentations $\mathcal{D}M_n$.

If a presentation $(F : V, X)$ is in $\mathcal{D}M_n$, X is called a *coordinate domain of M_n* and the coordinates $(v_1, ..., v_n)$ of a point $v \in V$ the *F-coordinates* of the point $F(v) \in X$.

As indicated earlier, *regular* manifolds M_n in E_r possess properties which must be postulated for the *abstract* manifolds of Part II. We shall define these properties and verify their possession by regular manifolds in E_r.

Compatibility. Let Γ_n be an arbitrary topological n-manifold and

$$\begin{aligned} v &\to F(v) : V \to X \qquad (V \text{ open in } E_n), \\ u &\to G(u) : U \to Y \qquad (U \text{ open in } E_n), \end{aligned} \tag{5.4}$$

be two homeomorphisms onto open subspaces X and Y of Γ_n. If $X \cap Y \neq \varnothing$, let

$$\begin{aligned} v &\to \hat{F}(v) : \hat{V} \to X \cap Y \qquad (\hat{V} \subset V), \\ u &\to \hat{G}(u) : \hat{U} \to X \cap Y \qquad (\hat{U} \subset U), \end{aligned} \tag{5.5}$$

be the unique restrictions of F and G, respectively, such that

$$\hat{F}(\hat{V}) = \hat{G}(\hat{U}) = X \cap Y.$$

Definition 5.4. *C^m-Compatibility, $m > 0$.* The homeomorphisms F and G introduced in (5.4) are termed *C^m-compatible* if $X \cap Y = \varnothing$, or if the "*transition*" homeomorphism

$$\lambda = \hat{G}^{-1} \circ \hat{F} : \hat{V} \to \hat{U} \qquad (\text{onto } \hat{U}) \tag{5.6}$$

is a C^m-diff.

We shall prove the following "compatibility" lemma.

Lemma 5.1. *Let the topological n-manifold Γ_n be a subspace of E_r which is the carrier of a regular C^m-manifold M_n in E_r, $0 < n \leqslant r$, and let F and G of (5.4) be presentations in $\mathscr{D}M_n$ of coordinate domains X and Y of M_n.*

Then F and G are C^m-compatible.

If $X \cap Y = \varnothing$, there is nothing to prove.

If $X \cap Y \neq \varnothing$, we refer to the "transition" homeomorphism $v \to \lambda(v)$ defined in (5.6) (assuming that $\Gamma_n = | M_n |$). Subject to the relation $u = \lambda(v)$, one has $G(u) = F(v)$ for $v \in \hat{V}$, or more explicitly

$$G_i(u) = F_i(v), \qquad i = 1,..., r; \quad v \in \hat{V}. \tag{5.7}$$

We shall show that both λ and its inverse are of class C^m, thereby implying that λ is a C^m-diff of $\hat{V}$ onto $\hat{U}$ (Definition 1.1).

If v^0 is given in $\hat{V}$, the point $u^0 = \lambda(v^0)$ is in $\hat{U}$. Since G is a C^m-embedding in E_r, there is a set ω of n of the integers $1,\ldots, r$ such that the n mappings

$$u \to G_\mu(u), \qquad \mu \in \omega, \quad u \in U, \tag{5.8}$$

have a nonvanishing Jacobian J_0 at $u = u^0$.

Now the n equations

$$G_\mu(u) = F_\mu(v), \qquad \mu \in \omega, \quad v \in \hat{V}, \quad u \in \hat{U}, \tag{5.9}$$

with v prescribed in $\hat{V}$, have the solution $u = \lambda(v)$ with $u^0 = \lambda(v^0)$ as an initial solution. Since the Jacobian $J_0 \neq 0$, it follows from the classical implicit function theory that the mapping $v \to \lambda(v)$ is of class C^m in some neighborhood of v^0 in $\hat{V}$. Since v^0 is arbitrary in $\hat{V}$, it follows that λ is of class C^m on $\hat{V}$.

One shows, similarly, that the homeomorphism λ^{-1} of $\hat{U}$ onto $\hat{V}$ is of class C^m. The compatibility lemma now follows from Definition 1.1.

The following theorem characterizes a regular C^m-manifold M_n in E_r in a manner which shows that regular C^m-manifolds in E_r are differentiable manifolds in the sense of the general definition in Part II.

Theorem 5.1. *Let M_n be a regular C^m-manifold in E_r, $0 < n \leqslant r$, and* (5.3) *a family of presentations in $\mathscr{D}M_n$ covering $| M_n |$. Then $\mathscr{D}M_n$ is the ensemble of those presentations $(F : V, X)$ of open subspaces X of $| M_n |$ which are C^m-compatible with each presentation in the given family* (5.3).

$\mathscr{D}M_n$ is given as the set of all regular C^m-presentations of open subspaces of $| M_n |$, while $| M_n |$ is itself given as a topological n-space which is a subspace of E_r.

Taking Lemma 5.1 into account, it is sufficient to show that a presentation $(F : V, X)$ of an open subspace X of $| M_n |$ such that F is C^m-compatible with each C^m-presentation of the family (5.3) is a regular C^m-presentation of X.

By hypothesis, for $v \in V, F(v)$ is a point

$$(F_1(v),\ldots, F_r(v)) \in | M_n | \subset E_r .$$

We must show that F is of class C^m on V, and that the corresponding functional matrix (5.2) has the rank n for $v \in V$.

Given $v^0 \in V$, set $x^0 = F(v^0)$. Let $(G : U, Y)$ be a regular C^m-presentation of a coordinate domain Y of M_n such that $x^0 \in Y$. Such a presentation will be taken from the given family (5.3). In the notation of Definition 5.4, with $\Gamma_n = | M_n |$ the transition homeomorphism λ given by (5.6) is, by hypothesis, a C^m-diff. Moreover, (5.6) implies that

$$\hat{F} = \hat{G} \circ \lambda. \tag{5.10}$$

Relation (5.10) shows that F is of class C^m on $\hat{V}$ and that the functional matrix (5.2) is of rank n for $v \in \hat{V}$.

Since v^0 is an arbitrary point of V, it follows that $(F : V, X)$ is a regular C^m-presentation of X, thereby establishing Theorem 5.1.

Covering Countability. A second property of regular C^m-manifolds M_n in E_r, which is postulated in the general definition of differentiable manifolds in Part II, is characterized in Theorem 5.2.

Theorem 5.2. *Corresponding to a regular C^m-manifold M_n in E_r, there always exists a "countable family* (5.3)*" of presentations in $\mathscr{D}M_n$ covering $| M_n |$.*

Let ω be the set of all positive integers. In the countable ensemble of open r-balls in E_r with rational radii and centers with rational coordinates let $(B_i)_{i\in\omega}$ be the subensemble of those r-balls whose intersections with $| M_n |$ are included in some coordinate domain of M_n. Corresponding to B_i, let $(F_i : V_i, X_i)$ then be a presentation in $\mathscr{D}M_n$ such that $X_i \supset B_i \cap | M_n |$. It is clear that the family

$$(F_i : V_i, X_i)_{i\in\omega}$$

satisfies the theorem.

Differentiable Monge Presentations. Let E_n be a coordinate n-plane in a Euclidean r-space E_r, with $0 < n < r$, and let $v_1, \ldots, v_n$ be the subset of the coordinates $x_1, \ldots, x_r$ of E_r which vary on E_n. A continuously differentiable homeomorphism

$$v \to F(v) : V \to X \qquad (V \text{ open in } E_n)$$

of V onto a subspace X of E_r, such that F^{-1} is the restriction to X of the orthogonal projection of E_r onto E_n, is called a *differentiable Monge presentation of* X. It is clear that such a presentation is regular.

Monge Coverings. Let Γ be a subspace of E_r which is a topological n-manifold. If Γ is the carrier of a regular n-manifold M_n, one sees that there exists a family $(F)_\Gamma$ of differentiable Monge presentations of open subsets of Γ whose union is Γ. We term $(F)_\Gamma$ a *differentiable Monge covering* of Γ.

The Differentiability Index m_Γ. If a presentation $(F : V, X)$ is of class C^m, we term m a *differentiability index* of F. If Γ admits a differentiable Monge covering $(F)_\Gamma$, let m_Γ be the minimum of the "differentiability indices" of presentations $F \in (F)_\Gamma$. One sees that m_Γ is independent of the particular family $(F)_\Gamma$ covering Γ. Taking account of Theorem 5.1, one readily verifies the following theorem:

Theorem 5.3. *If Γ is a subspace of E_r which is a topological n-manifold for which a family $(F)_\Gamma$ of differentiable Monge presentations covering Γ exists, the following is true: Corresponding to each integer m such that $1 \leqslant m \leqslant m_\Gamma$, there exists a unique regular C^m-manifold $M_n^{(m)}$ with carrier Γ. For each such m $(F)_\Gamma$ is a proper subset of the set of regular C^m-presentations in $\mathscr{D}M_n^{(m)}$.*

In verifying this theorem, it should be understood that two regular C^m-manifolds in E_r are identical if and only if they have the same carriers and identical sets of regular C^m-presentations.

Submanifolds of Regular Manifolds. If M_n is a regular C^m-manifold in E_r and Γ_n a nonempty open subspace of $| M_n |$, then Γ_n is the carrier of a unique regular C^m-manifold in E_r, termed a *submanifold* of M_n. The set of Monge presentations of open subsets of Γ_n is a subset of the set of presentations $\mathscr{D}M_n$. These Monge presentations cover Γ_n.

Sources of Regular n-Manifolds in E_r. Let W be a nonempty open subset of E_r. With $0 < s < r$ let

$$x \to H_j(x) : W \to R, \qquad j = 1, \ldots, s,$$

be s real-valued functions of class C^m, $m > 0$, and let Γ be the subspace of E_r of points $x \in E_r$ such that

$$0 = H_1(x) = H_2(x) = \cdots = H_s(x).$$

Set $n = r - s$. If at each point of Γ the functional matrix of the s functions H_i has the rank s, the classical theorems of implicit function theory imply that Γ is a topological n-manifold admitting a family $(F)_\Gamma$ of differentiable Monge presentations of open subsets of Γ whose union is Γ. For each m such that $1 \leqslant m \leqslant m_\Gamma$ there then exists a unique regular C^m-manifold $M_n^{(m)}$ determined, in accord with Theorem 5.3, by the family $(F)_\Gamma$.

Functions on M_n. Let M_n be a regular C^m-manifold in E_r, $0 < n \leqslant r$. Let

$$p \to f(p) : | M_n | \to R$$

be a real-valued function on $| M_n |$ and $(F : V, X)$ a presentation in $\mathscr{D}M_n$. The composite function (for composition in the extended sense, see Appendix I)

$$v \to (f \circ F)(v) : V \to R \tag{5.11}$$

will be called the *representation* of $f | X$ in F-coordinates v. The representations $f \circ F$ are used as follows to define properties of f.

Definition 5.5a. *Differentiability of f on M_n.* We say that f is of *class C^μ* on M_n, if for each presentation $(F : V, X)$ in $\mathscr{D}M_n$, $f \circ F$ is of class C^μ.

Definition 5.5b. *Critical points of f on M_n.* Suppose that f is of class C^μ, $\mu > 0$, on M_n. A point p^0 in a coordinate domain X of a presentation $(F : V, X)$ in $\mathscr{D}M_n$ is termed a *critical point* of f if and only if $u^0 = F^{-1}(p^0)$ is a critical point of $f \circ F$ on V.

Definition 5.5c. *Nullity and Index of a Critical Point.* Suppose that M_n and f are of class C^m, $m > 1$, and that $p^0 \in M_n$ is a critical point of f. Let $(F : U, X)$ be a presentation in $\mathscr{D}M_n$ such that $p_0 \in X$. Then p^0 is said to have the *nullity* ν and the *index* k if and only if the critical point $u^0 = F^{-1}(p^0)$ of $f \circ F$ has the nullity ν and index k.

Invariance of the above Properties of f. That the properties of f and of its critical points characterized in Definition 5.5 are independent of the representations $f \bar{\circ} F$ used to characterize these properties follows from Lemma 5.2 and Lemma 4.2.

The critical point p_0 is termed ND *if its nullity is zero.*

Lemma 5.2. *Given presentations F and G in $\mathscr{D}M_n$ of coordinate domains in M_n with a nonempty intersection X, let $\hat{F}$ and $\hat{G}$ be restrictions of F and G, respectively, which present X. Then the "transition" homeomorphism onto*

$$\lambda = \hat{G}^{-1} \circ \hat{F} : F^{-1}(X) \to G^{-1}(X) \tag{5.12}$$

[see (5.6)] *is a C^m-diff in the sense of* § 1.

Proof. That λ is a C^m-diff follows from the C^m-compatibility of F and G.

To apply Lemma 5.2, note that

$$\hat{G} \circ \lambda = \hat{F} \tag{5.13}$$

as a consequence of (5.12), so that if f is any real-valued function mapping $| M_n |$ into R,

$$(f \bar{\circ} \hat{G}) \circ \lambda = (f \bar{\circ} \hat{F}). \tag{5.14}$$

This relation between the "representation" $f \bar{\circ} \hat{G}$ and $f \bar{\circ} \hat{F}$ of f, taken with Lemma 4.2, implies the "invariance" of the properties of f and its critical points characterized in Definition 5.5.

Induced Mappings of Regular Manifolds. Diffs of abstract differentiable manifolds onto other differentiable manifolds will be studied in Part II. The mappings of regular manifolds M_n in E_r onto other such manifolds in E_r which occur most naturally are those induced by diffs of open sets of E_r onto open sets of E_r. The following theorem is concerned with such an "induced" mapping:

Theorem 5.4. *Let M_n be a regular C^m-manifold in E_r, $0 < n \leqslant r$, and*

$$x \to \varphi(x) : Z \to W \qquad (Z \text{ open in } E_r) \tag{5.15}$$

a C^m-diff of Z onto an open subset W of E_r.

If $|M_n| \subset Z$, *then* $\varphi(|M_n|)$ *is the carrier of a regular* C^m*-manifold* N_n *in* E_r *whose presentations* $(G : V, Y)$ *correspond biuniquely to the presentations in* $\mathscr{D}M_n$, *with* $(F : V, X)$ *in* $\mathscr{D}M_n$ *corresponding to the presentation*

$$(G : V, Y) = (\varphi \,\bar{\circ}\, F : V, \varphi(X)) \tag{5.16}$$

in $\mathscr{D}N_n$. *Thus* $G^{-1} \circ (\varphi \,\bar{\circ}\, F)$ *is the identity mapping of* V *onto* V.

By hypothesis the functional matrix of F has the maximum rank n at each point $v \in V$. That the functional matrix of $G = \varphi \,\bar{\circ}\, F$ has the rank n follows from the matrix equality

$$\left\| \frac{\partial G_i(v)}{\partial v_j} \right\| = \left\| \frac{\partial \varphi_i(x)}{\partial x_k} \right\| \left\| \frac{\partial F_k(v)}{\partial v_j} \right\|, \qquad v \in V,$$

where $x = F(v) = F(v_1, \ldots, v_n)$ and $G = (G_1, \ldots, G_r)$, $F = (F_1, \ldots, F_r)$, $\varphi = (\varphi_1, \ldots, \varphi_r)$.

The mapping of *presentations*,

$$(F : V, X) \to (\varphi \,\bar{\circ}\, F : V, \varphi(X)) : \mathscr{D}M_n \to \mathscr{D}N_n,$$

is clearly biunique. It is *onto* the ensemble of C^m-presentations in N_n, since the mapping of presentations,

$$(G : V, Y) \to (\varphi^{-1} \,\bar{\circ}\, G : V, \varphi^{-1}(Y)) : \mathscr{D}N_n \to \mathscr{D}M_n,$$

is its inverse.

Thus Theorem 5.4 is true.

We say that the manifold N_n of Theorem 5.4 is the *image* $\varphi(M_n)$ of M_n under φ.

Theorem 5.4 is supplemented by the following theorem.

Theorem 5.5. *On the manifolds* M_n *and* $N_n = \varphi(M_n)$ *of Theorem* 5.4 *let real-valued functions* f *and* g, *respectively, be given with*

$$f(p) = g(q); \qquad p \in M_n; \quad q \in N_n; \quad q = \varphi(p). \tag{5.17}$$

Then f *is of class* C^2 *on* M_n *if and only if* g *is of class* C^2 *on* N_n, *and if* f *and* g *are of class* C^2, *the critical points of* f *and* g *correspond biuniquely under* φ, *with preservation of nullity and indices.*

Let $(F : V, X)$ be a presentation $\mathscr{D}M_n$ and $(G : V, Y)$ the corresponding presentation (5.16) in $\mathscr{D}N_n$.

Then $f \mid X$ and $g \mid Y$ are, respectively, of class C^2 by definition if and only if $f \bar{\circ} F$ and $g \bar{\circ} G$ are of class C^2 on V. If ψ is the restriction of φ to X then, by definition of g and G

$$g \bar{\circ} G = (f \bar{\circ} \psi^{-1}) \circ (\psi \circ F) = f \bar{\circ} F \qquad (\text{on } V), \tag{5.18}$$

so that $f \mid X$ is of class C^2 if and only if $g \mid Y$ is of class C^2.

By definition the critical points of $f \mid X$ and $g \mid Y$ are, respectively, the F and G images of critical points v^0 of $f \bar{\circ} F$ and $g \bar{\circ} G$ on V, with nullity and indices determined by $f \bar{\circ} F$ and $g \bar{\circ} G$. Since (5.18) holds and since $G(v^0)$ is the image of $F(v^0)$ under φ, the lemma follows.

Exercice 5.1. Making use of Definition 5.4 of C^m-compatibility, verify the following local characterization of C^m-compatibility.

Let Γ_n be a topological n-manifold and

$$(F_1 : U_1, X_1) \qquad \text{and} \qquad (F_2 : U_2, X_2) \qquad (U_1, U_2 \text{ open in } E_n) \tag{5.19}$$

be presentations of intersecting open subsets X_1 and X_2 of Γ_n. If $X_1 \cap X_2 \neq \varnothing$, a necessary and sufficient condition (ξ) that F_1 and F_2 be C^m-compatible is as follows:

(ξ) *Each point $p_0 \in X_1 \cap X_2$ shall admit an open neighborhood N relative to $X_1 \cap X_2$ such that the restrictions*

$$(\hat{F}_1 : \hat{U}_1, N) \qquad \text{and} \qquad (\hat{F}_2 : \hat{U}_2, N), \tag{5.20}$$

respectively, of the presentations (5.19) *be presentations of N which are C^m-compatible.*

§ 6

FIRST THEOREMS ON NONDEGENERACY

The "Volume Ratio Method." Historical Review. Theorems and lemmas preceding Lemma 6.3 are stated without proof. They are presented to indicate the development of the theory from 1926 to 1932, the year Morse's Colloquium Lectures [13] were given.

The Role of Null Jordan Content. A bounded set A in E_n has a *null Jordan content* if, corresponding to a prescribed positive constant e, there exists a finite set of n-rectangles in E_n whose union includes A and whose total volume is less than e. A set of null Jordan content has a null Lebesgue measure $\mathbf{m}A = 0$, but the converse is not true. A set of null Jordan content is nowhere dense in E_n, a property not implied by null Lebesgue measure. The union

$$A = A_1 \cup A_2 \cup \cdots \tag{6.1}$$

of a countable ensemble of bounded disjoint sets each with null Jordan content has a Lebesgue measure $\mathbf{m}A = 0$. However, to say that $\mathbf{m}A = 0$ does not imply that each set A_i has a null Jordan content.

The first theorem in the 1927 abstract (Morse [2], p. 814) concerned an arbitrary compact regular manifold Σ_n of class C^2 in E_{n+1} and was stated as follows.

Theorem 6.0. *There exist on Σ_n infinitely many* ND *functions of class C^2.*

A first proof of this theorem in 1926 was by way of an elementary principle which we shall term the *gradient lemma*. In this lemma we are concerned with an open subset X of E_n and a closed n-cube Q included in X. There is given a real-valued function f of class C^2 defined on X. The Hessian $H(x)$ of f at a point $x \in X$ is the Jacobian at x of grad f. That is,

$$H(x) = \frac{D(f_{x1}, \ldots, f_{x_n})}{D(x_1, \ldots, x_n)}(x), \qquad x \in X. \tag{6.2}$$

We shall refer to points $\mathbf{a} = (a_1, \ldots, a_n)$ in a Euclidean space A_n. The "*gradient lemma*" follows.

Lemma 6.0. *Under the preceding conditions on f the subset*

$$\{\mathbf{a} \in A_n \mid (\operatorname{grad} f)(x) = \mathbf{a} \quad \text{for some} \quad x \in Q \quad \text{with} \quad H(x) = 0\} \tag{6.3}$$

of A_n has the Jordan content zero in A_n.

The following lemma was recognized as an immediate consequence of Lemma 6.0:

Lemma 6.1. *Under the preceding conditions on f the subset*

$$\{\mathbf{a} \in A_n \mid (\operatorname{grad} f)(x) = \mathbf{a} \quad \text{for some} \quad x \in X \quad \text{with} \quad H(x) = 0\} \tag{6.4}$$

of A_n has the Lebesgue measure zero in E_n.

Lemma 6.1 is obviously equivalent to the following:

Lemma 6.2. *Under the preceding conditions on f, for almost all values of $\mathbf{a} \in A_n$ the mapping*

$$x \to f(x) - (a_1x_1 + \cdots + a_nx_n) : X \to R \tag{6.5}$$

is ND.

Lemma 6.2 led at once to a proof of Theorem 6.0. However, the truth of Theorem 6.0 was confirmed around 1930 in another way. The Colloquium Lectures by Morse [12], given in 1932, contained a general theorem, Theorem 14.1, p. 243, on focal points of extremals transverse to a manifold. This theorem has the following special corollary:

Theorem 6.1. *For $0 < r < n$ let M_r be a regular C^2-manifold in E_n. For almost all points $a \in E_n - M_r$ the C^2-function*

$$x \to f(x) = \| x - a \|, \qquad x \in M_r, \tag{6.6}$$

is ND *on M_r.*

Theorem 6.1 will be extended in § 15 without reference to variational theory. It implies Theorem 6.0.

Although no explicit proof of the gradient lemma was published by Morse, a proof of an analogous lemma on conjugate points (Lemma 16.1 by Morse [3], p. 625) gives, except for trivial changes of terminology, a proof of Lemma 6.0. The essence of the proof, both in the case of conjugate points and in the case of ordinary critical points, is the use of infinitesimal "volume ratios."

In 1931 it was noted that the proof of the gradient lemma by the use of "volume ratios" led, with at most trivial changes, to another theorem in which grad f and the Jacobian of grad f were replaced by f and the Jacobian of f. We shall formulate this modification of Lemma 6.0 as Lemma 6.3. Lemma 6.3 concerns a C^1-mapping,

$$x \to F(x) = (F_1(x), \ldots, F_n(x)) \tag{6.7}$$

of an open subset X of E_n into E_n. The points $x \in X$ at which

$$J(x) = \frac{D(F_1, \ldots, F_n)}{D(x_1, \ldots, x_n)}(x) = 0 \tag{6.8}$$

are called *singular* points of F.

Lemma 6.3. *If the open subset X of E_n includes a closed unit n-cube Q and if σ is the subset of singular points of F in Q, then $F(\sigma)$ has the Jordan content zero in E_n.*

It was found very recently that a lemma similar to Lemma 6.3 was proved by Knopp and Schmidt [1] in 1926. A proof in case $n = 2$ was regarded as sufficient. Around 1932 proof of Lemma 6.3 by the volume ratio method was suggested to Sard as a starting point of a Harvard thesis on measure aspects of general differentiable mappings of open subsets E_n into E_m. Sard's results were published in 1942 (see Sard [1]).

We shall resume our exposition by giving a detailed proof of Lemma 6.3 by the "volume ratio method."

Proof of Lemma 6.3. For each integer $r > 0$ let Q be divided by a network of r^n congruent nonoverlapping n-cubes with edges of length $1/r$. If Z_r is such a cube and ρ the length of its diagonal, its volume is found to be

$$V(z_r) = c_n \rho^n, \qquad c_n = n^{-n/2}. \tag{6.9}$$

We shall apply Corollary 2.2, setting $K = Q$ in Corollary 2.2. Let z_r^* be any one of the cubes z_r which contains a point $u \in \sigma$. Since $J(u) = 0$, by hypothesis, there is at least one unit vector $\lambda = (\lambda_1, \ldots, \lambda_n)$ such that u is a critical point of the dot product $\lambda \cdot F$.

Let S_x be the system of coordinate x_i-axes. Let S_y be a system of coordinate y_i-axes obtained from S_x by a translation of the origin in S_x to the point $F(u)$, followed by a rotation of the x_n-axis into the y_n-axis, taken as the direction $(\lambda_1, \ldots, \lambda_n)$ in S_x.

Suppose that the point $F(x)$ with its coordinates $(F_1(x), \ldots, F_n(x))$ in S_x is represented in S_y by the point $(y_1, \ldots, y_n)$. If $x \in z_r^*$, the orthogonal projection of the vector $F(x) - F(u)$ onto a sensed straight line with the direction λ has an algebraic value $y_n = \lambda \cdot (F(x) - F(u))$ such that, in terms of φ of Corollary 2.2 and in accord with (2.20) and (2.25), $|y_n| \leqslant \rho\varphi(\rho)$. Recall that the mapping $t \to \varphi(t)$ depends only on the values of F in an open convex neighborhood of Q and not upon the choice of $u \in \sigma$.

The coordinate y_i, $i = 1, \ldots, n-1$, in S_y of the point $F(x)$ is similarly obtained by projecting the vector $F(x) - F(u)$ orthogonally onto the y_i-axis. If the direction of the y_i-axis has the components $\alpha_1, \ldots, \alpha_n$ relative to S_x, then

$$y_i = \alpha \cdot (F(x) - F(u)), \qquad i = 1, \ldots, n-1, \tag{6.10}$$

and by virtue of (2.21) $|y_i| \leqslant \rho H_Q$, where H_Q is a constant independent of the choice of $u \in \sigma$.

The image $F(z_r^*)$ in E_n of the cube z_r^* is thus contained in a rectangle w_r with center at the point $F(u)$ in S_x and with volume

$$V(w_r) \leqslant 2\rho\varphi(\rho)(2\rho H_Q)^{n-1}. \tag{6.11}$$

Recall that φ is continuous and that $\varphi(0) = 0$. A comparison of (6.11) and (6.9) shows that the "*volume ratio*," $V(w_r)$ to $V(z_r^*)$, tends to zero as ρ tends to zero, independently of the choice of $u \in \sigma$.

The sum of the volumes of the n-cubes z_r^* which contain points $u \in \sigma$ is at most the volume 1 of Q. The union of the corresponding rectangles w_r covers $F(\sigma)$. Let $e > 0$ be prescribed. It follows from (6.11) that if r is sufficiently large, $F(\sigma)$ is coverable by a finite set of rectangles w_r whose total volume is less than e. Hence $F(\sigma)$ has null Jordan content.

Thus Lemma 6.3 is true.

We state the following extension of Lemma 6.3:

Extended Lemma 6.3. *Let $x \to F(x) : X \to E_n$ be a C^1-mapping of an open subset X of E_n into R_n.*

If σ is the subset of singular points of F in X, and Y any relatively compact subset of X, then

$$\text{Jordan content}\, F(\sigma \cap Y) = 0.$$

The proof of the lemma depends upon the fact that the relatively compact subset Y of X can be covered by a finite set of n-cubes each included in X (see Appendix I).

ND Functions in Prescribed Families. Lemma 6.2 does not guarantee the existence of a ND function in a prescribed family of differentiable functions. If f is a function of the family, there may be no function f' of the form

$$x \to f(x) - (a_1x_1 + \cdots + a_nx_n) = f'(x)$$

other than f in the given family. The following family of electrostatic potentials f has this property:

Electrostatic Potentials. In E_3 let there be given $\mu > 0$ points

$$p^{(1)}, \ldots, p^{(\mu)} \tag{6.12}$$

and ν points

$$q^{(1)}, \ldots, q^{(\nu)}, \tag{6.13}$$

all distinct. Let $x \in E_3$ be distinct from each of the points (6.12) and (6.13). Let $\eta_1, \ldots, \eta_\mu$ be positive numbers and $\zeta_1, \ldots, \zeta_\nu$ negative numbers representing the magnitudes of electrostatic charges at the respective points (6.12) and (6.13). Set

$$r_i = \| p^{(i)} - x \|, \qquad \rho_j = \| q^{(j)} - x \|; \qquad i = 1, \ldots, \mu; \quad j = 1, \ldots, \nu. \tag{6.14}$$

For fixed points (6.12) and (6.13) and corresponding charges the function[†]

$$x \to V_{pq}(x) = \frac{\eta_1}{r_1} + \cdots + \frac{\eta_\mu}{r_\mu} + \frac{\zeta_1}{\rho_1} + \cdots + \frac{\zeta_\nu}{\rho_\nu} \tag{6.15}$$

is the associated electrostatic potential.

The parameters of this family are the points $p^{(i)}$, $q^{(j)}$ and the corresponding charges. The question arises: For fixed charges and fixed integers μ and ν, are ND potentials of the form (6.15) dense among all such potentials? We shall answer this question affirmatively by proving the following theorem:

Theorem 6.2.[‡] *If the $\mu + \nu$ charges are fixed, as well as the points $p^{(i)}$ and $q^{(j)}$, excepting one of these points, say $p^{(1)} = (a_1, a_2, a_3)$, then the resulting potential*

$$x \to V(x_1, x_2, x_3; a_1, a_2, a_3) \tag{6.16}$$

is ND *for almost all points (a_1, a_2, a_3) in E_3.*

We shall state a general theorem, Theorem 6.3, that implies Theorem 6.2.

Notation for Theorem 6.3. Let A_m be a Euclidean space of points $a = (a_1, \ldots, a_m)$. Let W be an open nonempty subset of the product space $E_n \times A_m$. Let there be given a C^2-mapping

$$(x, a) \to U(x, a) : W \to R. \tag{6.17}$$

[†] The critical points of V_{pq} are the points of equilibrium in the corresponding field of electrostatic forces. In §3.2 of Part IV relations between these points of equilibrium will be obtained whenever V_{pq} is ND.

[‡] One could prove this theorem by having recourse to the general theory of real analytic functions. However, such recourse would not be possible for similar families of nonanalytic functions.

Given a subset X of $E_n \times A_m$, let $\mathrm{pr}_1 X$ and $\mathrm{pr}_2 X$ denote the orthogonal projections of X into E_n and A_m, respectively.

The Partial Mappings U^a. For each point $a \in \mathrm{pr}_2 W$ we introduce the open subset

$$W(a) = \{x \in E_n \mid (x, a) \in W\}$$

of $\mathrm{pr}_1 W$ and the partial mapping

$$x \to U(x, a) = U^a(x) : W(a) \to R. \tag{6.18}$$

We regard U as defining a *family of mappings* U^a.

The x-Critical Ensemble Ω of U. Let Ω denote the set of points $(x, a) \in W$ such that

$$0 = U_{x_1}(x, a) = U_{x_2}(x, a) = \cdots = U_{x_n}(x, a).$$

A point (x, a) is in Ω if and only if $x \in E_n$ is a critical point of the function U^a. The set Ω may be empty.

The x-Critical Matrix $\mathbf{H}(U)$. Setting $U_{x_i} = U_i$ for $i = 1,.., n$, let the n by $n + m$ functional matrix of the n functions $U_1, ..., U_n$ with respect to the $n + m$ variables $x_1, ..., x_n$; $a_1, ..., a_m$, evaluated at (x, a) be denoted by

$$\| U_{ij}(x, a) \| = \mathbf{H}(U)(x, a). \tag{6.19}$$

We call $\mathbf{H}(U)$ the *x-critical matrix* of U.

Theorem 6.3. *Let W be an open nonempty subset of $E_n \times A_m$ and*

$$(x, a) \to U(x, a) : W \to R \tag{6.20}$$

a C^2-mapping such that for each pair (x, a) in the "x-critical ensemble" Ω of U

$$\mathrm{rank}\, \| U_{ij}(x, a) \| = n. \tag{6.21}$$

Then for almost all $a \in \mathrm{pr}_2 W$, the partial mappings

$$x \to U(x, a) = U^a(x) \tag{6.22}$$

are ND.

Theorem 6.3 will be established in § 7, together with Theorem 6.5. Theorems 6.3 and 6.5 will be given extensions in Part II. In these extensions the product space $E_n \times A_m$ will be replaced by the product $M_n \times A_m$ of the above Euclidean space A_m and an abstract differentiable manifold M_n. In the remainder of this section we shall clarify Theorem 6.3 by giving three applications.

First Application. Proof of Lemma 6.2. One sets

$$U(x, a) = f(x) - (a_1 x_1 + \cdots + a_n x_n), \qquad x \in X, \tag{6.23}$$

for each $a \in A_n$. For such a U the last n columns of the $n \times 2n$ matrix of (6.21) have the rank n, so that (6.21) holds. In this application $w = X \times A_n$ and $\mathrm{pr}_2 W = A_n$. Lemma 6.2 follows from Theorem 6.3.

Second Application. Proof of Theorem 6.2. We apply Theorem 6.3 taking $n = m = 3$, identifying U of Theorem 6.3 with V of Theorem 6.2. One finds that

$$\frac{\partial V}{\partial a_1} = \eta_1 \frac{(a_1 - x_1)}{\| x - a \|^3}, \qquad \frac{\partial V}{\partial a_2} = \eta_1 \frac{(a_2 - x_2)}{\| x - a \|^3}, \qquad \frac{\partial V}{\partial a_3} = \eta_1 \frac{(a_3 - x_3)}{\| x - a \|^3}.$$

A simple calculation shows that the 3-square determinant

$$\left| \frac{\partial^2 V(x, a)}{\partial x_i \, \partial a_j} \right| \neq 0$$

for x distinct from the charged points. The 3×6 matrix in (6.21) thus has the rank 3. Theorem 6.2 accordingly follows from Theorem 6.3.

Third Application. ND *boundary values.* Let Σ be a regular n-manifold in E_{n+1} of class C^μ, $\mu > 1$, bounding a compact subset Z of E_{n+1}. A real-valued function f of class C^m, $\mu \geqslant m > 1$, defined on an open neighborhood D_f of Z, will be termed *admissible relative to* Z on Σ if $f \mid Z$ is ND and if f is ordinary at each point of Σ.

The deepest global study of critical points of $f \mid Z$ is possible only when $f \mid \Sigma$ is ND (see § 10). That this is the general case is shown by the following theorem:

Theorem 6.4. *If f is admissible relative to Z and if e_f is a sufficiently small positive constant, then for almost all points $a = (a_1, ..., a_{n+1})$ in E_{n+1} such that $\| a \| < e_f$, the function*

$$x \to f(x) - a_j x_j = g^a(x), \qquad x \in D_f, \tag{6.24}$$

summing with respect to j has the following properties:

(i) *The critical points of $g^a \mid Z$ are* ND *and may be made to correspond biuniquely to those of $f \mid Z$ with preservation of index, and so as to tend uniformly to those of $f \mid Z$ as $\| a \|$ tends to zero.*

(ii) *The function $g^a \mid \Sigma$ is* ND.

Proof of (i). For fixed points a the critical points of g^a are solutions $x \in E_{n+1}$ of the $n + 1$ equations $\partial g^a / \partial x_1 = \cdots = \partial g^a / \partial x_{n+1} = 0$. Now, $g^a = f$ when $a = \mathbf{0}$. By hypothesis

$$\frac{D(g_1{}^a, ..., g_{n+1}^a)}{D(x_1, ..., x_{n+1})}(x) \neq 0$$

when $\| a \| = 0$ and x is a critical point of $f \mid Z$. Statement (i) follows on making use of the principle: the index of a ND quadratic form Q with coefficients a_{ij} remains constant as the coefficients a_{ij} vary continuously provided the determinant $| a_{ij} |$ remains non-null (cf. Theorem 3.4).

Proof of (ii). Let $(F : V, X)$ be a presentation of a coordinate domain of Σ. Then $g^a \mid X$ has the representation

$$u \to U(u, a) = f(F(u)) - a_j F_j(u),$$

summing with respect to j, for j on the range $1, ..., n + 1$ and $(u, a) \in V \times E_{n+1}$. We shall apply Theorem 6.3, identifying $(u_1, ..., u_n)$ with $(x_1, ..., x_n)$ of Theorem 6.3. The "u-critical matrix" $\| U_{ij}(u, a) \|$ of (6.21) is then n by $2n + 1$ and has as submatrix the functional matrix of $F_1, ..., F_{n+1}$ with respect to the variables $u_1, ..., u_n$, and so has rank n, since the presentation $(F : V, X)$ is regular. Hence by Theorem 6.3 $g^a \mid X$ is ND for almost all points $a \in E_{n+1}$.

Now, Σ can be covered by a finite set of coordinate domains X. Hence (ii) and Theorem 6.4 follow.

The following theorem is needed in § 16, where its counterpart for an abstract differentiable manifold M_n is established.

Theorem 6.4′. *Let Σ be a regular n-manifold of class C^μ, $\mu > 1$, in E_r, $0 < n < r$. There exist r real-valued functions*

$$p \to \varphi_j(p) : \Sigma \to R$$

of class C^μ on Σ such that the following is true: If $p \to f(p)$ is a real-valued function of class C^m, $\mu \geqslant m > 1$, on Σ, the function

$$p \to f(p) + \sum_{j=1}^{r} a_j \varphi_j(p) = f^a(p) \tag{6.25}$$

is ND *on Σ for almost all points $(a_1, ..., a_r) \in E_r$.*

Proof. We take φ_j as the restriction $x_j \mid \Sigma$, where $x_1, ..., x_r$ are coordinates in E_r. The proof is completed by using Theorem 6.3 as in the proof of Theorem 6.4(ii). If $(F : V, X)$ is one a of countable set of presentations covering $\Sigma, f^a \mid X$ has a representation

$$U(x, a) = f^a(F(u)), \qquad (x, a) \in X \times E_r,$$

to which Theorem 6.3 applies, showing that $f^a \mid X$ is ND for almost all points $a \in E_r$. Since the union of a countable set of subsets of E_r of zero measure is a set of zero measure, Theorem 6.4′ follows.

The proof in § 7 of Theorem 6.3 will include a proof of Theorem 6.5 below. Theorem 6.5 implies Theorem 6.3, as we shall see in § 7.

Theorem 6.5. *If under the hypotheses of Theorem* 6.3 *W_0 is an arbitrary relatively compact subset of W of Theorem* 6.3, *the subsets of points $a \in \mathrm{pr}_2 W_0$ such that the partial mappings U^a fail to be* ND *have a Jordan content in A_m which is zero.*

In particular, if W in Theorem 6.3 is a relatively compact subset of $E_n \times A_m$ such that U of Theorem 6.3 admits an extension of class C^2 over an open neighborhood of $\overline{W}$ in $E_n \times A_m$, then the conclusion of Theorem 6.3 can be given the form: the partial maps U^a which fail to be ND have parameters a in a set of Jordan content zero. This exceptional set is nowhere dense in A_m.

Historical Note. In preparing "Functional topology and abstract variational theory," published in France in 1939, the following conjecture was made by Morse around 1932:

A C^r-mapping of an open subset of E_n, $n > 1$, into E_1 for which $r = n$ is constant on any connected subset σ of its critical set.

This theorem was not needed in treating the ND case, nor in the abstract theory, but was desirable in treating degenerate mappings from E_n to E_m when $n > m$.

The problem was made known to interested graduate students and colleagues at Harvard from 1932 on. A first response was the counter-example of Whitney [1], which showed that f need not be constant on σ if $r < n$. The main problem remained unsolved until M. Morse communicated it to A. Morse at Princeton in 1938. A. Morse [1] verified the conjecture.

EXERCISE 6.1. Suppose that in Theorem 6.3 $n = 1$, $m = 2$, $W = E_1 \times A_2$, and

$$U(x, a) = \tfrac{1}{3}(x_1 - 1)^3 + a_1^2 x_1 - a_2^2 x_1 - x_1, \qquad (x, a) \in E_1 \times A_2.$$

Verify the following: The set Ω is the quadric in $E_3 = E_1 \times A_2$ on which

$$U_{x_1}(x, a) = (x_1 - 1)^2 + a_1^2 - a_2^2 - 1 = 0.$$

The matrix condition (6.21) is satisfied. The subsets Ω^* of points $(x, a) \in \Omega$ such that x is a degenerate x-critical point of U^a is the plane hyperbola on which $x_1 = 1$, $a_1^2 - a_2^2 = 1$, and meas $\mathrm{pr}_2\, \Omega^* = 0$. Theorem 6.5 holds for each open relatively compact subset W_0 of W.

§ 7

NONDEGENERACY: THEOREM 7.1

We shall show that Theorems 6.3 and 6.5 are implied by the following more geometric theorem. This theorem is of general character, with other applications. Following a statement of the theorem an essential geometric interpretation of it will be given.

Theorem 7.1. *Let n C^1-mappings*

$$(x, a) \to U_i(x, a) : W \to R, \qquad i = 1,\dots, n, \tag{7.1}$$

of an open subset W of $E_n \times A_m$ be given, and let Ω be the subset of points $(x, a) \in W$ such that

$$U_1(x, a) = U_2(x, a) = \cdots = U_n(x, a) = 0. \tag{7.2}$$

Suppose that at each point $(x, a) \in \Omega$ the n by $n + m$ functional matrix of $U_1, \dots, U_n$ has the maximum rank n. Then the subset Ω^* of points $(x, a) \in \Omega$ at which

$$\frac{D(U_1, \dots, U_n)}{D(x_1, \dots, x_n)}(x, a) = 0 \tag{7.3}$$

(termed x-singular points of Ω) has the following properties [of which (η) implies (ζ)]:

(η) *There exists an open neighborhood N_p in W of each point $p \in W$ so small that*

$$\text{Jordan content } \mathrm{pr}_2(N_p \cap \Omega^*) = 0. \tag{7.4}$$

(ζ) *The* meas $\mathrm{pr}_2\, \Omega^* = 0$ *in* A_m.

Geometric Interpretation of Theorem 7.1. The set Ω can be regarded as the carrier of a regular m-manifold in the Euclidean $(n + m)$-space $E_n \times A_m$ of rectangular coordinates $x_1, \ldots, x_n ; a_1, \ldots, a_m$. We denote this manifold as well as its carrier by Ω.

We regard E_n and A_m as coordinate n- and m-planes of $E_n \times A_m$. On E_n, $a_1 = \cdots = a_m = 0$, while on A_m, $x_1 = \cdots = x_n = 0$. A point (x^0, a^0) of Ω is in Ω^* if and only if there is at least one ray tangent to Ω at (x^0, a^0) and orthogonal to the coordinate m-plane, A_m.

Choice of x_i and a_j Axes, Both the hypotheses and conclusions of Theorem 7.1 will be unaffected if the x_i-axes are permuted among themselves and the a_j-axes are similarly permuted. It is understood that under any such change of coordinates the functions $U_1, \ldots, U_n$ are to remain invariant in value. We thus understand that the manifold Ω, as well as its subset Ω^*, remains invariant.

Statement (η) of Theorem 7.1, and not (ζ), is the principal conclusion of Theorem 7.1. Statement (η) implies (ζ) and has other implications of importance for us.

Statement (A) will now be verified:

(A) *Theorem* 7.1 (η) *implies Theorem* 7.1 (ζ).

Proof of (A). Granting the truth of (η), there exists a sequence $\eta_1, \eta_2, \ldots$ of open subsets of W covering W and such that the following is true: For each positive integer k the set $w_k = \text{pr}_2(\eta_k \cap \Omega^*)$ has a null measure $\mathbf{m}w_k$ in A_m. Moreover,

$$\text{pr}_2\, \Omega^* \subset \underset{k>0}{\text{Union}}\; \text{pr}_2(\eta_k \cap \Omega^*), \tag{7.5}$$

since the sets η_k cover W. Hence

$$\mathbf{m}(\text{pr}_2\, \Omega^*) \leqslant \mathbf{m}w_1 + \mathbf{m}w_2 + \cdots = 0, \tag{7.6}$$

so that (ζ) is true.

Proof of Theorem 7.1 (η). We begin the proof of (η) by verifying the truth of (η) in the trivial case in which the point $p \in W$ is not in Ω^*. Since Ω^* is a closed subset of W, there exists an open neighborhood N_p of p in W such that $N_p \cap \Omega^* = \varnothing$ whenever $p \notin \Omega^*$. In this case (η) is accordingly true.

The Case $p = (x^0, a^0) \in \Omega^*$. In this case there exist by hypothesis integers $r \geqslant 0$ and $s > 0$ such that $r + s = n$ and the Jacobian of $U_1, \ldots, U_n$ with respect to r of the coordinates $x_1, \ldots, x_n$ and with respect to s of the coordinates $a_1, \ldots, a_m$ does not vanish at (x^0, a^0). If use is made of our freedom to renumber the coordinates x_i and the coordinates a_j, we can suppose that

$$\frac{D(U_1, \ldots, U_n)}{D(x_1, \ldots, x_r; a_1, \ldots, a_s)}(x^0, a^0) \neq 0, \qquad s > 0. \tag{7.7}$$

As a special case r can vanish and $s = n$.

A Monge Representation of Ω *near* (x^0, a^0). By hypothesis $U_i(x^0, a^0) = 0$ for $i = 1, \ldots, n$, and (7.7) holds. The classical implicit function theorem accordingly implies the existence of solutions (x, a) of the equations (7.2) near (x^0, a^0). The notation involved in presenting the implicit function theorem requires definition.

Notation for Lemma 7.1 Of the coordinates $x_1, \ldots, x_n; a_1, \ldots, a_m$ of a point $(x, a) \in E_n \times A_m$ the subset

$$x_1, \ldots, x_r; a_1, \ldots, a_s, \qquad r + s = n, \quad s > 0, \tag{7.8}$$

has appeared in (7.7). There remains the complementary set

$$x_{r+1}, \ldots, x_n; a_{s+1}, \ldots, a_m, \qquad 0 \leqslant r < n, \tag{7.9}$$

of m coordinates of (x, a). Let Π denote the coordinate m-plane of $E_n \times A_m$ on which the coordinates (7.9) are variable and the coordinates (7.8) vanish. Let π project $E_n \times A_m$ orthogonally onto Π, so that

$$\pi(x, a) = (x_{r+1}, \ldots, x_n; a_{s+1}, \ldots, a_m) = z, \tag{7.10}$$

introducing z.

The classical implicit function theorem, supplemented in (i) by continuity considerations, gives the following:

Lemma 7.1. *Let* $(x^0, a^0) \in \Omega^*$ *be such that* (7.7) *holds. Corresponding to any sufficiently small open neighborhood* ω *of* $\pi(x^0, a^0)$ *relative to the* *coordinate m-plane Π there exists an ϵ-neighborhood $N(\epsilon)$ in W of (x^0, a^0) so small that the following are true:*

(*i*) $\pi(N(\epsilon)) \subset \omega$ *and*

$$\frac{D(U_1, ..., U_n)}{D(x_1, ..., x_r; a_1, ..., a_s)}(\bar{x}, \bar{a}) \neq 0, \qquad (\bar{x}, \bar{a}) \in N(\epsilon), \tag{7.11}$$

(ii) *The m-manifold Ω includes an open subset Ω_ω which projects homeomorphically under π onto ω and has a Monge representation on ω of class C^1 giving the inverse of $\pi | \Omega_\omega$,*

(iii) $N(\epsilon) \cap \Omega = N(\epsilon) \cap \Omega_\omega$.

By virtue of (ii) the points $(x, a) \in \Omega_\omega$ have coordinates (x, a), uniquely determined by their projections,

$$\pi(x, a) = (x_{r+1}, ..., x_n; a_{s+1}, ..., a_m) = z \tag{7.12}$$

in ω and by equations

$$\begin{matrix} x_1 = \varphi_1(z) & a_1 = \psi_1(z) \\ \vdots \quad \vdots & \vdots \quad \vdots \\ x_r = \varphi_r(z) & a_s = \psi_s(z) \end{matrix} \qquad z \in \omega \tag{7.13}$$

where z is arbitrary in ω and the functions $\varphi_1, ..., \varphi_r$ and $\psi_1, ..., \psi_s$ are of class C^1 on ω.

Two additional lemmas are required to establish (η) of Theorem 7.1. In Lemma 7.2 use is made of a neighborhood $N(\epsilon)$ of a point (x^0, a^0), as given in Ω^*, with $N(\epsilon)$ conditioned as in Lemma 7.1. Reference is made to the Monge representation of $N(\epsilon) \cap \Omega$, given by (7.13). Recall that $r + s = n$ and $0 \leqslant r < n$.

Lemma 7.2. *A point $(\bar{x}, \bar{a}) \in N(\epsilon) \cap \Omega$ such that*

$$\frac{D(\psi_1, ..., \psi_s)}{D(x_{r+1}, ..., x_n)}(\pi(\bar{x}, \bar{a})) \neq 0 \tag{7.14}$$

is not in Ω^*, the set of "x-singular" points of Ω.

Suppose the lemma false and that $(\bar{x}, \bar{a}) \in \Omega^*$, that is, that $(\bar{x}, \bar{a}) \in \Omega$ and

$$\frac{D(U_1, ..., U_n)}{D(x_1, ..., x_n)}(\bar{x}, \bar{a}) = 0. \tag{7.15}$$

There then exists a ray λ tangent to Ω at $(\bar{x}, \bar{a})$ and orthogonal to the coordinate m-plane A_m. Let g be a regular arc,

$$t \to (x_1(t),..., x_n(t); a_1(t),..., a_m(t)), \qquad -d < t < d, \tag{7.16}$$

in $N(\epsilon) \cap \Omega$ meeting the point $(\bar{x}, \bar{a})$ when $t = 0$, and having the direction of λ when $t = 0$. Since λ is orthogonal to A_m,

$$a_1'(0) = a_2'(0) = \cdots = a_m'(0) = 0. \tag{7.17}$$

The coordinates of points on g satisfy the relations (7.13) for each $t \in (-d, d)$. Hence

$$| x_{r+1}'(0)| + | x_{r+2}'(0)| + \cdots + | x_n'(0)| \neq 0. \tag{7.18}$$

Otherwise (7.17) and (7.13) would imply that

$$| x_1'(0)| + | x_2'(0)| + \cdots + | x_r'(0)| = 0, \qquad 0 \leqslant r < n,$$

and g would not be regular.

According to (7.13) for $-d < t < d$

$$a_k(t) = \psi_k(x_{r+1}(t),..., x_n(t); a_{s+1}(t),..., a_m(t)), \qquad k = 1,..., s.$$

We infer that when $t = 0$, $z = \pi(\bar{x}, \bar{a})$, and (7.17) holds

$$0 = \frac{\partial \psi_k(z)}{\partial x_{r+1}} x_{r+1}'(0) + \cdots + \frac{\partial \psi_k(z)}{\partial x_n} x_n'(0), \qquad k = 1,..., s. \tag{7.19}$$

The relations (7.19) are impossible when (7.18) and (7.14) hold.

We infer the truth of the lemma.

Lemma 7.2 should be paired with Lemma 7.3:

Lemma 7.3. *Let $N(\epsilon)$ be a neighborhood relative to W of the above point (x^0, a^0) with $N(\epsilon)$ conditioned as in Lemma 7.1, and let e be a constant such that $0 < e < \epsilon$.*

Then the subset $Z(e)$ of points $(x, a) \in N(e) \cap \Omega$ such that

$$\frac{D(\psi_1, ..., \psi_s)}{D(x_{r+1}, ..., x_n)} (\pi(x, a)) = 0 \tag{7.20}$$

has the property that in A_m

$$\text{Jordan content } \mathrm{pr}_2\, Z(e) = 0. \tag{7.21}$$

Proof of Lemma 7.3. Set $Y(e) = N(e) \cap \Omega$. Then $Z(e) \subset Y(e)$. We shall define a C^1-mapping

$$z \to \Theta(z) : \pi(Y(e)) \to A_m$$

onto $\mathrm{pr}_2\, Y(e)$, recalling that $\pi(Y(e))$ is an *open* neighborhood in Π of $\pi(x^0, a^0)$ by virtue of Lemma 7.1 (i) and (ii). We shall apply the Extended Lemma 6.3 to Θ. Set $Y(\epsilon) = N(\epsilon) \cap \Omega$.

The Mapping Θ. Under Θ a point

$$z = (x_{r+1}, \ldots, x_n\,;\, a_{s+1}, \ldots, a_m) \in \pi(Y(e)) \tag{7.22}$$

shall go into the point

$$(\Theta_1(z), \ldots, \Theta_m(z)) = (\psi_1(z), \ldots, \psi_s(z);\, a_{s+1}, \ldots, a_m) \in A_m\,. \tag{7.23}$$

In the m-plane Π the domain $\pi(Y(e))$ is an open relatively compact subset of $\pi(Y(\epsilon))$. Moreover, Θ admits a C^1-extension over $\pi(Y(\epsilon))$ defined by (7.23). The Jacobian of $\Theta_1, \ldots, \Theta_m$ with respect to the m coordinates of z reduces to

$$\frac{D(\psi_1, \ldots, \psi_s)}{D(x_{r+1}, \ldots, x_n)}(z), \qquad z \in \pi Y(\epsilon),$$

by virtue of (7.23), and vanishes by hypothesis when $z = \pi(x, a)$ and $(x, a) \in Z(e)$. It follows from Extended Lemma 6.3 that

$$\text{Jordan content } \Theta(\pi Z(e)) = 0. \tag{7.24}$$

Relation (7.21) will follow once we have verified the equation

$$\mathrm{pr}_2(x, a) = \Theta(\pi(x, a)), \qquad (x, a) \in Y(e). \tag{7.25}$$

Now, (7.25) is true since both sides of (7.25) are equal to $(a_1, \ldots, a_m)$. The left side of (7.25) is equal to $(a_1, \ldots, a_m)$ by virtue of the definition of pr_2. The right side of (7.25) is equal to $(a_1, \ldots, a_m)$ by virtue of (7.23) and the relations of (7.13)

$$a_1 = \psi_1(z), \ldots, a_s = \psi_s(z)$$

[when $z = \pi(x, a)$ and $(x, a) \in Y(e)$]. From (7.25) we infer that

$$\Theta(\pi(Z(e))) = \mathrm{pr}_2\, Z(e),$$

[since $Z(e) \subset Y(e)$] so that (7.21) follows from (7.24).

Thus Lemma 7.3 is true.

Completion of Proof of (η). We have seen that (η) is true if $p \notin \Omega^*$. If $p = (x^0, a^0) \in \Omega^*$, let $N_p = N(e)$, where $N(e)$ is conditioned as in Lemma 7.3. According to Lemma 7.2, $N_p \cap \Omega^* \subset$ the subset $Z(e)$ of $N(e) \cap \Omega$ introduced in Lemma 7.3. Lemma 7.3 implies that (7.4) holds.

This completes the proof of Theorem 7.1.

Implications of Theorem 7.1. A theorem which is equivalent to Theorem 7.1 (η) is equivalent to Theorem 7.1. Such a theorem follows.

Theorem 7.2. *If, under the hypotheses of Theorem* 7.1, W_0 *is an arbitrary, open, relatively compact subset of* W, *then*

$$\text{Jordan content } \mathrm{pr}_2(W_0 \cap \Omega^*) = 0, \tag{7.26}$$

where Ω^* *is the set of "x-singular" points of* Ω.

Theorem 7.2 implies Theorem 7.1(η), since for each $p \in W$ there clearly exists a relatively compact open neighborhood of $p \in W$.

Conversely, Theorem 7.1 (η) implies Theorem 7.2, since Cl W_0 can be covered by a finite ensemble of neighborhoods $N_p \subset W$ for each of which (7.4) holds, thereby implying (7.26).

Theorem 6.3 *as Corollary of Theorem* 7.1. To infer Theorem 6.3 from (ζ) of Theorem 7.1, we first identify the domain W of U of Theorem 6.3 with the domain W of the functions $U_1, \ldots, U_n$ of Theorem 7.1. We then identify the functions $U_1, \ldots, U_n$ of Theorem 7.1 with the partial derivatives $U_{x_1}, \ldots, U_{x_n}$ of Theorem 6.3. The set Ω of points (x, a) in W of Theorem 7.1 at which the functions $U_1, \ldots, U_n$ vanish is identified with the set Ω of points at which the partial derivatives $U_{x_1}, \ldots, U_{x_n}$ vanish. Theorems 6.3 and 7.1 have the respective hypotheses that the functional matrices of the functions

$U_1, \ldots, U_n$ and $U_{x_1}, \ldots, U_{x_n}$ have rank n at points of their respective Ω. It follows from Theorem 7.1 (ζ) that for almost all $a \in \mathrm{pr}_2 \Omega$ of Theorem 6.3, the Hessian of U with respect to the variables $x_1, \ldots, x_n$ fails to vanish, implying Theorem 6.3.

The Meaning of Ω^*. In the context of Theorem 7.1 Ω^* is the subset of x-singular points of the regular manifold Ω, while in the context of Theorem 6.3 Ω^* is the subset of Ω of degenerate x-critical points of U.

Theorem 6.5 *as Corollary of Theorem* 7.2. One makes the identifications of the preceding paragraphs and in addition identifies the relatively compact sets W_0 of Theorem 6.5 and Theorem 7.2. It follows from (7.26) that the parameters a of partial mappings $x \to U^a(x)$ which fail to be ND are contained in a subset of $\mathrm{pr}_2 W_0$ of A_m of Jordan content 0, implying Theorem 6.5.

§ 8

Σ-NORMAL φ-COORDINATES

The Family of Normals ν_p to Σ. Let Σ be a regular compact n-manifold of class C^∞ in E_{n+1}. The strong hypothesis that Σ be of class C^∞ is made for simplicity of exposition in dealing with the family of normals to Σ at points p. The results presented here are effectively supplemented by the theorem on "elevating manifold differentiability" in Appendix II.

We make a useful definition:

Definition 8.1. *Strongly Extensible Presentations φ.* A presentation $(\varphi : U, X)$ in $\mathscr{D}\Sigma$ of a coordinate domain X of Σ will be called *strongly extensible* if φ admits an extension as a presentation $(\varphi' : U', X')$ in $\mathscr{D}\Sigma$ such that U is relatively compact in U' or, equivalently, X is relatively compact in X'.

Since Σ is compact, it can be covered by a finite set of coordinate domains X each given by a strongly extensible presentation in $\mathscr{D}\Sigma$. A presentation $(\varphi : U, X)$ will be *understood to be strongly extensible* unless otherwise stated.

Normals ν_p to Σ. Let p be an arbitrary point in Σ with coordinates $p_1, \ldots, p_{n+1}$ in E_{n+1}. Let Σ be the boundary of a compact subset Z of E_{n+1}. Let

$$\lambda(p) = (\lambda_1(p), \ldots, \lambda_{n+1}(p)) \tag{8.0}$$

be a sensed unit vector normal to Σ at p, leaving Z at p. Let ν_p be the normal to Σ at p. Then ν_p is the ensemble of points

$$x = p + s\lambda(p), \qquad s \in R. \tag{8.1}$$

The mapping

$$(p, s) \to p + s\lambda(p) : \Sigma \times R \to E_{n+1} \tag{8.2}$$

is continuous but not biunique.

Notation. Let e be a positive constant and set $I(e) = (-e, e)$. Let $\nu_p{}^e$ be the subarc of the normal ν_p on which $s \in I(e)$. Given an arbitrary subset W of $|\Sigma|$, set

$$W^e = \bigcup_{p \in W} \nu_p{}^e. \tag{8.3}$$

In particular, $|\Sigma|^e$ is written as Σ^e. This notation is used throughout Part I.

We shall prove the following:

Lemma 8.1. *Let $(\varphi : U, X)$ be a presentation in $\mathscr{D}\Sigma$ of a coordinate domain X of Σ. If e is a sufficiently small positive constant, the mapping*

$$(u, s) \to \varphi(u) + s\lambda(\varphi(u)) = \psi(u, s) : U \times I(e) \to X^e \tag{8.4}$$

is a diffeomorphism of class C^∞, onto X^e in E_{n+1}.

Let u^0 be a point of U. We shall show that the Jacobian

$$\frac{D(\psi_1, \ldots, \psi_{n+1})}{D(u_1, \ldots, u_n, s)}(u^0, s^0) \neq 0 \qquad \text{when} \quad s^0 = 0. \tag{8.5}$$

To verify (8.5), recall that the first n columns of the determinant (8.5) represent n vectors tangent to Σ at the point $p^0 = \varphi(u^0)$. These vectors are linearly independent, since the functional matrix of $\varphi_1, \ldots, \varphi_{n+1}$ has rank n at u^0, by hypothesis. The last column of the determinant (8.5) represents the vector $\lambda(p^0)$. Since $\lambda(p^0)$ is orthogonal to Σ at p^0, (8.5) follows.

With each point $u^0 \in U$ there can accordingly be associated a constant $e > 0$ and an open neighborhood $V \subset U$ of u^0 with e and V so small that $\psi \mid (V \times I(e))$ is a diff onto Y^e and Y^e is open in E_{n+1}.

A similar result holds if one replaces $(\varphi : U, X)$ by a strong extension $(\varphi' : U', X')$ and u^0 by any point in U', including points $u^0 \in \bar{U}$. Since $\bar{U}$ is compact, we infer that if e is sufficiently small, two arcs $\nu_{p'}^e$, $\nu_{p''}^e$ normal to $\bar{X}$ at distinct points p' and p'' are disjoint and the mapping (8.4) accordingly biunique. The mapping (8.4) is locally a diff onto open subsets of E_{n+1}. It is accordingly a diff onto X^e.

The lemma follows.

Objectives of § 8. We are concerned with relations between sets of coordinates of points near $| \Sigma |$: the ordinary rectangular coordinates $x_1, \ldots, x_{n+1}$ of E_{n+1}, the pairs (p, s) introduced in (8.2) and termed *Σ-normal parameters*, and the pairs (u, s) introduced in Lemma 8.1 and termed Σ-normal φ-coordinates. All conventions concerning the existence, nondegeneracy, and indices of critical points are ultimately to be referred to Σ-normal φ-coordinates. Σ-normal parameters (p, s) are used primarily to define functions globally near $| \Sigma |$.

Definition 8.2. *Σ-Normal φ-Coordinates* (u, s). Under the conditions of Lemma 8.1 a pair $(u, s) \in U \times I(e)$ will be called Σ-normal φ-coordinates of the point

$$x = \varphi(u) + s\lambda(\varphi(u)) = \psi(u, s). \tag{8.6}$$

The set X^e will be called the *domain* of points x represented by these Σ-normal φ-coordinates.

A Representation of g. If $x \to g(x)$ is a real-valued function whose domain in E_{n+1} includes the set X^e of Lemma 8.1, the mapping

$$(u, s) \to g(\varphi(u) + s\lambda(\varphi(u))) = h(u, s), \qquad (u, s) \in U \times I(e) \tag{8.7}$$

is called the *representation* of $g \mid X^e$ in Σ-normal φ-coordinates (u, s).

Since ψ is a diff of class C^∞, $g \mid X^e$ is of class C^μ, if and only if h is of class C^μ on its domain in (8.7). By Lemma 4.2 a point x^0 with Σ-normal φ-coordinates (u^0, s^0) will be *critical* and ND *with index k* relative to g if and only if (u^0, s^0) is respectively critical and ND with index k relative to h.

Σ-Normal Parameters (p, s). These parameters enter by way of the following lemma, a corollary of Lemma 8.1.

Lemma 8.2. *If ϵ is a sufficiently small positive constant, the mapping*

$$(p, s) \to p + s\lambda(p) : \Sigma \times I(\epsilon) \to \Sigma^\epsilon \tag{8.8}$$

is a homeomorphism onto Σ^ϵ, and Σ^ϵ is the union of a finite set of domains X^ϵ of Σ-normal φ-coordinates.

In the remainder of Part I the number ϵ of Σ^ϵ will always be so small that Σ^ϵ satisfies Lemma 8.2.

Definition 8.3. (i) *Σ-Normal Parameters.* Under the conditions of Lemma 8.2 a pair $(p, s) \in \Sigma \times I(\epsilon)$ will be called *Σ-normal parameters* of the point $x = p + s\lambda(p) \in \Sigma^\epsilon$.

(ii) The set Σ^ϵ will be called the *domain of points x represented* by these Σ-normal parameters.

(iii) Given a presentation $(\varphi : U, X)$ in $\mathscr{D}\Sigma$ and $(u, s) \in U \times I(\epsilon)$, the coordinates (u, s) will be called *Σ-normal φ-coordinates* of the Σ-normal parameters $(p, s) = (\varphi(u), s)$.

Mappings $(p, s) \to H(p, s)$. Let there be given a real-valued mapping

$$(p, s) \to H(p, s) : \Sigma \times I(\epsilon) \to R. \tag{8.9}$$

The mapping

$$(u, s) \to H(\varphi(u), s) = G(u, s), \qquad (u, s) \in U \times I(\epsilon), \tag{8.10}$$

is called the *representation* of H in Σ-normal φ-coordinates (u, s).

Definition 8.4. *Conventions Concerning H:*

(i) *The class C^μ of H.* If $(\varphi : U, X)$ is a presentation in $\mathscr{D}\Sigma$ of a coordinate domain X of Σ and if H is defined as in (8.9), the restriction $H\,|(X \times I(\epsilon))$ of H will be said to be of *class C^μ* if and only if G of (8.10) is of class C^μ. The unrestricted mapping H will be said to be of class C^μ if and only if each such restriction of H is of class C^μ.

(ii) *The Critical Points of H.* A point $(p^\circ, s^\circ) \in X \times I(\epsilon)$ with Σ-normal φ-coordinates (u°, s°) will be said to be *critical* and to be ND with *index k* relative to H if and only if (u°, s°) is, respectively, critical and ND with index k relative to G.

Invariance Principle. The entities defined in the preceding two paragraphs are independent of the particular presentations ($\varphi : U, X$) entering into these definitions. This follows from the "compatibility" of presentations of coordinate domains of Σ and from Lemma 4.2 (see Lemma 5.2).

We state two lemmas concerning the use of Σ-normal parameters:

Lemma 8.3. (i) *Let $x \to g(x)$ and $(p, s) \to H(p, s)$ be real-valued functions on Σ^ϵ and $\Sigma \times I(\epsilon)$, respectively. If*

$$g(p + s\lambda(p)) = H(p, s), \qquad (p, s) \in \Sigma \times I(\epsilon), \tag{8.11}$$

then g is of class C^μ on its domain if and only if H is of class C^μ on its domain.

(ii) *If $x^\circ \in \Sigma^\epsilon$ has Σ-normal parameters (p°, s°), then x° is critical and* ND *with index k relative to g if and only if (p°, s°) is, respectively, critical and* ND *with index k relative to H.*

Proof. Suppose that x° is in a domain X^ϵ of points x represented by Σ-normal φ-coordinates (u, s). Reference to (8.7) and to (8.10) shows that the *representation* both of g and H in terms of these coordinates is the mapping

$$(u, s) \to g(\varphi(u) + s\lambda(\varphi(u))) = H(\varphi(u), s), \qquad (u, s) \in U \times I(\epsilon),$$

where the equality is a consequence of (8.11). Lemma 8.3 is now a trivial consequence of the conventions on H preceding the lemma.

The Partial Mappings H^s. Given a mapping

$$(p, s) \to H(p, s) : \Sigma \times I(\epsilon) \to R, \tag{8.12}$$

we shall refer to the *partial mapping*

$$p \to H(p, s) = H^s(p) : \Sigma \to R \tag{8.13}$$

and prove the following fundamental lemma:

Lemma 8.4. *If the mapping H of* (8.12) *is of class C^1, a point $(p^\circ, s^\circ) \in \Sigma \times I(\epsilon)$ is a critical point of H if and only if $H_s = 0$ at (p°, s°), and p° is a critical point of H^{s° on Σ.*

Let $(\varphi : U, X)$ be a presentation in $\mathscr{D}\Sigma$ of a coordinate domain X which contains p°. Then (p°, s°) is in $X \times I(\epsilon)$ and $H \,|(X \times I(\epsilon))$ has a "representation" [see (8.10)]

$$(u, s) \to H(\varphi(u), s) = G(u, s), \qquad (u, s) \in U \times I(\epsilon) \tag{8.14}$$

in Σ-normal φ-coordinates (u, s). By virtue of Definition 8.4 (p°, s°) is a critical point of H, if and only if (u°, s°) is a critical point of G, that is, if and only if $G_s = 0$ at (u°, s°) and u° is a critical point of the partial mapping $u \to G(u, s^\circ)$. Since the critical points of H^s on Σ are understood in the sense of Definition 5.5, the lemma follows.

§ 9

NONDEGENERATE FUNCTIONS UNDER BOUNDARY CONDITIONS A

Introduction. Global critical point theory begins with the study of ND functions f on regular domains Z of a Euclidean space E_{n+1} under the simplest boundary conditions on f, namely *boundary conditions* A. These conditions will now be defined.

Definition 9.1. *Regular C^{μ}-Domains Z in E_{n+1}.* By such a domain we mean a compact subset Z of E_{n+1} bounded by the carrier $|\Sigma|$ of a regular n-manifold Σ of class C^{μ}, $\mu > 1$. By abuse of language we refer to Σ as the *boundary* of Z. The *interior* of Z is denoted by $\mathring{Z}$.

We do not assume that $|\Sigma|$ or Z is connected. The number of components of Z or $|\Sigma|$ is finite.

Definition 9.2. *Functions f Admissible Relative to a Regular Domain Z.* We shall *admit* real-valued functions of class C^m, $m > 1$, on some open neighborhood D_f of Z, ND on Z and without critical points on Σ.

A non-null vector V at a point $p \in \Sigma$ is termed *entrant* or *emergent* if V is not tangent to Σ at p and is directed into the interior or exterior, respectively, of Z at p.

Definition 9.3. *Boundary Conditions* A. We shall say that a function f which is admissible relative to a regular domain Z of E_{n+1} satisfies *boundary conditions* A, if grad f is emergent at each point of Σ.

Definition 9.4. *The Type Numbers of* $f \mid Z$. By the kth type number m_k of $f \mid Z$, $k = 0, 1, \ldots, n + 1$, we mean the number of critical points of $f \mid Z$ of index k.

The Connectivities of Z. Let $\mathscr{K}$ be any commutative field (see Appendix I). In Part III we shall review the foundations of singular homology theory, taken in the sense of Eilenberg [1] (see also Eilenberg and Steenrod [1] pp. 185–211). The connectivities

$$R_k(Z, \mathscr{K}), \qquad k = 0, 1, \ldots, n + 1, \tag{9.1}$$

are the dimensions of the homology groups of singular k-cycles on Z, taking the coefficients from $\mathscr{K}$. When $\mathscr{K}$ is fixed, as is the case in Part I, we shall denote the connectivities (9.1) by

$$R_0, R_1, \ldots, R_{n+1}, \tag{9.2}$$

understanding that $R_{n+1} = 0$, as will be shown to be the case in Part III.

Theorem 9.1 below is a fundamental theorem. The theorem remains true if the n-manifold Σ bounding Z is merely of class C^1 instead of class C^∞ (see Appendix II for proof).

Theorem 9.1. *Let* f *be a function of class* C^2 *admissible relative to a regular* C^∞*-domain* Z *of* E_{n+1}. *If* $m_0, m_1, \ldots, m_{n+1}$ *are the type numbers of* $f \mid Z$ *and*

$$R_0, R_1, \ldots, R_{n+1}, \qquad R_{n+1} = 0, \tag{9.3}$$

the connectivities of Z, *then under boundary conditions* A *on* f

$$\begin{aligned}
&(m_0 - R_0) \geqslant 0,\\
&(m_1 - R_1) - (m_0 - R_0) \geqslant 0,\\
&(m_2 - R_2) - (m_1 - R_1) + (m_0 - R_0) \geqslant 0,\\
&\cdots\\
&(m_{n+1} - R_{n+1}) - (m_n - R_n) + \cdots (-1)^{n+1}(m_0 - R_0) = 0.
\end{aligned} \tag{9.4}$$

Proof of Theorem 9.1. The proof of Theorem 9.1 is in two parts. The first part involves homotopy considerations based on a proper use in § 21 of f-trajectories of the level manifolds of f. It is postponed until Riemannian geometry has been introduced in Part II because the complexities of "homotopy equivalences" involved can be dispensed with quickly by a suitable global change of Riemannian metric which modifies this metric near each critical point of f. This device was introduced by Morse, [9], § 6, in 1960 and has not yet been widely understood.

The second part of the proof of Theorem 9.1 involves homology theory as well as homotopy theory, and will be presented in Part III, § 30.

Boundary Conditions B. The statement of Theorem 9.1 has not been postponed until Part III because its validity under boundary conditions A implies Theorem 10.2, in which the condition that grad f be emergent on Σ is relaxed. Theorem 10.2 is concerned with *boundary conditions* B, conditions which are topologically consistent with much more general distributions of critical points on Z than are boundary conditions A.

Theorem 9.1 is stated here because the proof that Theorem 9.1 implies Theorem 10.2 involves neither homotopy nor homology, but rather the redefinition of f near $|\Sigma|$ so as to reduce an f satisfying boundary conditions B to an f satisfying boundary conditions A.

Boundary conditions B will be defined in § 10 and Theorem 10.1 will be related to Theorem 9.1 in an explicit way in § 11 by a construction of $\hat{f}$ of Theorem 10.1.

We shall devote the remainder of this section to remarks about Theorem 9.1, beginning with two readily verified corollaries:

Corollary 9.1. *Under the conditions of Theorem* 9.1

$$m_k \geqslant R_k, \qquad k = 0, 1, \ldots, n + 1. \tag{9.5}$$

To state Corollary 9.2 we introduce the *excess numbers* $\mathscr{E}_k = m_k - R_k$, $k = 0, 1, \ldots, n + 1$ and note the truth of the following:

Corollary 9.2. *Under the conditions of Theorem* 9.1

$$\mathscr{E}_{k+1} + \mathscr{E}_{k-1} \geqslant \mathscr{E}_k, \qquad k = 1, \ldots, n. \tag{9.6}$$

We state an extension, Theorem 9.1′ of Theorem 9.1:

Theorem 9.1′ *reads as does Theorem* 9.1, *replacing* C^∞ *by* C^2.

In Appendix II Theorem 9.1′ is shown to be a consequence of Theorem 9.1 and of Theorem A on "elevating manifold differentiability" (see Morse [9]).

In general, theorems involving regular domains Z in E_{n+1} will be stated first under the assumption that the boundary Σ of Z is of class C^∞. So stated, the proofs are simpler than would be the case if Σ were assumed to be in the minimum differentiability class for which the proposed theorem is true. Theorem A on "elevating manifold differentiability" enables one, in most cases, to lower the differentiability condition on Σ as far as possible. What the final differentiability assumption is will depend on the theorem (see Theorems 10.2, 10.3, and 12.1′).

The Completeness of the Relations (9.4). The question arises: Under the conditions of Theorem 9.1, are there relations other than the relations (9.4) or implied by the relations (9.4) which always hold between the type numbers and connectivities? That the answer is *no* is shown by the following theorem:

Theorem 9.2. *Corresponding to nonnegative integers* $R_0, R_1, \ldots, R_n, R_{n+1}$, *of which* $R_0 = 1$ *and* $R_{n+1} = 0$, *and to nonnegative integers* $m_0, m_1, \ldots, m_{n+1}$ *satisfying relations* (9.4) *there exists a regular* C^∞-*domain* Z *in* E_{n+1} *with the integers* R_i *as connectivities and a function* f *of class* C^∞, *admissible relative to* Z, *satisfying boundary conditions* A *with the integers* m_i *as type numbers of* $f \mid Z$.

A proof of Theorem 9.2 can be obtained by carrying out the following exercises:

Exercise 1. Show that if the relations (9.4) hold, integers $e_0, e_1, \ldots, e_n$ are uniquely defined by the equations $R_0 + e_0 = m_0$ and

$$R_k + e_{k-1} + e_k = m_k, \qquad k = 1, \ldots, n, \tag{9.7}$$

and are nonnegative. Moreover, $e_n = m_{n+1}$.

EXERCISE 2. Given a ND function $x \to \varphi(x)$ of class C^∞ on an open domain D of E_{n+1} and an integer such that $0 \leqslant k < n$, it is possible to modify φ on an arbitrary small neighborhood N of an ordinary point of φ in D so as to obtain a new function of class C^∞ on D identical with φ except on N and possessing just two critical points in N with indices k and $k + 1$ (cf. John [1]).

EXERCISE 3. Given integers R_i, as in Theorem 9.2, there exists a regular C^∞-domain Z in E_{n+1} with the integers R_i as connectivities and, relative to Z, an admissible function φ of class C^∞, without critical points on Σ, assuming a value on Σ exceeding each value of φ on $\mathring{Z}$, and such that the type numbers of $\varphi \mid Z$ are numbers

$$m_0^* = R_0, \qquad m_1^* = R_1, \ldots, m_n^* = R_n, \qquad m_{n+1}^* = 0. \tag{9.8}$$

EXERCISE 4. If the relations (9.4) hold with $R_0 = 1$ and $R_{n+1} = 0$, if φ is defined as in Exercise 3, and if integers $e_0, e_1, \ldots, e_n$ are chosen as in Exercise 1, it is possible to modify φ successively on $\mathring{Z}$ as in Exercise 2 so as to add e_k critical points of index k and e_k critical points of index $k + 1$ for each k on the range 0, 1,..., n and thereby obtain a function f with type numbers m_i satisfying relations (9.4) with the prescribed connectivities R_i.

§ 10

NONDEGENERATE FUNCTIONS UNDER BOUNDARY CONDITIONS B

Let Z be a regular C^∞-domain of E_{n+1} and let f be of class C^m, $m > 1$, and admissible relative to Z in the sense of Definition 9.2. By hypothesis f has no critical point on the boundary Σ of Z. Consequently, a point $p \in \Sigma$ is a critical point of $f \mid \Sigma$ if and only if grad f is orthogonal to Σ at p.

The following example shows how the critical points of $f \mid \Sigma$ can condition the critical points of $f \mid Z$.

EXAMPLE 10.1. Let $n + 1 = 2$ and let Z be the disk in E_2 on which $x_1{}^2 + x_2{}^2 \leqslant 1$. The function f with values $f(x) = x_2{}^2 - x_1{}^2$ does not satisfy boundary conditions A but, as we shall see, satisfies boundary conditions B. On Z, f has a critical point of index 1. The existence of this critical point is implied by the connectivities of the disk Z and the nature of $f \mid \Sigma$, as Theorem 10.2 will show.

Objectives. We aim to reveal the conditions on the critical points of $f \mid Z$ implied by the connectivities of Z and the nature of $f \mid \Sigma$. In the absence of any simplifying assumption as to the nature of $f \mid \Sigma$ our problem is prohibitively complex.

Definition 10.1. *Boundary Conditions* B. We shall say that a function f which is admissible relative to Z (Definition 9.2) satisfies boundary conditions B if $f \mid \Sigma$ is ND.

In case $f \mid \Sigma$ is degenerate, Theorem 6.4 shows how f can be approximated arbitrarily closely so as to obtain a function g which satisfies boundary conditions B while remaining admissible relative to Z and having critical points on Z which differ arbitrarily little in position from those of f and not at all in index. Theorem 6.4 shows in what sense the functions admissible relative to Z which satisfy boundary conditions B represent the "general" case.

To present Theorem 10.1 two definitions are needed:

Definition 10.2. *The Entrant and Emergent Portals of Z.* If f is admissible relative to a regular domain Z in E_{n+1}, the open subsets of $|\Sigma|$ on which grad f is entrant and emergent, respectively, will be denoted by $|\Sigma|_-$ and $|\Sigma|_+$. One of these sets may be empty, and their union is in general not $|\Sigma|$. Let Σ_- and Σ_+ denote the submanifolds (see § 5) of Σ with carriers $|\Sigma|_-$ and $|\Sigma|_+$. We term Σ_- and Σ_+ the *entrant* and *emergent portals*, respectively, of Z.

Definition 10.3. *The Augmented Type Numbers of $f \mid Z$.* If f is admissible relative to Z and satisfies *boundary conditions* B, we shall denote the type numbers of $f \mid \Sigma_-$ by

$$\mu_0, \mu_1, ..., \mu_n. \tag{10.1}$$

If the type numbers of $f \mid Z$ are denoted by

$$m_0, m_1, ..., m_{n+1} \tag{10.2}$$

as previously, the $n + 1$ integers $m_0', m_1', ..., m_{n+1}'$ defined by

$$m_0 + \mu_0, m_1 + \mu_1, ..., m_n + \mu_n, m_{n+1} \tag{10.3}$$

will be called the *augmented type numbers* of $f \mid Z$.

The following theorem shows that a function f admissible relative to Z and satisfying boundary conditions B can be modified in a well-defined way so as to satisfy boundary conditions A. Theorem 10.1 is basic.

Theorem 10.1. *Let f be of class C^m, $m > 1$, and admissible relative to a regular C^∞-domain Z of E_{n+1}; suppose that $f \mid \Sigma$ is* ND *and let N_f be an open neighborhood of Σ which contains no critical points of $f \mid Z$. There then exists a function $\acute{f}$ of class C^m on a neighborhood of Z,* *admissible relative to Z, with* grad $\hat{f}$ *now emergent on Σ, identical with f on $Z - N_f$ and such that the ordinary type numbers of $\hat{f} \mid Z$ are equal to the augmented type numbers of $f \mid Z$.*

In this section we present some of the implications of this theorem, deferring a proof until § 11.

EXAMPLE 10.2. We refer to Example 10.1, where Z is the unit disk in E_2 with center at the origin. If $f(x) = x_2{}^2 - x_1{}^2$ and if one sets

$$Z_- = \{x \in Z \mid f(x) < 0\}, \tag{10.4}$$

the f-entrant portal Σ_- of Z has the carrier $Z_- \cap |\Sigma|$. The critical points of $f \mid \Sigma_-$ are the points $(\pm 1, 0)$, both of index zero. The augmented type numbers of $f \mid Z$ are 2, 1, 0.

Theorems 9.1 and 10.1 imply Theorem 10.2. Theorem 10.2 and its extension, Theorem 10.2′ below, are the principal theorems of § 10.

Theorem 10.2. *If f is of class C^2 and admissible relative to a regular C^∞-domain Z in E_{n+1} and satisfies boundary conditions* B, *the augmented type numbers*

$$m_0', \; m_1', \ldots, \; m_{n+1}' \tag{10.5}$$

of $f \mid Z$ satisfy the relations (9.4) *of Theorem* 9.1, *m_i' replacing m_i for each i.*

In the notation of Definition 10.3 the relations of Theorem 9.1 can be put in the more explicit form

$$\begin{aligned} m_k - m_{k-1} + m_{k-2} - \cdots (-1)^k m_0 \\ \geqslant (R_k - \mu_k) - (R_{k-1} - \mu_{k-1}) + \cdots (-1)^k (R_0 - \mu_0), \end{aligned} \tag{10.6}$$

where k has the range $0, 1, \ldots, n + 1$ and the equality prevails in (10.6) when $k = n + 1$.

Theorem 10.2 implies Theorem 10.3. Theorem 10.3 in turn implies Theorem 9.1.

Theorem 10.3. *If f is of class C^2 and admissible relative to a regular C^∞-domain Z in E_{n+1} and if* grad f *is never both entrant and normal to Σ*

at points of Σ, *then the type numbers of* $f \mid Z$ *satisfy the relations* (9.4) *of Theorem* 9.1.

Proof of Theorem 10.3. Observe first that the minimum angle between inner normals to Σ and grad f at a point p of Σ is bounded from zero for $p \in \Sigma$. It follows then from Theorem 6.4 that there exists a real-valued function g of class C^2 on a neighborhood of Z such that the following are true:

(A_0) grad g approximates grad f so closely at points of Σ that grad g is never both entrant and normal to Σ.

(A_1) g is admissible relative to Z.

(A_2) The type numbers of $g \mid Z$ and $f \mid Z$ are equal.

(A_3) g satisfies boundary conditions B.

(A_4) Because of (A_0), $g \mid \Sigma_-$ has no critical points.

(A_5) The augmented type numbers of $g \mid Z$ are equal to the type numbers of $g \mid Z$ and hence those of $f \mid Z$.

(A_6) Theorem 10.2 is applicable to g, implying that the type numbers of $f \mid Z$ satisfy the relations (9.4).

This establishes Theorem 10.3.

We state extensions of Theorems 10.2 and 10.3:

Theorem 10.2′. *This reads as does Theorem* 10.2 *with* C^∞ *replaced by* C^2.

Theorem 10.3′. *This reads as does Theorem* 10.3 *with* C^∞ *replaced by* C^2.

Theorems 10.2′ *and* 10.3′ *follow, respectively, from Theorems* 10.2 *and* 10.3 *with the aid of Theorem* A *of Appendix* II.

We have established Theorem 10.2′ in Appendix II. We leave to the reader the relatively simple task of establishing Theorem 10.3′ using Theorem 10.3 and Theorem *A* of Appendix II.

The proof of Theorem 10.3′ is nearer to the proof of Theorem 9.1′ than to the proof of Theorem 10.2′, in that in the proof of Theorem 10.3′, as in the proof of Theorem 9.1′, it is sufficient to use Theorem A to modify Z and Σ but to leave f unmodified.

Theorems 10.2 and 10.3 have had industrial applications. They can be applied to determine the types of instability and the distribution of points of equilibrium of "conservative fields of force" defined by ND "functions of forces."

The following is one of several applications to harmonic functions of two variables (see Morse [6], p. 48):

Corollary 10.1. *If f is a harmonic function of two variables admissible relative to a regular C^2-domain Z in E_2, of the topological type of a circular disk, with f satisfying boundary conditions B, then, in the notation of Definition* 10.3,

$$\mu_0 \geqslant 1, \qquad \mu_0 - \mu_1 = m_1 + 1. \tag{10.7}$$

For harmonic functions of three variables the following corollary of Theorem 10.2 is more novel. It is presented here for the first time.

Corollary 10.2. *If f is a harmonic function of three variables, admissible relative to a regular C^2-domain Z in E_3 of the topological type of a solid 3-ball, with f satisfying boundary conditions B, then, in the notation of Definition* 10.3,

$$\mu_0 \geqslant 1, \qquad \mu_0 - \mu_1 \leqslant m_1 + 1, \qquad \mu_0 - \mu_1 + \mu_2 = m_1 - m_2 + 1. \tag{10.8}$$

Proof. Taking account of the nonexistence of isolated interior points of relative maximum or minimum of harmonic functions, relations (10.8) follow from relations (10.6), with $R_0 = 1$ and $R_1 = R_2 = R_3 = 0$ therein.

EXERCISE 10.1 Suppose that the harmonic function f of Corollary 10.2 has a finite number of isolated singularities in Z, at m_0 of which f "becomes negatively infinite" and at m_3 of which f "becomes positively infinite" (see Kellogg [1], p.271). Find the relations which replace the relations (10.8) and involve the integers m_0 and m_3 as well as the integers already appearing in (10.8).

§ 11

PROOF OF THEOREM 10.1 UNDER BOUNDARY CONDITIONS B

Notation Recalled. To establish Theorem 10.1, we must define the function $\hat{f}$ which is to replace f in Theorem 10.1. Let $D_f \supset Z$ be the open domain of f in E_{n+1}, Σ the regular boundary of Z, and Σ^ϵ the neighborhood of Σ, as defined in Lemma 8.2. The constant ϵ shall be so small that $\Sigma^\epsilon \subset D$, and Σ^ϵ contains no critical points of f. We shall modify f at most on Σ^ϵ. (see Morse and Van Schaack [1]).

To that end, we turn to Σ-normal parameters (p, s) on Σ^ϵ and set

$$f(p + s\lambda(p)) = H(p, s), \qquad (p, s) \in \Sigma \times I(\epsilon). \tag{11.1}$$

So defined, H is of class C^m, $m > 1$, with f, and has no critical points on $\Sigma \times I(\epsilon)$ (see Lemma 8.3). Moreover, the partial mapping

$$p \to H^0(p) = H(p, 0) = f(p) : \Sigma \to R \tag{11.2}$$

is ND, by the hypotheses of Theorem 10.1, and so has at most a finite set [denoted by $(q)^0$] of critical points $q \in \Sigma$.

Definition 11.1. *H^s-Critical Arcs on Σ.* A C^1-mapping

$$s \to p(s) : [-\sigma, \sigma] \to \Sigma, \qquad 0 < \sigma < \epsilon, \tag{11.3}$$

such that each point $p(s)$ is a critical point of the corresponding partial mapping H^s will be called an *H^s-critical arc* on Σ.

We state a fundamental lemma concerning the existence of such arcs.

Lemma 11.1. *If $\sigma < \epsilon$ is a sufficiently small positive constant, the following are true*:

(A$_1$) *Corresponding to each point* $q \in (q)^0$ *there is a unique* H^s*-critical arc* γ^q *on* Σ *of form*

$$s \to p^q(s) : [-\sigma, \sigma] \to \Sigma \qquad \text{with} \quad p^q(0) = q. \tag{11.4}$$

(A$_2$) *The* H^s*-critical arcs* γ^q *are disjoint for distinct points* $q \in (q)^0$.

(A$_3$) *For* $s \in [-\sigma, \sigma]$ *each critical point of* H^s *in* Σ *is a point* $p^q(s)$ *in one of the arcs* γ^q *of* (A$_1$).

(A$_4$) *At each point* $q \in (q)^0$, $H_s(q, 0) \neq 0$. *If one sets*

$$H_s(p^q(s), s) = h_q(s), \qquad -\sigma \leqslant s \leqslant \sigma, \tag{11.5}$$

then $h_q(s)$ *is negative or positive depending on whether* $q \in \Sigma$ *is in the f-entrant or f-emergent portal of* Z (Definition 10.2).

(A$_5$) *As a critical point of* H^s, $p^q(s)$ *is* ND *and has an index which does not vary with* $s \in [-\sigma, \sigma]$.

Proof of (A$_1$). A critical point q of H^0 in Σ is ND by hypothesis. Let $(\varphi^q : U, X)$ be a presentation in $\mathscr{D}\Sigma$ of a coordinate domain X of Σ, such that $q \in X$ and $\varphi^q(u^0) = q$ for some $u^0 \in U$. Set

$$H(\varphi^q(u), s) = G^q(u, s), \qquad (u, s) \in U \times I(\epsilon). \tag{11.6}$$

For s fixed in $I(\epsilon)$ a point $p = \varphi^q(u) \in X$ is a critical point of H^s if and only if $u \in U$ is a solution of the n equations

$$G^q_{u_i}(u, s) = 0, \qquad i = 1, \ldots, n. \tag{11.7}$$

Since q is a ND critical point of H^0, the n-square determinant

$$| G^q_{u_i u_j}(u, s)| \neq 0, \qquad \text{when} \quad (u, s) = (u^0, 0). \tag{11.8}$$

If $\sigma > 0$ is sufficiently small, there accordingly exists a solution

$$s \to u^q(s) : [-\sigma, \sigma] \to U, \qquad u^q(0) = u^0, \tag{11.9}$$

of (11.7). The C^{m-1}-mapping

$$s \to p^q(s) = \varphi^q(u^q(s)) : [-\sigma, \sigma] \to \Sigma \tag{11.10}$$

defines an H^s-critical arc γ^q of form (11.4).

Proof of (A_2). For distinct points $q \in (q)^0$ the corresponding arcs γ^q will be disjoint if σ is sufficiently small, since $p^q(0) = q$ and the mappings $s \to p^q(s)$ are continuous.

Proof of (A_3). To establish (A_3), one uses appropriate Σ-normal φ-coordinates (u, s) to prove the following: There exists a covering of Σ by coordinate domains X^q containing the respective points $q \in (q)^0$, and a finite set of other coordinate domains $Y^1, \ldots, Y^r$, together with a value of σ so small that the following is true: There are no Σ-normal parameters (p, s) in $Y^i \times \bar{I}(\sigma)$, $i = 1, \ldots, r$, such that p is a critical point of H^s in Y^i. When Σ-normal parameters (p, s) in $X^q \times \bar{I}(s)$ are such that p is a critical point of H^s then $p = p^q(s)$.

Proof of (A_4). Since H has no critical point in Σ, and each point $q \in (q)^0$ is a critical point of H^0, it follows from Lemma 8.4 that $H_s(q, 0) \neq 0$ for $q \in (q)^0$. Hence $h_q(0) \neq 0$, so that $h_q(s)$ has one sign on $[-\sigma, \sigma]$ if σ is sufficiently small. Since $H_s(q, 0)$ is positive or negative depending on whether grad f at q is emergent or entrant, statement (A_4) follows.

Proof of (A_5). The critical point $p^q(s)$ of H^s in Σ is ND if and only if the determinant

$$| G^q_{u_i u_j} |^s \neq 0, \tag{11.11}$$

where the *superscript* s indicates evaluation for $(u, s) = (u^q(s), s)$. Now, (11.11) holds for $s = 0$, by virtue of (11.8). Hence (11.11) holds for $s \in [-\sigma, \sigma]$ if σ is sufficiently small.

Finally, the index k of the point $p^q(s) \in \Sigma$, as a critical point of H^s, varies continuously with $s \in [-\sigma, \sigma]$ provided (11.11) holds, since k is the index of the quadratic form whose coefficients are the elements of the determinant (11.11), and this index cannot change as s varies on $[-\sigma, \sigma]$, since (11.11) holds. This follows from Theorem 3.4.

The Constant σ. We suppose $\sigma < \epsilon$ chosen so that Lemma 11.1 is satisfied and fix σ for the remainder of this section. We set $\sigma = 2\rho$.

Constants Involved in Defining the Replacement $\hat{f}$ of f. To define the required modification $\hat{f}$ of f, we need to specify two constants M and η associated with H. Of these constants, M depends in part on the Hessian of G^q (written Hess G^q) as q ranges over the set $(q)^0$.

Hess G^q is obtained by bordering the n-square determinant $| G^q_{u_i u_j} |$ by an $(n + 1)$th column $G^q_{u_1 s}, \ldots, G^q_{u_n s}, G^q_{ss}$ and an $(n + 1)$th row $G^q_{su_1}, \ldots, G^q_{su_n}, G^q_{ss}$ with the element G^q_{ss} in common. Each element in Hess G^q will be evaluated for $(u, s) = (u^q(s), s)$, and this evaluation will be indicated by a superscript s as in (11.11).

The Constant M_q. Let the element G^q_{ss} in Hess G^q be replaced by $G^q_{ss} + M$ and the resulting $(n + 1)$-determinant be denoted by $\text{Hess}^M G^q$. If $M > M_q > 0$ and M_q is sufficiently large,

$$\operatorname{sign}(\text{Hess}^M G^q)^s = \operatorname{sign}| G^q_{u_i u_j} |^s, \qquad -\sigma \leqslant s \leqslant \sigma. \tag{11.12}$$

We suppose M_q so chosen.

A Constant M. Let M be a positive constant such that with $\sigma = 2\rho$ the conditions

$$M > \max_{q \in (q)^0} M_q, \tag{11.13}$$

$$M\rho > \max_{p \in \Sigma} | H_s(p, 0)|, \tag{11.14}$$

$$M\rho > \max_{q \in (q)^0} [\max_{s \in \bar{I}(\sigma)} (| h_q(0)| + \rho | h_q'(s)|)], \tag{11.15}$$

[see (11.5)] are satisfied. Here $\bar{I}(\sigma) = [-\sigma, \sigma]$.

A Constant η. A positive constant η such that

$$0 < \eta < \min_{q \in (q)^0} [\min_{s \in \bar{I}(\sigma)} | h_q(s)|] \tag{11.16}$$

exists by virtue of Lemma 11.1 (A_4).

An Auxiliary Function ζ. Given the constants ρ, η, and M, there exists a monotone increasing C^∞-mapping $s \to \zeta(s)$ of R into R such that

$$\zeta(s) = 0, \qquad s \leqslant -2\rho, \tag{11.17}$$

$$\zeta'(s) < \eta, \qquad -2\rho \leqslant s \leqslant -\rho, \tag{11.18}$$

$$\zeta''(s) = M, \qquad -\rho \leqslant s. \tag{11.19}$$

The Existence of ζ. Let $e < \rho$ be a positive constant. A conventional use of exponentials (Munkres [1], p. 6) suffices to define a monotone increasing C^∞-mapping $t \to \mu(t)$ of R into R such that

$$\mu(t) = 0, \qquad t < -e - \rho$$

$$\mu(t) = M \qquad -\rho \leqslant t.$$

If e is so small that $eM < \eta$, a mapping ζ such that

$$\zeta'(s) = \int_{-2\rho}^{s} \mu(t)\, dt$$

satisfies (11.18) and (11.19), and if $\zeta(-2\rho) = 0$, satisfies (11.17).

Lemma 11.2 presents essential properties of ζ.

Lemma 11.2. *If for each* $q \in (q)^0$ *one sets*

$$\hat{h}_q(s) = h_q(s) + \zeta'(s), \qquad -\sigma \leqslant s \leqslant \sigma, \tag{11.20}$$

then (i) *and* (ii) *are true*:

(i) *If* $q \in \Sigma_+$, $\hat{h}_q(s)$ *vanishes for no value of* s *admitted in* (11.20).

(ii) *If* $q \in \Sigma_-$, $\hat{h}_q(s) = 0$ *for some value* $s = s_q$ *in* $(-\rho, 0)$ *and for no other value of* s *admitted in* (11.20).

Proof of (i). Under the hypotheses of (i) $h_q(s) > 0$ in (11.20) by Lemma 11.1 (A_4). Since $\zeta'(s) \geqslant 0$ for every s, (i) follows from (11.20).

Proof of (ii). Under the hypothesis of (ii) we shall show that

$$\hat{h}_q(s) < 0, \qquad -2\rho \leqslant s \leqslant -\rho, \tag{11.21}$$

$$\hat{h}_q(0) > 0, \tag{11.22}$$

$$\hat{h}_q'(s) > 0, \qquad -\rho \leqslant s \leqslant 2\rho. \tag{11.23}$$

These three relations trivially imply (ii).

Verification of (11.21). By (11.18) and (11.16)

$$\operatorname{sign}(h_q(s) + \zeta'(s)) = \operatorname{sign}(h_q(s) + \eta) = \operatorname{sign} h_q(s)$$

for s as in (11.21). For such an s and for $q \in \Sigma_-$, $h_q(s) < 0$ by Lemma 11.1 (A_4), implying (11.21).

Verification of (11.22). From (11.19) we infer that

$$\zeta'(0) = \zeta'(-\rho) + \int_{-\rho}^{0} M\,dt > M\rho > |\,h_q(0)|, \tag{11.24}$$

since $\zeta'(-\rho) > 0$ and (11.15) holds. Hence

$$0 < \zeta'(0) + h_q(0) = \hat{h}_q(0).$$

Verification of (11.23). By (11.20), (11.19), and (11.15),

$$\hat{h}_q'(s) = h_q'(s) + \zeta_q''(s) = h_q'(s) + M > 0, \qquad -\rho \leqslant s \leqslant 2\rho.$$

Thus (ii) is true and Lemma 11.2 is established.

Definition 11.2. *H-s-arcs* Γ^q. Corresponding to each H^s-critical arc $s \to p^q(s)$ on Σ we introduce the arc

$$s \to (p^q(s), s) \in \Sigma \times \bar{I}(\sigma), \tag{11.25}$$

calling this arc the *H-s-arc* Γ^q.

The H^s-critical arcs γ^q in Σ may be regarded as the "projections" into Σ of the corresponding *H-s-arcs* Γ^q.

The Replacement $\hat{f}$ of f. We shall modify f on Σ^ϵ by replacing the representation $(p, s) \to H(p, s)$ of $f \mid \Sigma^\epsilon$, as defined in (11.1), by the mapping

$$(p, s) \to \hat{H}(p, s) = H(p, s) + \zeta(s), \qquad (p, s) \in \Sigma \times I(\epsilon). \tag{11.26}$$

So defined, $\hat{H}$ is of class C^m with H. Since $\zeta(s) = 0$ for $s \leqslant -\sigma$,

$$\hat{H}(p, s) = H(p, s), \qquad -\epsilon < s \leqslant -\sigma, p \in \Sigma, \tag{11.27}$$

a relation to which we shall return.

$\hat{H}$-s-arcs. Replacing H by $\hat{H}$, partial mappings $\hat{H}^s$ are defined as in (8.13), $\hat{H}^s$-critical arcs as in Definition 11.1, and $\hat{H}$-s-arcs as in Definition 11.2.

The critical points of $\hat{H}^0$ in Σ are the critical points $q \in (q)^0$ of H^0. For $-\sigma \leqslant s \leqslant \sigma$, $\hat{H}^s$-critical arcs and $\hat{H}$-s-arcs exist and are identical, respectively, with H^s-critical arcs γ^q and H-s-arcs Γ^q.

Let an H-s-arc Γ^q for which q is in Σ_- be denoted by Γ_-^q.

Lemma 11.3. *$\hat{H}$ has the following properties*:

(i) $\hat{H}_s(p, 0) > 0, \quad p \in \Sigma$.

(ii) *The set of critical points (p, s) of $\hat{H}$ for which $-\sigma \leqslant s \leqslant 0$ includes a point $(p^q(s), s)$ on each arc Γ_-^q, and no other critical points. At the critical point of $\hat{H}$ on Γ_-^q, $-\rho < s < 0$.*

(iii) *The critical point $(p^q(s), s)$ of $\hat{H}$ on Γ_-^q is* ND *and has as index k, the index of q as a critical point of H^0.*

Proof of (i). By (11.26), (11.14), and (11.24)

$$\hat{H}_s(p, 0) = H_s(p, 0) + \zeta'(0) > -M\rho + M\rho = 0, \qquad p \in \Sigma.$$

Proof of (ii). Recall that the $\hat{H}^s$-critical arcs are the H^s-critical arcs $s \to p^q(s)$, one for each $q \in (q)^0$. By (11.26), (11.5), and (11.20)

$$\hat{H}_s(p^q(s), s) = h_q(s) + \zeta'(s) = \hat{h}_q(s), \qquad -2\rho \leqslant s \leqslant 2\rho. \quad (11.28)$$

According to Lemma 8.4, (p, s) is a critical point of $\hat{H}$ if and only if $p \in \Sigma$ is a critical point of $\hat{H}^s$ and $\hat{H}_s(p, s) = 0$. When s is restricted to the interval $[-\sigma, \sigma]$, (p, s) accordingly is a critical point of $\hat{H}$ if and only if $p = p^q(s)$ (Lemma 11.1) and $\hat{h}_q(s) = 0$ [(11.28)]. Statement (ii) now follows from Lemma 11.2.

Proof of (iii). To establish (iii), we must use suitable Σ-normal φ-coordinates.

Notation. Following the proof of (A_1) of Lemma 11.1, let $(\varphi^q : U, X)$ be a presentation of a coordinate domain X of $\mathcal{D}\Sigma$ such that $q \in X$. For $(u, s) \in U \times I(\epsilon)$ set

$$H(\varphi^q(u), s) = G^q(u, s), \qquad \hat{H}(\varphi^q(u), s) = \hat{G}^q(u, s),$$

[as in (11.6)] so that [using (11.26)]

$$\hat{G}^q(u, s) = G^q(u, s) + \zeta(s). \quad (11.29)$$

Since (11.10) holds, $u^q(s)$ is the "Σ-normal φ^q-coordinate" of the point $p^q(s) \in \Sigma$. By virtue of (11.11)

$$| G^q_{u_i u_j} |^s \neq 0, \qquad -\sigma \leqslant s \leqslant \sigma, \tag{11.30}$$

where the superscript s has the same meaning as in (11.11) and (11.12).

The Index k. It follows from (11.29) and (11.19) that (with $\sigma = 2\rho$)

$$\hat{G}^q_{ss}(u, s) = G^q_{ss}(u, s) + \zeta''(s) = G^q_{ss}(u, s) + M, \qquad -\rho \leqslant s \leqslant 0,$$

so that, with the notation of (11.12),

$$(\text{Hess } \hat{G}^q)^s = (\text{Hess}^M G^q)^s, \qquad -\rho \leqslant s \leqslant 0.$$

Since $M > M_q$ by (11.13), we infer from (11.12) that

$$\text{sign}(\text{Hess } \hat{G}^q)^s = \text{sign}| G^q_{u_i u_j} |^s, \qquad -\rho \leqslant s \leqslant 0. \tag{11.31}$$

According to (ii), there is a critical point $(p^q(s), s)$ of $\hat{H}$ in $\Gamma\_^q$ with $-\rho < s < 0$. By virtue of (11.31) and (11.30) $(\text{Hess } \hat{G}^q)^s \neq 0$, so that this critical point is ND.

By virtue of (11.31) and the Kronecker Corollary 3.2 the index k of $(p^q(s), s)$, as a critical point of $\hat{H}$, is equal to the index of $p^q(s) \in \Sigma$, as a critical point of H^s. By (A_5) of Lemma 11.1 k is the index of $p^q(0) = q$ as a critical point of H^0.

This establishes (iii) and completes the proof of Lemma 11.3.

Final Definition of $\hat{f}$ and Proof of Theorem 10.1. The domain of definition of $\hat{f}$ shall be the subset $D_{\hat{f}} = Z \cup \Sigma^\epsilon$ of E_{n+1}. We shall define $\hat{f}$ separately on the overlapping open subsets

$$Z - \text{Cl } \Sigma^\sigma, \quad \Sigma^\epsilon \tag{11.32}$$

of $D_{\hat{f}}$, noting that the union of these two sets in $D_{\hat{f}}$. To this end we set

$$\hat{f}(x) = f(x), \qquad x \in Z - \text{Cl } \Sigma^\sigma \tag{11.33}$$

$$\hat{f}(p + s\lambda(p)) = \hat{H}(p, s), \qquad (p, s) \in \Sigma \times I(\epsilon). \tag{11.34}$$

The definitions (11.33) and (11.34) are consistent on the intersection of the sets (11.32), as one infers from (11.27) and (11.1). So defined,

$\hat{f}$ is of class C^m on each of the sets (11.32). Since these sets are open, $\hat{f}$ is of class C^m on $D_{\hat{f}}$.

That $\hat{f}$ satisfies Theorem 10.1 follows if we show that $\hat{f}$ has the following additional properties:

Properties (α), (β), and (γ) *of* $\hat{f}$.

(α) *Restricted to* $Z - \mathrm{Cl}\,\Sigma^\sigma$, f *and* $\hat{f}$ *are identical and have in common all critical points of* $f \mid Z$.

(β) Grad $\hat{f}$ *is emergent on* Σ, *and* $\hat{f} \mid Z$ *is* ND.

(γ) *The critical points of* $\hat{f} \mid \mathring{Z}$ *in excess of those of* $f \mid \mathring{Z}$ *correspond biuniquely to critical points of* $f \mid \Sigma_-$, *with preservation of index.*

Verification of (α). By choice of ϵ, Σ^ϵ, and hence Σ^σ, excludes all critical points of $f \mid Z$. Since (11.33) holds, (α) follows.

Verification of (β). Grad $\hat{f}$ is emergent on Σ because of (i) of Lemma 11.3. Moreover, $f \mid Z$ is ND by hypothesis, and $\hat{H}$ ND at critical points (p, s) for which $s \leqslant 0$, in accord with Lemma 11.3. Hence $\hat{f} \mid Z$, as defined by (11.33) and (11.34), is ND.

Verification of (γ). The critical points of $\hat{f} \mid \mathring{Z}$ in excess of those of $f \mid \mathring{Z}$ are those represented by critical points of $\hat{H}$ at which $s < 0$. According to Lemma 11.3 the critical points (p, s) of $\hat{H}$ at which $s < 0$ correspond biuniquely to the critical points of $f \mid \Sigma_-$ with preservation of indices.

The properties of $\hat{f}$ are such that $\hat{f}$ is "admissible" in the sense of Definition 9.2 relative to the regular subset Z of E_{n+1} because $\hat{f}$ is of class C^m on $D_{\hat{f}}$, ND on Z, and grad $\hat{f}$ is emergent on Σ.

The function $\hat{f}$ *thus satisfies Theorem* 10.1 *as a replacement of* f.

In § 12 we shall sketch two extensions of Theorem 10.1 fundamental in embedding and in cobordism. To this end, we give another definition of $H(p, s)$, which was previously defined in (11.1). This new definition is equivalent to the definition (11.1) in the context of § 11, but in the context of Theorems 12.3 and 12.4 it is much more general.

Definition 11.2. *The value of* $H(p, s)$ in (11.1) at the point $(p, s) \in \Sigma \times I(\epsilon)$ is the value of f at the point $x \in \Sigma^\epsilon$ represented as in § 8 by Σ-normal parameters (p, s).

§ 12

f-LEVEL BOUNDARIES

The proof of Theorem 9.1 is simplest if the boundary Σ of the given regular C^∞-domain Z is a level manifold of f, that is, a manifold on which f is constant. Theorem 12.2 shows that this is the only case that need be considered in proving Theorem 9.1 (see final proof in § 30).

We shall conclude this section with a proof of the following theorem. The proof makes essential use of Theorems 6.4 and 10.1 and follows a proof of Theorem 12.2.

Theorem 12.1. *Let Σ be an arbitrary, regular, compact, differentiable n-manifold in E_{n+1} of class C^∞ bounding a compact subset Z of E_{n+1}. There then exists a* ND *function g of class C^∞ on a neighborhood of the compact domain Z bounded by Σ such that Σ is a level manifold of g on which g is ordinary and assumes a value exceeding each value of g on $\mathring{Z}$.*

We state an ultimate extension of Theorem 12.1.

Theorem 12.1′. *This reads as does Theorem* 12.1 *on replacing C^∞ by C^m, $m > 1$.*

To prove Theorem 12.1′, use Theorem 12.1 and Theorem A of Appendix II.

In formulating Theorem 12.2 the following definition is needed.

Definition 12.1. *Functions f and g critically equivalent on Z.* Two real-valued functions f and g of the same class, admissible relative to some regular domain Z in the sense of Definition 9.2, will be said to be *critically equivalent* on Z if the critical points of f and g on Z correspond biuniquely with preservation of indices.

If f and g are critically equivalent on Z, the type numbers of $f \mid Z$ equal those of $g \mid Z$. If in addition grad f and grad g are emergent on Σ, to verify Theorem 9.1 for f it is accordingly sufficient to verify Theorem 9.1 for g. If g satisfies the conditions of the following theorem, a verification of Theorem 9.1 will be simpler for g than for f.

Theorem 12.2. *Let f be of class C^m, $m > 1$, and admissible* (Definition 9.2) *relative to a regular C^∞-domain Z in E_{n+1}*, grad f *emergent on Σ, the boundary of Z. There then exists a real-valued function g of class C^m on an open neighborhood D_g of Z "critically equivalent" to f on Z and such that the boundary Σ of Z is a level manifold of g on which g is ordinary and assumes a value exceeding the value of g at each point $x \in \mathring{Z}$.*

We shall define a function g which satisfies Theorem 12.2 by altering f in a neighborhood Σ^ϵ of $\mid \Sigma \mid$, where ϵ is so small that the conclusions of Lemma 8.2 are valid, that Σ^ϵ is included in the domain D_f of f and excludes the critical points of $f \mid Z$.

In the notation of Lemma 8.2, set

$$f(p + s\lambda(p)) = H(p, s), \qquad (p, s) \in \Sigma \times I(\epsilon), \tag{12.1}$$

thereby defining a function H of class C^m. Since grad f is emergent on Σ by hypothesis, $H_s(p, 0) > 0$ for $p \in \Sigma$.

The Constants ω and M. Let $\omega < \epsilon$ be so small a positive constant that

$$H_s(p, s) > 0, \qquad (p, s) \in \Sigma \times \bar{I}(\omega), \tag{12.2}$$

and let M be a positive constant such that

$$M > \max_{x \in Z} f(x). \tag{12.3}$$

A Modification K of H. We shall make use of a C^∞-mapping $s \to \xi(s)$ of R into R such that $\xi(1) = 1$, $\xi(s) = 0$ for $s \leq 0$, and $\xi'(s) > 0$ for $s > 0$. We then introduce a modification K of H with values

$$K(p, s) = H(p, s) + (M - H(p, 0))\, \xi\left(\frac{s + \omega}{\omega}\right), \qquad (p, s) \in \Sigma \times I(\epsilon). \tag{12.4}$$

It is clear that K is of class C^m on its domain of definition, that

$$K(p, s) = H(p, s), \qquad -\epsilon < s < -\omega, \quad p \in \Sigma \tag{12.5}$$

$$K(p, 0) = M, \qquad p \in \Sigma, \tag{12.6}$$

and, making use of (12.2), that

$$K_s(p, s) > 0, \qquad (p, s) \in \Sigma \times \bar{I}(\omega). \tag{12.7}$$

Definition of g. The domain of g shall be the open subset $D_g = Z \cup \Sigma^\epsilon$ of E_{n+1}. We shall define g separately but consistently on the overlapping open subsets,

$$Z - \mathrm{Cl}\, \Sigma^\omega, \quad \Sigma^\epsilon \tag{12.8}$$

of D_g, setting

$$g(x) = f(x), \qquad x \in Z - \mathrm{Cl}\, \Sigma^\omega, \tag{12.9}$$

$$g(p + s\lambda(p)) = K(p, s), \qquad (p, s) \in \Sigma \times I(\epsilon). \tag{12.10}$$

That (12.9) and (12.10) give the same value to $g(x)$ when x is in the intersection of the two sets (12.8) follows from (12.5) and (12.1). Since f and K are of class C^m, we infer that g is of class C^m.

We shall verify the following additional properties of g:

(i) $g(p) = M > g(x)$, $(x \in \mathring{Z}, p \in \Sigma)$.

(ii) On $Z \cap \Sigma^\epsilon$, g is ordinary.

(iii) On Z, f and g are critically equivalent.

Verification of (i). For $p \in \Sigma$, $g(p) = M$ by virtue of (12.6) and (12.10). M is larger than any value of g on $\mathring{Z} \cap \Sigma^\omega$ by virtue of (12.6), (12.7), and (12.10), and larger than any value of g on $\mathring{Z} - \Sigma^\omega$ by virtue of (12.3) and (12.9).

Verification of (ii). g is ordinary on $Z \cap \Sigma^\omega$ because of (12.7) and (12.10) and at the remaining points of $Z \cap \Sigma^\epsilon$ because (12.9) holds and f is ordinary on Σ^ϵ.

Verification of (iii). On $Z \cap \Sigma^\epsilon$ there are no critical points of g by (ii), and of f by virtue of the choice of ϵ. Since (12.9) holds, (iii) follows.

The function g so defined on D_g satisfies Theorem 12.2.

Proof of Theorem 12.1. Let $x \to h(x) : E_{n+1} \to R$ be any ND function of class C^∞ without critical points on Σ. The restriction $h \mid \Sigma$ may fail to be ND. However, one can infer from Theorem 6.4 that there exists a function f of class C^∞, "critically equivalent" to h on Z, and so admissible relative to Z, and in addition such that $f \mid \Sigma$ is ND. According to Theorem 10.1, f can be altered near Σ so as to yield a function $\hat{f}$ of class C^∞, admissible relative to Z, but with grad $\hat{f}$ emergent on Σ. Finally, given $\hat{f}$, Theorem 12.2 implies that there is a ND function g of class C^∞, on a neighborhood of Z, satisfying Theorem 12.1.

Non-Euclidean Extensions. Theorem 12.1 has an extension for abstract, compact, differentiable C^∞-manifolds M_{n+1} (Def 13.1) which can be readily proved with the aid of extensions of theorems in Part I. This extension can be used effectively in the theory of embedding of differentiable manifolds in Euclidean spaces. We shall state it, and associate with it the theorems of Part I, upon which its proof depends. The reader will find Part II essential.

Introduction to Theorem 12.3. Let there be given a compact subset Z^* of M_{n+1}. Here Z^* is to generalize the regular compact domain Z of E_{n+1} introduced in Definition 9.1; one supposes that Z^* is bounded on M_{n+1} by a compact manifold Σ^*, C^∞-embedded in M_{n+1} (Definition 16.2). Note that $\mathring{Z}^*$ is an *open* subset of M_{n+1}.

Theorem 12.3 *extending Theorem* 12.1. *Let* M_{n+1}, Z^* *and its boundary* Σ^* *be given as in the preceding paragraph. We affirm that there exists a real-valued function g which is of class* C^∞ *and* ND (*see* § 13) *on some open neighborhood* N *of* Z^* *relative to* M_{n+1} *and such that* Σ^* *is a level manifold on* N *of g on which g is ordinary and assumes a value exceeding each value of g on* $\mathring{Z}^*$.

We shall indicate a mode of proof of Theorem 12.3. This proof is modeled on the proof of Theorem 12.1.

The concepts and theorems of Part I and their extensions essential to the proof of Theorem 12.3 are as follows.

Σ*-normal* φ*-coordinates of* § 8. One must define Σ^**-normal* φ*-coordinates* on M_{n+1} near Σ^*, the boundary of Z^*. A simple generalization of Definition 8.2 suffices. In this extension short geodesic arcs on M_{n+1} normal to Σ^* replace the straight arcs in E_{n+1} normal to Σ as given in § 8.

Theorem 6.4. Theorem 6.4 is used in the concluding paragraph of the proof of Theorem 12.1. It is easily generalized for a C^∞-function f^* defined on an open neighborhood N of Z^* relative to M_{n+1} under the assumption that f^* is ND on N and ordinary at each point of Σ^*. The proof of Theorem 6.4 made use of Theorem 6.3, while the proof of an extension of Theorem 6.4 makes similar use of Theorem 14.1; an explicit formulation of an extension of Theorem 6.4 is given as Theorem 9.1 of Morse [15].

Theorem 10.1. Theorem 10.1 is used in the concluding paragraph of the proof of Theorem 12.1. Theorem 10.1 has an obvious extension to the case of a ND f^* of class C^∞ on an open neighborhood of Z^* relative to M_{n+1}, assuming that there are no critical points of f^* on Σ^*. A proof of Theorem 10.1 is given in § 11; the proof of the extension, Theorem 10.1′, of Theorem 10.1 is similar. One begins by replacing the Σ-normal φ-coordinates and parametric points $p \in E_{n+1}$ near Σ, as introduced in § 8, by geodesically defined Σ^*-normal φ-coordinates and points $p \in M_{n+1}$ near Σ^*. Definition 11.2, rather than (11.1), is used to define $H(p, s)$ when Σ^*-normal parameters (p, s) replace the Σ-normal parameters (p, s) of § 8. A proof of Theorem 10.1′, then follows the proof of Theorem 10.1 in § 11 in exact fashion.

With these generalizations of Theorems 6.4 and 10.1 at one's disposal, a proof of Theorem 12.3 can be trivially modeled on the proof of Theorem 12.1.

"Cobordism." Suppose that there is given a compact connected subset Z^* of a C^∞-manifold M_n, bounded on M_n by two disjoint $(n-1)$-manifolds Σ' and Σ'', "C^∞-embedded" in M_n (see § 16). Then Σ' and Σ'' are termed "*cobordant*."

Let compact $(n-1)$-manifolds M'_{n-1} and M''_{n-1} of class C^∞ be given, with no mention of M_n. Let (M'_{n-1}) and (M''_{n-1}) denote the classes of compact $(n-1)$-C^∞-manifolds respectively diffeomorphic to M'_{n-1} and M''_{n-1}. A basic problem has been studied by Thom [1]:

Under what conditions is there a pair of $(n - 1)$-manifolds $\Sigma' \in (M'_{n-1})$, $\Sigma'' \in (M''_{n-1})$ which are cobordant on some C^∞-manifold M_n?

Theorem 12.4 gives a first necessary condition that Σ' and Σ'' be cobordant on some C^∞-manifold M_n (see Wallace [1] and Milnor [3]).

Theorem 12.4. *Let Σ' and Σ'' be compact disjoint $(n - 1)$-manifolds of class C^∞,C^∞-embedded in a C^∞-manifold M_n and bounding a compact connected subset Z^* of M_n. There then exists a real-valued function g of class C^∞ on some open neighborhood N of Z^* relative to M_n and such that Σ' and Σ'' are level manifolds on N of g on which g is ordinary and assumes values* 1 *and* 0, *respectively, values greater than and less than each value of g on $\mathring{Z}^*$.*

Our proof of Theorem 12.4 is a simple extension of the proof of Theorem 12.3, making use of the extensions described above of Theorem 6.4 and Theorem 10.1. One focuses first on Σ' and then on Σ''.

For a more complete study of the "cobordism" theory see the papers of Thom and particularly the recently published book by Milnor [3].

In Part II we return to details.

PART II

ABSTRACT DIFFERENTIABLE MANIFOLDS

§ 13

THE MANIFOLDS DEFINED

Abstract differentiable manifolds, as we shall define them, include "regular" manifolds in a Euclidean space E_r as special cases. Among the characteristic properties of regular manifolds in E_r are the following: A regular manifold M_n in E_r is a topological n-manifold Γ_n given as a "subspace" of E_r provided with a *differentiable structure* defined by the set of all regular presentations of open subspaces of Γ_n (Definition 5.3). These presentations satisfy a *compatibility* condition (Lemma 5.1). The carrier Γ_n of M_n admits a *countable* covering by regular presentations.

In defining an abstract differentiable manifold M_n a topological n-manifold Γ_n is again given, but in general not as a "subspace" of a Euclidean space E_r. "Presentations" of open subspaces of Γ_n are defined as before, but no definition of "regular" presentation is required. M_n is again a topological manifold Γ_n provided with a differentiable structure defined by a set of presentations of open subspaces of Γ_n, but these presentations are now required (formerly proved) to be *C^m-compatible* for some m (Definition 5.4) and to include a countable subset *covering* Γ_n (Definition 5.2). The precise definition of M_n follows.

Definition 13.1. *Abstract Differentiable C^m-Manifolds M_n. Let Γ_n be a topological n-manifold coverable by a countable set*

$$(F_k : V_k , X_k)_{k \in \omega} \tag{13.1}$$

of C^m-compatible presentations (Definition 5.4). *Then the set $\mathscr{D}$ of all presentations of open subspaces of Γ_n which are C^m-compatible with the* *presentations of the covering* (13.1) *of* Γ_n *is a set of pairwise* C^m*-compatible presentations of open subspaces of* Γ_n.

Granted $\mathscr{D}$ *exists,* Γ_n, *associated with* $\mathscr{D}$, *is called a* C^m*-manifold* M_n *with carrier* $| M_n | = \Gamma_n$ *and set of presentations* $\mathscr{D}M_n = \mathscr{D}$.

This definition requires the following theorem [a consequence of (A_4) below]:

Theorem 13.1. *Any two presentations in the set* $\mathscr{D}$ *introduced in Definition* 13.1 *are* C^m*-compatible presentations of open subspaces of* Γ_n.

Two C^m-manifolds M_n and N_n are regarded as *identical* if $| M_n | = | N_n |$ and $\mathscr{D}M_n = \mathscr{D}N_n$. A C^m-manifold M_n and a C^μ-manifold N_n such that $\mu \neq m$ and $| M_n | = | N_n |$ can never have identical sets $\mathscr{D}M_n$ and $\mathscr{D}N_n$.

The proof of Theorem 13.1 will follow readily once we have verified Propositions (A_1)–(A_4) on C^m-compatibility as stated below.

C^m-Compatibility. Let Γ_n be a topological manifold. Given a presentation $(F : U, X)$ of an open subspace of Γ_n, a restriction $(\hat{F} : \hat{U}, \hat{X})$ of F which is a presentation of an *open* subspace $\hat{X}$ of Γ_n will be called an *open restriction* of F.

Proposition (A_1). Let there be given two presentations

$$(F_1 : U_1, X_1), \qquad (F_2 : U_2, X_2) \tag{13.2}$$

of open subspaces X_1 and X_2 of Γ_n. If $X_1 \cap X_2 = X \neq \varnothing$, F_1 and F_2 will be C^m-compatible in the sense of Definition 5.4 if and only if the unique open restrictions, respectively of F_1 and F_2 with range X are C^m-compatible.

Proposition (A_2). If $X_1 = X_2 = X \neq \varnothing$ in (13.2), the presentations F_1 and F_2 are C^m-compatible if and only if $F_2^{-1} \circ F_1 = \lambda$ and $F_1^{-1} \circ F_2 = \lambda^{-1}$, where λ is a C^m-diff of U_1 onto U_2.

Proposition (A_3). If the presentations F_1 and F_2 of (13.2) are C^m-compatible, arbitrary open restrictions $(\hat{F}_1 : \hat{U}_1, \hat{X}_1)$ and $(\hat{F}_2 : \hat{U}_2, \hat{X}_2)$ of F_1 and F_2, respectively are C^m-compatible.

Proof of (A_3). If $\hat{X}_1 \cap \hat{X}_2 = \varnothing$, ($A_3$) is trivially true. To verify (A_3) in case $\hat{X}_1 \cap \hat{X}_2 \neq \varnothing$, it will be sufficient to verify (A_3) in case $X_1 = X_2 = X \neq \varnothing$ and $\hat{X}_1 = \hat{X}_2 = \hat{X}$ is a nonempty open subspace of X. In this special case it is a hypothesis that $F_2^{-1} \circ F_1$ is a C^m-diff λ of U_1 onto U_2. It follows that λ maps $\hat{U}_1$ onto $\hat{U}_2$ and

$$\lambda \mid \hat{U}_1 = (F_2 \mid \hat{U}_2)^{-1} \circ (F_1 \mid \hat{U}_1) = \hat{F}_2^{-1} \circ \hat{F}_1 . \tag{13.3}$$

Hence (A_3) is true.

Proposition (A_4) is put in italics because of its special importance:

PROPOSITION (A_4). *Let K be a set of presentations of open subspaces of Γ_n that cover Γ_n. If the presentations F_1 and F_2, given in* (13.2), *are C^m-compatible with each presentation in K, then F_1 and F_2 are C^m-compatible.*

Proof of (A_4). If $X_1 \cap X_2 = \varnothing$, then ($A_4$) is trivially true. It follows from (A_1)–(A_3) that (A_4) will be true in any other case if true when $X_1 = X_2$. Assuming then that $X_1 = X_2 = X \neq \varnothing$, we wish to show that $F_2^{-1} \circ F_1$ is a C^m-diff of U_1 onto U_2.

To that end, let u_0 be an arbitrary point in U_1, and set $p_0 = F_1(u_0)$. There exists an open restriction $(G : V, Y)$ of some presentation in K such that $p_0 \in Y$ and $Y \subset X$. By hypothesis G and F_i, $i = 1, 2$, are C^m-compatible. Given $Y \subset X$, there exist open restrictions $(\hat{F}_1 : \hat{U}_1, Y)$ and $(\hat{F}_2 : \hat{U}_2, Y)$, respectively, of F_1 and F_2.

Then, by (A_3) G is C^m-compatible with $\hat{F}_i$, $i = 1, 2$. Hence the homeomorphisms

$$\left.\begin{aligned} G^{-1} \circ \hat{F}_1 &= \lambda_1 : \hat{U}_1 \to V \\ G^{-1} \circ \hat{F}_2 &= \lambda_2 : \hat{U}_2 \to V \end{aligned}\right\} \quad (\text{onto } V)$$

are C^m-diffs. It follows that

$$\hat{F}_2^{-1} \circ \hat{F}_1 = \lambda_2^{-1} \circ \lambda_1 : \hat{U}_1 \to \hat{U}_2$$

is a C^m-diff of $\hat{U}_1$ onto $\hat{U}_2$. Moreover,

$$(F_2^{-1} \circ F_1) \mid \hat{U}_1 = \hat{F}_2^{-1} \circ \hat{F}_1 . \tag{13.4}$$

Now, $\hat{U}_1$ is an open neighborhood of the point u_0 prescribed in U_1, and the restriction (13.4) is a C^m-diff. Since $F_2^{-1} \circ F_1$ is known *a priori* to be homeomorphism, we conclude that $F_2^{-1} \circ F_1$ is a C^m-diff of U_1 onto U_2.

This establishes (A_4).

Theorem 13.1 *is a consequence of* (A_4).

We state a theorem.

Theorem 13.2. *A regular C^m-manifold M_n in E_r, $0 < n \leqslant r$, is a C^m-manifold in the sense of Definition* 13.1.

Theorem 13.2 is a simple consequence of Theorems 5.1 and 5.2.

Terminology. Let M_n be a C^m-manifold. If $(F : U, X) \in \mathscr{D}M_n$, the range X of F is called a *coordinate domain* of the manifold M_n, and the coordinates $(u^1, \ldots, u^n)$ of a point $u \in U$ are called *F-coordinates* of the point $F(u) \in X$ or, more loosely, *local coordinates on* X. The subset U of E_n is called the *Euclidean domain* of F.

It is to be noted that the indices of the coordinates of u are taken here as superscripts, while in Part I they were taken as subscripts. This is in conformity with the conventions of Riemannian geometry which we shall presently follow.

Real-Valued Functions on M_n. As in § 5, we are concerned with real-valued functions $p \to f(p) : |M_n| \to R$. If $(F : V, X) \in \mathscr{D}M_n$, the composite function $v \to (f \bar{\circ} F)(v) : V \to R$ [cf. (5.11)] is called a *representation* of $f \mid X$ in F-coordinates $v^1, \ldots, v^n$, as in § 5.

Definition 5.5 *Recalled.* Using representations $f \bar{\circ} F$, $f \bar{\circ} G$, etc. of f, we take over Definition 5.5 verbatim, noting that there is no reference in Definition 5.5 to the "regularity" of the presentations. Definition 5.5 characterizes the following: (i) *The differentiability of f on M_n.* (ii) *Critical points of f on M_n.* (iii) *Nondegeneracy of f, and indices of critical points.* Since an open subset Y of $|M_n|$ may be included in the coordinate domain X of infinitely many presentations $(F{:}V, X) \in \mathscr{D}M_n$, there may be infinitely many representations $f \bar{\circ} F$ which serve to define the characteristics (i)–(iii) of $f \mid Y$. That these characteristics of $f \mid Y$ are independent of the special representations $f \bar{\circ} F$ which serve to define them follows, as in § 5, from Lemmas 5.2 and 4.2.

Notation. The class of all real-valued C^m-functions on M_n will be denoted hereafter by $C^m(M_n)$.

Products of Differentiable Manifolds. Let M_n and N_r be two C^m-manifolds. The Cartesian product $| M_n | \times | N_r |$ of the carriers of these manifolds with the usual product topology (Bourbaki [3], p. 47) is a topological $(n + r)$-manifold Γ_{n+r}. We shall assign Γ_{n+r} a "*product*" differentiable structure, termed *induced* by M_n and N_r. With this product differentiable structure Γ_{n+r} becomes the carrier of a C^m-manifold M_{n+r} denoted by $M_n \times N_r$.

Product Presentations. Let presentations

$$(F : U, X) \in \mathscr{D}M_n, \qquad (G : V, Y) \in \mathscr{D}N_r \tag{13.5}$$

be given. The homeomorphism

$$(u, v) \to (F(u), G(v)) : U \times V \to | M_n | \times | N_r | \tag{13.6}$$

into Γ_{n+r} is onto the open subset $X \times Y$ of Γ_{n+r}. A presentation

$$((F, G) : U \times V, X \times Y) = (F, G) \tag{13.7}$$

of an open subspace $X \times Y$ of $| M_n | \times | N_r |$, termed a *product* of presentations F and G, is thereby defined. As F and G range over the presentations of $\mathscr{D}M_n$ and $\mathscr{D}N_r$, respectively, the resultant presentations (13.7) of open subspaces of Γ_{n+r} are C^m-compatible.

To verify this C^m-compatibility, it is sufficient, in accord with Proposition (A_2), to verify the C^m-compatibility of a presentation (F, G) of form (13.7) with an arbitrary presentation (F', G') of similar type induced by presentations $(F' : U', X) \in \mathscr{D}M_n$ and $(G' : V', Y) \in \mathscr{D}N_n$, where X and Y are identical with the subspaces X and Y presented in (13.7). The transition homeomorphism of the presentations (F', G') and (F, G) is a mapping [see (5.12)] (onto)

$$(u, v) \to ((F^{-1} \circ F')(u), \quad (G^{-1} \circ G')(v)) : U' \times V' \to U \times V. \tag{13.8}$$

By virtue of the C^m-compatibility of G' and G and that of F' and F, the homeomorphism (13.8) is a C^m-diff of $U' \times V'$ onto $U \times V$. Thus presentations in $\mathscr{D}\Gamma_{n+r}$ of form (13.7) are C^m-compatible.

Definition 13.2. *Product Differentiable Structures.* Since these product presentations cover Γ_{n+r} and are C^m-compatible by Definition 13.1, Γ_{n+r} admits a differentiable structure defined by the set of all presentations of open subspaces of Γ_{n+r} C^m-compatible with the product presentations of open subspaces of Γ_{n+r}. Hence Γ_{n+r} taken with this structure is a C^m-manifold M_{n+r}. We denote this C^m-manifold by $M_n \times N_r$ and say that it has a "product" differentiable structure.

Countable Coverings. The hypothesis that the set (13.1) of presentations covering the topological manifold Γ_n is countable implies that Γ_n is coverable by a countable ensemble of open topological n-balls. It follows that in any ensemble K of open subsets Γ_n there exists a countable subset K_0 of sets of K such that $\bigcup K = \bigcup K_0$. If Γ_s, $0 < s \leqslant n$, is any topological manifold which is a *subspace* of Γ_n, then Γ_s is coverable by a countable set of open topological s-balls, and hence in any ensemble H of open subsets of Γ_s, there exists a countable subset H_0 of sets of H such that $\bigcup H = \bigcup H_0$.

Submanifolds N_n of M_n. Given a C^m-manifold M_n, by a C^m-submanifold of M_n we mean a manifold N_n whose carrier $| N_n |$ is an open nonempty subset of $| N_n |$ and for which $\mathscr{D}N_n$ is the set of restrictions of the presentations in $\mathscr{D}M_n$ whose coordinate domains are open subsets of $| N_n |$. That a C^m-manifold N_n is thereby defined is readily verified.

Carrier Problems. The "*simple carrier problem*," as we understand it, is to determine whether or not a prescribed topological manifold Γ_n is the carrier of a C^m-manifold M_n for a prescribed $m > 0$. In § 16 we shall be concerned with a variant of this "simple carrier problem" in which Γ_n is given as a subspace of a differentiable manifold Q_r for which $0 < n < r$ and one requires not only that $| M_n | = \Gamma_n$, but that M_n be "C^m-embedded" in Q_r in a sense there defined.

In an approach to carrier problems the following notation is useful:

Definition 13.3. $\mathscr{D}^0\Gamma_n$. If Γ_n is a topological n-manifold, the set of all presentations $(H : U, Z)$ of open subsets Z of Γ_n will be denoted by $\mathscr{D}^0\Gamma_n$.

A presentation in $\mathscr{D}^0\Gamma_n$ is subject to no differentiability condition, as distinguished from a presentation in $\mathscr{D}M_n$, where M_n is a C^m-manifold.

Existing differential topology can not solve the simple carrier problem in any generality. However, there are cases in which the simple carrier problem can be given a useful solution. One such is indicated in the following lemma.

Lemma 13.1. *If Γ_n is a topological n-manifold, if $(H : U, Z)$ is given in $\mathscr{D}^0\Gamma_n$ and if a positive integer m is prescribed, the subset Z of Γ_n is the carrier of a C^m-manifold $\mathbf{Z}_n$ for which $\mathscr{D}\mathbf{Z}_n$ consists of the presentation $(H : U, Z)$ and of all presentations of open subspaces of Z which are C^m-compatible with $(H : U, Z)$.*

This lemma is a direct consequence of Definition 13.1 on taking Γ_n of Definition 13.1 as Z and the set of presentations (13.1) as the presentation $(H : U, Z)$ and a countable set of its restrictions.

Definition 13.4. We shall term the C^m-manifold $\mathbf{Z}_n$, with carrier Z and presentations characterized in Lemma 13.1, *the prime C^m-manifold $\mathbf{Z}_n$ differentiability structured by $(H : U, Z)$ and m.*

Definition 13.5. *Euclidean Differentiable Structures.* Let X be an open nonempty subset of E_n. There exists a presentation of X of the form $(I : U, X)$, in which $U = X$ and I is the identity mapping of U onto X. We regard X as the carrier of a manifold $\mathbf{X}$ with a differentiable C^∞-structure determined by the presentation $(I : U, X)$ in accord with Definition 13.1. We say then that $\mathbf{X}$ has a *Euclidean* differentiable structure.

Exercise 13.1. Let Γ_n be a topological n-manifold and $\mathscr{D}^0\Gamma_n$ the set of presentations of open subspaces of Γ_n. Establish the following principle of *conditioned transitivity* of C^m-compatibility:

(A) *If F_1, F_2, and F_3 are in $\mathscr{D}^0\Gamma_n$, if F_1 and F_2 are C^m-compatible, and if F_2 and F_3 are C^m-compatible, then if range $F_1 \subset$ range F_2 or if range $F_2 \supset$ range F_3, F_1 and F_3 are C^m-compatible.*

In the absence of any condition on the ranges of F_1, F_2, and F_3, (A) is *false*.

§ 14

FAMILIES OF DIFFERENTIABLE FUNCTIONS ON M_n

The manifold M_n shall be an abstract differentiable manifold of class C^r, $r > 1$, as defined in § 13. We shall extend Theorems 6.3 and 6.5 on the existence of ND functions in a prescribed family **G** of differentiable functions. As in § 6, the parameters of the family shall be the coordinates $a^1,..., a^m$ of a point $a \in A_m$, an m-dimensional Euclidean space.

The Family **G**. The functions of the family will be the partial mappings (14.1) for fixed $a \in A_m$ of a real-valued C^μ-mapping, $1 < \mu \leqslant r$,

$$(p, a) \to \mathbf{G}(p, a) : Z \to R$$

defined on an open subset Z of the product $M_n \times A_m$. The first theorem, Theorem 14.1, is an easy consequence of Theorem 6.3 or 6.5 and has numerous applications. A difficulty not met in § 6 is that the "rank condition" in Theorem 14.1 is defined in terms of **G** and of presentations in M_n. This rank condition should accordingly be shown to be independent of the choice of presentations in M_n used to define it, in so far as this choice is arbitrary. This is done in Lemma 14.1.

Theorem 14.1. *Introduction* (a). Whenever we refer to a product $X \times A$ of two spaces X and A of points p and a, respectively, we shall understand that the "projections" pr_1 and pr_2 of $X \times A$ onto X

and A, respectively, are mappings such that, for arbitrary $(p, a) \in X \times A$

$$pr_1(p, a) = p \qquad pr_2(p, a) = a.$$

(b) **G** *as a Family of Mappings* $\mathbf{G}^a$. *The sets* $\Omega(\mathbf{G})$, $\Omega^{\text{deg}}(\mathbf{G})$. The domain Z of **G** is open in $M_n \times A_m$. For fixed $a \in \text{pr}_2 Z$ introduce the open subset

$$Z(a) = \{p \in M_n \mid (p, a) \in Z\}$$

of M_n. The partial mapping

$$p \to \mathbf{G}(p, a) = \mathbf{G}^a(p) : Z(a) \to R \tag{14.1}$$

is well-defined. The mappings **G** will be regarded as the *family* of *mappings* $\mathbf{G}^a$. Let $\Omega(\mathbf{G})$ denote the set of pairs $(p, a) \in Z$ such that p is a critical point on M_n of $\mathbf{G}^a$. Let $\Omega^{\text{deg}}(\mathbf{G})$ be the subset of pairs $(p, a) \in \Omega(\mathbf{G})$ such that p is a degenerate critical point of $\mathbf{G}^a$.

Theorem 14.1 is concerned with the measure of $\text{pr}_2\, \Omega^{\text{deg}}(\mathbf{G})$ in A_m.

(c) *Canonical Neighborhoods in Z.* Each point $(p_0, a_0) \in Z$ has a neighborhood $X \times N$ relative to Z, where N is an open neighborhood of a_0 relative to A_m and X is a coordinate domain of M_n given by a presentation $(F : V, X) \in \mathscr{D}M_n$. A neighborhood of (p_0, a_0) of this character which is a relatively compact subset of another neighborhood of this character, will be called *canonical.* We see that each point $(p_0, a_0) \in Z$ has an arbitrarily small canonical neighborhood. Since M_n is coverable by a countable set of presentations in $\mathscr{D}M_n$, it is clear that if ω is the set of positive integers, there exists a set

$$(X_k \times N_k)_{k \in \omega} \tag{14.2}$$

of canonical neighborhoods of points of Z whose union is Z.

(d) *Local Representations U of* **G**. *The sets* $\Gamma(U)$, $\Gamma^{\text{deg}}(U)$. The mapping **G** of (14.1) is assumed of "class C^μ" on Z in the following sense. Corresponding to a prescribed point $(p_0, a_0) \in Z$ there exists a "canonical" neighborhood $X \times N$ of (p_0, a_0) in Z such that the *local representation*

$$(v, a) \to \mathbf{G}(F(v), a) = U(v, a) : V \times N \to R \tag{14.3}$$

of **G** is of class C^μ on $V \times N$ for some presentation $(F : V, X) \in \mathscr{D}M_n$.

Let $\Gamma(U)$ be the set of pairs $(v, a) \in (V \times N)$ such that

$$0 = U_{v^1}(v, a) = \cdots = U_{v^n}(v, a) = 0. \tag{14.4}$$

A pair (v, a) is in $\Gamma(U)$ if and only if v is a critical point of the partial mapping $v \to U(v, a)$. Let the subset of pairs $(v, a) \in \Gamma(U)$ for which v is a degenerate critical point of the partial mapping $v \to U(v, a)$ be denoted by $\Gamma^{\text{deg}}(U)$.

(e) *The Local Rank Condition.* Given the "local representation" U of $\mathbf{G}$ as in (14.3), we proceed as in § 6, setting $U_{v^i} = U_i$, $i = 1,..., n$, and let $\| U_{ij}(v, a)\|$ be the n by $n + m$ functional matrix of $U_1,..., U_n$ with respect to the $n + m$ variables

$$v^1,..., v^n; a^1,..., a^m, \qquad (v, a) \in V \times N. \tag{14.5}$$

The rank condition on $\| U_{ij} \|$ requires that

$$\text{rank}\| U_{ij}(v, a)\| = n, \qquad (v, a) \in \Gamma(U). \tag{14.6}$$

We shall refer to $\Gamma(U)$ as the *v-critical ensemble* of U and to

$$\| U_{ij} \| = \mathbf{H}(U) \tag{14.7}$$

as the *v-critical matrix* of U.

Theorem 14.1 extends Theorem 6.3. It concerns a manifold M_n of class C^r and presentations $(F : V, X) \in \mathscr{D}M_n$.

Theorem 14.1. *Let Z be a nonempty open subset of $M_n \times A_m$ and $(p, a) \to \mathbf{G}(p, a)$ a real-valued C^μ-function on Z, $1 < \mu \leqslant r$, such that for each "local representation"*

$$(v, a) \to \mathbf{G}(F(v), a) = U(v, a) : V \times N \to R \tag{14.8}$$

of $\mathbf{G}$ the corresponding v-critical matrix $\mathbf{H}(U)$ satisfies the rank condition

$$\text{rank}\| U_{ij}(v, a)\| = n, \qquad (v, a) \in \Gamma(U). \tag{14.9}$$

Then meas $\text{pr}_2\, \Omega^{\text{deg}}(\mathbf{G}) = 0$.

Proof of Theorem 14.1. As we have seen, there exists a countable set (14.2) of "canonical" neighborhoods $X_k \times N_k$ in Z with union Z.

Because the neighborhood $X_k \times N_k$ is canonical, it is the relatively compact subset of another canonical neighborhood $\hat{X}_k \times \hat{N}_k$ in Z such that X_k and $\hat{X}_k$ are coordinate domains of M_n given by presentations $(F_k : V_k, X_k) \in \mathscr{D}M_n$, and $(\hat{F}_k : \hat{V}_k, \hat{X}_k) \in \mathscr{D}M_n$ of which $\hat{F}_k$ is an extension of F_k.

For F_k and $\hat{F}_k$ we introduce the corresponding local representations

$$(v, a) \to \mathbf{G}(F_k(v), a) = U^k(v, a), \qquad (v, a) \in V_k \times N_k, \tag{14.10}$$

$$(v, a) \to \mathbf{G}(\hat{F}_k(v), a) = \hat{U}^k(v, a), \qquad (v, a) \in \hat{V}_k \times \hat{N}_k \tag{14.11}$$

of $\mathbf{G}$. By hypothesis the critical matrices $\mathbf{H}(U^k)$ and $\mathbf{H}(\hat{U}^k)$ have the rank n at each point (p, a) in $\Gamma(U^k)$ and $\Gamma(\hat{U}^k)$, respectively. We infer from Theorem 6.5 that for each $k \in \omega$

$$\text{Jordan content } \mathrm{pr}_2\, \Gamma^{\deg}(U^k) = 0. \tag{14.12}$$

There accordingly exists a subset σ_k of N_k of measure zero in A_m such that for $a \in N_k - \sigma_k$ the partial mappings $v \to U^k(v, a)$, $v \in V_k$, $k \in \omega$, are ND. By definition of U^k we have

$$U^k(v, a) = \mathbf{G}^a(F_k(v)), \quad v \in V_k,$$

so that for $a \in N_k - \sigma_k$, $\mathbf{G}^a \mid X_k$ is ND. Now the domain $Z(a)$ of $\mathbf{G}^a$ is the union of a subset of the domains X_k. If then $\sigma = \bigcup_{k \in \omega} \sigma_k$, a function $\mathbf{G}^a$ of the family $\mathbf{G}$ is ND on $Z(a)$ for each $a \in \mathrm{pr}_2 Z - \sigma$, that is, for almost all $a \in \mathrm{pr}_2 Z$.

This establishes Theorem 14.1.

Theorem 14.1 is implied by Theorem 14.2 below, as Theorem 6.3 was implied by Theorem 6.5.

Theorem 14.2. *If, under the hypothesis of Theorem* 14.1 Z_0 *is an arbitrary, open, relatively compact subset of the set* Z *of Theorem* 14.1, *then*

$$\text{Jordan content } \mathrm{pr}_2(Z_0 \cap \Omega^{\deg}(\mathbf{G})) = 0. \tag{14.13}$$

Proof. The set Z_0 is covered by a finite subset of the canonical neighborhoods (14.2) covering Z. For each $k \in \omega$ (14.12) holds, or, equivalently, (14.13) holds with Z_0 replaced by $X_k \cap N_k$. Hence (14.13) holds as stated.

Theorem 14.1 follows from Theorem 14.2; Theorem 14.1 is a consequence of the validity of (14.12) for each k, and thus a consequence of Theorem 14.2.

The Invariance of the Rank Conditions (14.6). The principal condition of Theorem 14.1 is on the rank of the critical matrices $\mathbf{H}(U)$ of local representations (14.3) of the family $\mathbf{G}$. Because this condition bears on *each* such local representation U, it would be difficult to verify without the following lemma on the invariance of the rank condition (14.6) under a change of local coordinates of M_n.

Lemma 14.1. *If the critical matrix* $\mathbf{H}(U)$ *of a local representation* U *of* $\mathbf{G}$ *of the form* (14.3) *satisfies the associated rank condition* (14.6), *then the critical matrix* $\mathbf{H}(U')$ *of a local representation* U' *of* $\mathbf{G}$ *of the form*

$$(v', a) \to \mathbf{G}(F'(v'), a) = U'(v', a) : V' \times N \to R \qquad (14.13')$$

also satisfies the associated rank condition (14.6) *whenever* $(F : V, X) \in \mathscr{D}M_n$ *and* $(F' : V', X') \in \mathscr{D}M_n$ *present the same domain* $X = X'$ *in* M_n.

Proof of Lemma 14.1. *Notation.* By virtue of the C^r-compatibility of the presentations

$$(F : V, X), \qquad (F' : V', X'), \qquad X = X', \qquad (14.14)$$

in $\mathscr{D}M_n$ there exists a C^r-diff,

$$v' \to \psi(v') = v, \qquad (14.15)$$

of V' onto V, where

$$\psi(v') = (\psi^1(v'),\ldots, \psi^n(v')) = F^{-1}(F'(v')), \qquad v' \in V'. \qquad (14.16)$$

Let the partial derivatives of U and U' with respect to $v^1,\ldots, v^n$ and $v'^1,\ldots, v'^n$ be denoted, respectively, by

$$U_1,\ldots, U_n \qquad \text{and} \qquad U'_1,\ldots, U'_n. \qquad (14.17)$$

By definition of U and U' we have $U(v, a) = U'(v', a)$, subject to

(14.15), and, with integers s and t on the range $1,..., n$, we have (summing with respect to t)

$$U'_s(v', a) = U_t(v, a) \frac{\partial \psi^t(v')}{\partial v'^s}. \tag{14.18}$$

From (14.18) we infer that the pair $(v, a) \in V \times N$ is in the "critical ensemble" $\Gamma(U)$ of U if and only if, subject to (14.15), the corresponding pair (v', a) is in the critical ensemble $\Gamma(U')$ of U'.

Method of Proof of Lemma 14.1. Let (v_0, a_0) be an arbitrary pair in $\Gamma(U)$ and (v'_0, a_0) the corresponding pair in $\Gamma(U')$. Our object is to prove the following:

The condition

$$\text{rank}\| U_{ij}(v_0, a_0)\| = n \tag{14.19}$$

implies the condition

$$\text{rank}\| U'_{ij}(v'_0, a_0)\| = n. \tag{14.20}$$

To this end, for $r = n + 1, n + 2,..., n + m$ let linear functions U_r in the variables $v^1,..., v^n; a^1,..., a^m$ be added to the set of n functions $U_1,..., U_n$, choosing U_r in such a way that each function U_r vanishes at (v_0, a_0) and

$$\frac{D(U_1, U_2,..., U_{n+m})}{D(v^1,..., v^n; a^1,..., a^m)}(v_0, a_0) \neq 0. \tag{14.21}$$

Such a choice of $U_{n+1},..., U_{n+m}$ is possible.

Let functions U'_r be defined on $V' \times N$ by the condition that

$$U_r(v, a) = U'_r(v', a), \qquad r = n + 1,..., n + m, \tag{14.22}$$

subject to (14.15). To show that (14.19) implies (14.20), it is sufficient to show that (14.21) implies

$$\frac{D(U'_1, U'_2,..., U'_{n+m})}{D(v'^1,..., v'^n; a^1,..., a^m)}(v'_0, a_0) \neq 0, \tag{14.23}$$

since the matrices in (14.19) and (14.20) are composed of the first n rows of the determinants (14.21) and (14.23), respectively.

Change of Notation. To establish (14.23) as a consequence of (14.21) in a simple manner, we shall set

$$(v^1,\ldots, v^n; a^1,\ldots, a^m) = (x^1\ x^2,\ldots, x^{n+m}) = x, \tag{14.24}$$

$$(v'^1,\ldots, v'^n; a^1,\ldots, a^m) = (y^1, y^2,\ldots, y^{n+m}) = y, \tag{14.25}$$

restricting x and y to the domains defined, respectively, by (14.24) and (14.25) when $(v, a) \in V \times N$ and $(v', a) \in V' \times N$. With this understood the diff (14.15) gives rise to a diff

$$y \to (\varphi^1(y),\ldots, \varphi^{n+m}(y)) = x \tag{14.26}$$

of the domain of y onto the domain of x under which

$$\begin{aligned} \varphi^s(y) &= \psi^s(y^1,\ldots, y^n), \qquad s = 1,\ldots, n, \\ \varphi^r(y) &= y^r, \qquad r = n+1,\ldots, n+m. \end{aligned} \tag{14.27}$$

Subject to (14.24) and to (14.25), we write

$$\begin{aligned} U_\mu(v, a) &= \bar{U}_\mu(x), \qquad \mu = 1,\ldots, n+m, \\ U'_\mu(v', a) &= \bar{U}'_\mu(y), \qquad \mu = 1,\ldots, n+m, \\ x_0 = (v_0\,, a_0)&, \qquad y_0 = (v'_0\,, a_0). \end{aligned}$$

Proof of the Implication (14.21) $\Rightarrow$ (14.23). Let μ, ν, h, and k be indices on the range $1, 2,\ldots, n+m$. With x and y corresponding under the diff (14.26), Eqs. (14.18) take the form (summing with respect to h)

$$\bar{U}'_\mu(y) = \bar{U}_h(x) \frac{\partial \varphi^h}{\partial y^\mu}(y) \tag{14.28}$$

for $\mu = 1,\ldots, n$, while Eqs. (14.22) take the form (14.28) for $\mu = n+1,\ldots, n+m$. Taking account of the conditions [cf. (14.4)]

$$0 = \bar{U}_1(x_0) = \bar{U}_2(x_0) = \cdots = \bar{U}_{n+m}(x_0) \tag{14.29}$$

relations (14.28), subject to (14.26), imply that

$$\frac{\partial \bar{U}'_\mu(y)}{\partial y^\nu} = \frac{\partial \bar{U}_h(x)}{\partial x^k} \frac{\partial \varphi^h(y)}{\partial y^\mu} \frac{\partial \varphi^k(y)}{\partial y^\nu} + \text{remainder}, \qquad \mu, \nu = 1, 2,\ldots, n+m, \tag{14.30}$$

where the *remainder* vanishes when $x = x_0$ and $y = y_0$.

Let A and C be the $(n + m)$-square matrices

$$\| A_{hk} \| = \left\| \frac{\partial \bar{U}_h(x)}{\partial x^k} \right\|^{x=x_0}, \qquad \| C_{h\mu} \| = \left\| \frac{\partial \varphi^h(y)}{\partial y^\mu} \right\|^{y=y_0},$$

and A_1 similarly the matrix of the Jacobian (14.23) evaluated when $y = y_0$.

Then (14.30) with $x = x_0$ and $y = y_0$ implies the matrix equality $A_1 = C'AC$, where C' is the transpose of C. Hence the nonvanishing of the determinants $| C |$ and $| A |$ implies that $| A_1 | \neq 0$ or, equivalently, that (14.23) holds. Thus (14.21) implies (14.23), and hence (14.19) implies (14.20).

Lemma 14.1 follows.

Note. It follows from Lemma 14.1 that in verifying the rank conditions of Theorem 14.1 or 14.2 it is by no means necessary to examine *each* local representation U of **G**. It is sufficient to examine a subset of local representations U of **G** of the form (14.3) so chosen that the associated domains $F(V) \times N$ have Z as union. One can omit each local representation U whose critical ensemble $\Gamma(U)$ is empty.

§ 15

FOCAL POINTS OF REGULAR MANIFOLDS

Let M_n be a regular manifold of class C^2 in a Euclidean space E_m, $0 < n < m$. We are concerned in this section with two problems: the explicit *a priori* existence of ND functions on M_n and the density of focal points of M_n in E_m.

We shall begin with a proof of Theorem 6.1. Theorem 6.1 affirms that for almost all points $a \in E_m - M_n$ the function $x \to \| x - a \| \mid M_n$ is ND on M_n. Theorem 6.1 can be proved very simply by proving the following equivalent theorem:

Theorem 15.1. *If M_n is a regular C^2-manifold in E_m, $0 < n < m$ then for almost all points $a \in E_m - M_n$ the function*

$$x \to 1/\| x - a \| \tag{15.1}$$

restricted to M_n is ND *on M_n.*

We shall prove Theorem 15.1 by means of Theorem 14.1.

That Theorems 6.1 and 15.1 are equivalent is shown by the following lemma:

Lemma 15.1. *If $v \to f(v)$ is a real-valued nonvanishing* ND *function of class C^2 on an open subset V of E_n, then $1/f$ is* ND *on V.*

Proof of Lemma 15.1. The critical points of f and $1/f$ are clearly the same. A critical point v_0 of f is ND if the Jacobian

$$J(v_0) = \frac{D(f_{v^1}, \ldots, f_{v^n})}{D(v^1, \ldots, v^n)}(v_0) \neq 0.$$

As a critical point of $1/f$, v_0 is then ND, since

$$\frac{D\left(\dfrac{f_{v^1}}{f^2},\ldots,\dfrac{f_{v^n}}{f^2}\right)}{D(v^1,\ldots,v^n)}(v_0) = \frac{J(v_0)}{f^2(v_0)} \neq 0.$$

The lemma follows.

The following lemma leads to a proof of Theorem 15.1:

Lemma 15.2. *If for $x \neq a$ one sets $\varphi(x, a) = \| x - a \|^{-1}$, then the m-square determinant*

$$\left| \frac{\partial^2 \varphi(x, a)}{\partial x^\mu \partial a^\nu} \right| \neq 0, \qquad x \neq a. \tag{15.2}$$

Proof. For $\mu = 1,\ldots, m$ set

$$\varphi_\mu(x, a) = \frac{\partial \varphi}{\partial a^\mu}(x, a) = \frac{x^\mu - a^\mu}{\| x - a \|^3}, \qquad x \neq a.$$

We wish to show that

$$\frac{D(\varphi_1,\ldots,\varphi_m)}{D(x^1,\ldots,x^m)}(x, a) \neq 0, \qquad x \neq a. \tag{15.3}$$

To that end note that for a fixed point a the transformation $x \to y$, defined by setting

$$y^\mu = \varphi_\mu(x, a), \qquad x \neq a, \tag{15.4}$$

implies that $\| y \| = \| x - a \|^{-2}$, and hence that under the transformation (15.4)

$$x^\mu = a^\mu + y^\mu \| y \|^{-3/2}, \qquad \| y \| \neq 0. \tag{15.5}$$

Thus the transformation (15.4) has a continuously differentiable inverse and hence a nonvanishing Jacobian (15.3).

Thus (15.2) is true.

Proof of Theorem 15.1. We shall apply Theorem 14.1 to prove Theorem 15.1, taking $\mathbf{G}$ of Theorem 14.1 as the mapping

$$(p, a) \to 1/\| p - a \|, \qquad p \in M_n, \quad a \notin M_n, \tag{15.6}$$

identifying A_m of Theorem 14.1 with E_m of Theorem 15.1.

For fixed a_0 the partial mapping $p \to \mathbf{G}(p, a_0)$ has a point $p_0 \in M_n$ as critical point if and only if the vector $p_0 - a_0$ in E_m is orthogonal to M_n at p_0.

The truth or falsity of Theorem 15.1 is independent of the choice of the rectangular coordinate system in E_m among coordinate systems obtained one from the other by translations or orthogonal transformations of the coordinate axes $x^1,\ldots, x^m$. It is understood that the coordinates $a^1,\ldots, a^m$ are subject to the same transformations as the coordinates $x^1,\ldots, x^m$ so that $\| x - a \|$ is invariant.

The Rank Conditions of Theorem 14.1. Suppose that p_0 is a point of M_n and that the non-null vector $p_0 - a_0$ is orthogonal to M_n at p_0. Then p_0 is a critical point of $\mathbf{G}^{a_0}$. Suppose that the origin has been chosen as p_0 and the $x^1,\ldots, x^n$ axes as tangent to M_n at p_0. Let Π be the coordinate n-plane of the $x^1,\ldots, x^n$ axes. Let $(F : V, X)$ be a Monge presentation of a neighborhood X of p_0 relative to M_n, so that F is the inverse of the orthogonal projection of X into Π. If then $p_0 = F(v_0)$ is the origin,

$$\frac{\partial F^i}{\partial v^h}(v_0) = \delta_h^{\,i}, \qquad i = 1,\ldots, m; \quad h = 1,\ldots, n. \tag{15.7}$$

One introduces the local representation

$$(v, a) \to \varphi(F(v), a) = U(v, a), \qquad v \in V, \quad a \in N, \tag{15.8}$$

of φ, where N is an open neighborhood of a_0 not meeting M_n. For $h = 1,\ldots, n$ and i summed over the range $1,\ldots, m$ and for $v \in V$ and $a \in N$

$$\frac{\partial U(v, a)}{\partial v^h} = \frac{\partial \varphi(x, a)}{\partial x^i} \frac{\partial F^i(x)}{\partial v^h}, \qquad x = F(v). \tag{15.9}$$

For h and i as in (15.7), for k on the range $1,\ldots, m$, (15.9) implies that

$$\frac{\partial^2 U(v, a)}{\partial v^h\, \partial a^k} = \frac{\partial^2 \varphi(x, a)}{\partial x^i\, \partial a^k} \frac{\partial F^i(x)}{\partial v^h}, \qquad x = F(v). \tag{15.10}$$

We shall indicate evaluation of the terms of (15.10) when $v = v_0$, $a = a_0$, $x = p_0$ by adding a superscript 0. Taking account of (15.7), one arrives at the equality of the n by m matrices

$$\left\| \frac{\partial^2 U(v, a)}{\partial v^h\, \partial a^k} \right\|^0 = \left\| \frac{\partial^2 \varphi(x, a)}{\partial x^h\, \partial a^k} \right\|^0, \qquad h = 1, 2,\ldots, n; \quad k = 1, 2,\ldots, m. \tag{15.11}$$

Since (15.2) holds, the matrices (15.11) have the rank n. Thus the v-critical matrix $\mathbf{H}(U)$ has the rank n at (p_0, a_0).

It follows from the invariance of the rank condition (Lemma 14.1) under admissible change of local parameters on M_n that the rank condition of Theorem 14.1 is satisfied for *each* local representation U of $\mathbf{G}$.

Theorem 15.1 accordingly follows from Theorem 14.1, and Theorem 6.1 from Theorem 15.1.

Definition 15.1. **Focal points of M_n in E_m.** If a non-null vector $p_0 - a_0$ is orthogonal to M_n at p_0 and if p_0 is a degenerate critical point of the partial mapping

$$p \to \| p - a_0 \| : M_n \to R, \tag{15.12}$$

then a_0 is called a *focal point of M_n with base point p_0*.

The following was proved by Morse in lectures at the City University of New York in 1965 (see Morse [15]):

Theorem 15.2. *There exists a noncompact regular manifold M_n of class C^∞ in E_{n+1} whose focal points are everywhere dense in E_{n+1}.*

This theorem calls for the following supplement:

Theorem 15.3. *The focal points of a compact regular manifold M_n of class C^2 in E_m, $0 < n < m$, are nowhere dense in E_m.*

A proof of Theorem 15.3 was given by Morse [13], p. 243 in the analytic case without making use of properties of M_n other than differentiability. However, this theorem is implied by the following corollary of Theorem 14.2:

Lemma 15.3. *If M_n is a compact regular C^2-manifold in E_m, $0 < n < m$, and X a relatively compact open subset of E_m bounded from M_n the set of focal points of M_n in X has a Jordan content zero.*

As we have just seen, the mapping $(p, a) \to \mathbf{G}(p, a)$, defined by (15.6) on the open subdomain $Z = M_n \times (E_m - M_n)$ of $M_n \times E_m$, satisfies the conditions of Theorem 14.1 when $A_m = E_m$. An open relatively compact subset of Z is afforded by $Z_0 = M_n \times X$. It follows from Theorem 14.2 that

$$\text{Jordan content } \mathrm{pr}_2(Z_0 \cap \Omega^{\mathrm{deg}}(\mathbf{G})) = 0. \tag{15.13}$$

Lemma 15.1 implies that for fixed $a \notin M_n$, the mappings $p \to \| p - a \|$ and $p \to \| p - a \|^{-1}$, $p \in M_n$, are both ND if one is ND; Lemma 15.3 accordingly follows from (15.13).

§ 16

DIFFERENTIABLE MAPPINGS OF MANIFOLDS INTO MANIFOLDS

Let M_n and N_r be manifolds of class at least C^μ; suppose m an integer such that $0 < m \leqslant \mu$. We shall define differentiable mappings of M_n into N_r.

Definition 16.1. *A Mapping* $\varphi : M_n \to N_r$ *of Class* C^m. A mapping

$$p \to \varphi(p) : | M_n | \to | N_r | \tag{16.1}$$

will be said to *define* (or simply to *be*) a mapping of class C^m of M_n into N_r if corresponding to a prescribed point $p_0 \in M_n$ there exist presentations

$$(F : U, X) \in \mathscr{D}M_n \qquad \text{and} \qquad (G : V, Y) \in \mathscr{D}N_r \tag{16.2}$$

such that $p_0 \in X$, $\varphi(X) \subset Y$, and the mapping

$$G^{-1} \circ (\varphi \circ F) : U \to V \tag{16.3}$$

is of class C^m.

Note. The parentheses cannot be removed from $\varphi \circ F$ in (16.3), since the range of φ may not be included in the domain Y of G^{-1}.

Suppose that in the sense of Definition 16.1 φ is a C^m-mapping of M_n into N_r.

Definition 16.2. The mapping φ of (16.1) will be termed a C^m-*immersion* $M_n \to N_r$ if corresponding to each $p_0 \in M_n$, F and G in

(16.2) can be chosen so that $p_0 \in X$, $\varphi(X) \subset Y$, and the mapping (16.3) is a C^m-immersion in the sense of Definition 5.1. A C^m-immersion $\varphi : M_n \to N_r$ which is a homeomorphism into N_r is termed a C^m-*embedding* of M_n in N_r (cf. Munkres [1], p. 10).

An immersion $M_n \to N_r$ is possible at most if $n \leqslant r$. This is because dim $U = n$ and dim $V = r$.

Remark. If φ is a C^m-immersion (or embedding) $M_n \to N_r$, $\varphi \mid M_n'$ is a C^m-immersion (or embedding) $M_n' \to N_r$ of any C^m-submanifold M_n' of M_n. In fact, if $p_0 \in M_n'$, the condition of Definition 16.2 is satisfied by hypothesis by some $F \in \mathscr{D}M_n$, and hence is also satisfied by any restriction of F whose coordinate domain contains p_0.

The following lemma frees Definitions 16.1 and 16.2 of their dependence upon the choice of presentations $(F : U, X)$ such that $p_0 \in X$. However, one needs characterizations of immersions *both* in the form of Definition 16.2 and in the form of Lemma 16.1.

In Lemma 16.1 the inclusion of the phrase "(and an immersion)" in parentheses is meant to imply that Lemma 16.1 is true if this phrase is kept or deleted.

Lemma 16.1. *A mapping φ of M_n into N_r is of class C^m (and an immersion) if and only if the following is true: Corresponding to arbitrary presentations*

$$(F_1 : U_1, X_1) \in \mathscr{D}M_n \quad \text{and} \quad (G_1 : V_1, Y_1) \in \mathscr{D}N_r \tag{16.4}$$

such that $\varphi(X_1) \subset Y_1$, the mapping

$$G_1^{-1} \circ (\varphi \circ F_1) : U_1 \to V_1 \tag{16.5}$$

is of class C^m (and an immersion).

It is clear that φ is of class C^m (and an immersion) if the condition of the lemma is satisfied. We shall prove the converse (A):

(A) *If φ is of class C^m (and an immersion), each mapping* (16.5) *is of class C^m (and an immersion).*

It is sufficient to prove that if u_0 is a prescribed point of U_1, the restriction of $G_1^{-1} \circ (\varphi \circ F_1)$ to some neighborhood of u_0 in U_1 is of class C^m (and an immersion).

The point $p_0 = F_1(u_0)$ is in X_1 and $\varphi(p_0)$ is in Y_1. Since φ is of class C^m (and an immersion) by hypothesis, there exist presentations (16.2) such that

$$p_0 \in X \subset X_1, \qquad \varphi(X) \subset Y \subset Y_1, \tag{16.6}$$

and the mapping (16.3) is of class C^m (and an immersion). Corresponding to these sets X and Y there then exist unique presentations

$$(\hat{F} : \hat{U}, X) \in \mathscr{D}M_n \qquad \text{and} \qquad (\hat{G} : \hat{V}, Y) \in \mathscr{D}N_r \tag{16.7}$$

which are restrictions, respectively, of the presentations (16.4). Since $\varphi(X) \subset Y \subset Y_1$, the mapping

$$\hat{G}^{-1} \bar{\circ} (\varphi \bar{\circ} \hat{F}) : \hat{U} \to \hat{V} \tag{16.8}$$

is well-defined and a restriction of the mapping (16.5) to a neighborhood $\hat{U}$ in U of the point u_0 prescribed in U. The proof will accordingly be complete if we prove the following:

(B) *The mapping* (16.8) *is of class* C^m (*and an immersion*). By virtue of the C^μ-compatibility of F and $\hat{F}$ and of G and $\hat{G}$, there exists a C^μ-diff η of $\hat{U}$ onto U and a C^μ-diff ζ of V onto $\hat{V}$ such that

$$F = F \circ \eta, \qquad G = \hat{G} \circ \zeta. \tag{16.9}$$

It follows that

$$\hat{G}^{-1} \bar{\circ} (\varphi \bar{\circ} \hat{F}) = \zeta \bar{\circ} (G^{-1} \bar{\circ} (\varphi \bar{\circ} F)) \circ \eta. \tag{16.10}$$

Since $0 < m \leqslant \mu$, the mapping (16.8) is then of class C^m (and an immersion), with the mapping (16.3).

This completes the proof of (B) and thereby of the lemma.

Note. One can show that Lemma 16.1 is false if the condition $0 < m \leqslant \mu$ is omitted.

We state a theorem on the composition of C^m-mappings of manifolds.

Theorem 16.1. *For* $i = 1, 2, 3$, *let* M_{n_i} *be differentiable manifolds each of class* C^μ. *Suppose that* $0 < m \leqslant \mu$. *If*

$$p \to \varphi(p) : M_{n_1} \to M_{n_2} \qquad p \to \psi(p) : M_{n_2} \to M_{n_3} \tag{16.11}$$

$$n_1 \leqslant n_2 \leqslant n_3$$

are C^m-mappings (and immersions), the composite mapping $\Theta = \psi \circ \varphi : M_{n_1} \to M_{n_3}$ *is of class C^m (and an immersion).*

Proof. Let points $p_i \in M_{n_i}$ be defined by prescribing $p_1 \in M_{n_1}$ and setting $p_2 = \varphi(p_1)$ and $p_3 = \psi(p_2)$. Our hypotheses imply that presentations

$$(F_i ; U_i, X_i) \in \mathscr{D}M_{n_i}, \qquad i = 1, 2, 3, \tag{16.12}$$

can be chosen in the order $i = 3, 2, 1$ so that $p_i \in X_i$, the inclusions $X_3 \supset \psi(X_2)$ and $X_2 \supset \varphi(X_1)$ are valid, and the mappings

$$F_3^{-1} \bar{\circ} (\psi \bar{\circ} F_2) : U_2 \to U_3, \qquad F_2^{-1} \bar{\circ} (\varphi \bar{\circ} F_1) : U_1 \to U_2 \tag{16.13}$$

are of class C^m (and immersions).

It follows from the above inclusions that $X_3 \supset (\psi \circ \varphi)(X_1)$, so that the mapping

$$F_3^{-1} \bar{\circ} (\Theta \bar{\circ} F_1) = [F_3^{-1} \bar{\circ} (\psi \bar{\circ} F_2)] \circ [F_2^{-1} \bar{\circ} (\varphi \bar{\circ} F_1)] : U_1 \to U_3 \tag{16.14}$$

is well-defined and the equality (16.14) valid. The mapping (16.14) is of class C^m (and an immersion), with the mappings (16.13). Since p_1 is prescribed in M_{n_1} and is in X_1, $\Theta = \psi \circ \varphi$ is of class C^m (and an immersion) in accord with Definition 16.1.

This establishes Theorem 16.1.

Definition 16.3. Diffs φ. Let M_n and N_n be differentiable manifolds of at least class C^μ. Suppose that $0 < m \leqslant \mu$. We then term a homeomorphism φ of $| M_n |$ onto $| N_n |$ a C^m-diff of M_n onto N_n if both φ and φ^{-1} are mappings of class C^m.

By virtue of this definition φ is a C^m-diff of M_n onto N_n if and only if φ^{-1} is a C^m-diff of N_n onto M_n.

The following is a corollary of Theorem 16.1.

Corollary 16.1. *Let M_n, N_n, and Q_n be C^μ-manifolds. Suppose that $0 < m \leqslant \mu$. If φ is a C^m-diff of M_n onto N_n and if ψ is a C^m-diff of N_n onto Q_n, then the composite mapping $\psi \circ \varphi$ is a C^m-diff of M_n onto Q_n.*

We add two theorems.

Theorem 16.2a. *Let φ be a homeomorphism of $|M_n|$ onto $|N_n|$. A necessary and sufficient condition that φ be a C^m-diff of M_n onto N_n, $0 < m \leqslant \mu$, is that for arbitrary presentations*

$$(F : U, X) \in \mathscr{D}M_n \quad \text{and} \quad (G : V, Y) \in \mathscr{D}N_n \tag{16.15}$$

such that $\varphi(X) = Y$ the homeomorphism

$$G^{-1} \circ (\varphi \circ F) : U \to V \tag{16.16}$$

be a C^m-diff of U onto V.

A Convention. Presentations F and G in this order such that $\varphi(X) = Y$ will be called *φ-paired.* If ψ is the inverse of φ, presentations G and F are ψ-paired in that $X = \psi(Y)$.

Regardless of whether or not φ has any differentiability properties, it is seen that if p_0 is prescribed in M_n, there exist "φ-paired" presentations F and G such that $p_0 \in X$. We can now prove (i):

(i) *The condition of the theorem is sufficient that φ be a C^m-diff.*

Proof of (i). Given $p_0 \in M_n$, by hypothesis there exist φ-paired presentations F and G such that $p_0 \in X$ and the homeomorphism (16.16) is a C^m-diff. Hence by Definition 16.1 φ is of class C^m. When the mapping (16.16) is a diff of class C^m its inverse,

$$F^{-1} \circ (\psi \circ G) : V \to U, \tag{16.17}$$

is of class C^m according to Definition 1.1. It follows then from Definition 16.1 that ψ is of class C^m, so that by Definition 16.3 φ is a C^m-diff.

We now prove (ii):

(ii) *If φ is a C^m-diff, the condition of the theorem is satisfied.*

Since φ is of class C^m, it follows from Lemma 16.1 that the mapping (16.16) is of class C^m. Since ψ is of class C^m, it follows similarly from Lemma 16.1 that the mapping (16.17) is of class C^m. Since the mappings (16.16) and (16.17) are inverse homeomorphisms, both are diffs.

This establishes (ii) and completes the proof of the theorem.

Theorem 16.2b. *A homeomorphism φ of M_n onto N_n is a C^m-diff if and only if φ is a C^m-embedding $M_n \to N_n$.*

The inverse ψ of a C^m-embedding $\varphi : M_n \to N_n$ onto N_n, is both a homeomorphism and an immersion and in particular of class C^m. Hence a C^m-embedding φ is a C^m-diff. Conversely, a C^m-diff of M_n onto N_n is a C^m-embedding in the sense of Definition 16.2, as we now verify.

If φ is a C^m-diff of M_n onto N_n, then by Theorem 16.2a each homeomorphism (16.16) is a C^m-diff of U onto V and hence by Lemma 1.1 an immersion. By Definition 16.2 φ is then a C^m-embedding $M_n \to N_n$.

Theorem 16.3a. *If M_n and N_n are C^m-manifolds and φ a C^m-diff of M_n onto N_n, the implications*

$$(F : U, X) \in \mathscr{D}M_n \Rightarrow (\varphi \bar{\circ} F : U, \varphi(X)) \in \mathscr{D}N_n, \tag{16.18}$$

$$(G : V, Y) \in \mathscr{D}N_n \Rightarrow (\varphi^{-1} \bar{\circ} G : V, \varphi^{-1}(Y)) \in \mathscr{D}M_n \tag{16.19}$$

are valid for arbitrary $F \in \mathscr{D}M_n$ and $G \in \mathscr{D}N_n$.

Moreover, the biunique mapping of $\mathscr{D}M_n$ into $\mathscr{D}N_n$ implied by (16.18) *is onto $\mathscr{D}N_n$, and equivalently the inverse biunique mapping of $\mathscr{D}N_n$ into $\mathscr{D}M_n$ implied by* (16.19) *is onto $\mathscr{D}M_n$.*

Proof of (16.18). The homeomorphism

$$\varphi \bar{\circ} F : U \to \varphi(X) \tag{16.20}$$

is onto the open subspace $\varphi(X)$ of $|N_n|$. This mapping is a presentation in $\mathscr{D}N_n$ by Definition 13.1 if and only if the presentations

$$(G : V, Y) \in \mathscr{D}N_n \qquad \text{and} \qquad (\varphi \bar{\circ} F : U, \varphi(X)) \tag{16.21}$$

of $\mathscr{D}^0 | N_n |$ are C^m-compatible for arbitrary choice of $G \in \mathscr{D}N_n$.

If the presentations (16.21) are nonoverlapping, they are trivially C^m-compatible. In any other case set $\hat{Y} = (Y \cap \varphi(X))$ and set $\hat{X} = \varphi^{-1}(\hat{Y})$. Then, in accord with (A_1) and (A_2) of § 13, the presentations (16.21) are C^m-compatible if restrictions $(\hat{G} : \hat{V}, \hat{Y})$ and $(\varphi \bar{\circ} \hat{F} : \hat{U}, \varphi(\hat{X}))$ of the respective presentations (16.21) are C^m-compatible, that is, if the homeomorphism

$$\hat{G}^{-1} \bar{\circ} (\varphi \bar{\circ} \hat{F}) : \hat{U} \to \hat{V}, \qquad \hat{Y} = \varphi(\hat{X}), \tag{16.22}$$

of $\hat{U}$ onto $\hat{V}$ is a C^m-diff. On applying Theorem 16.2a, given the diff φ and the presentations

$$(\hat{F} : \hat{U}, \hat{X}) \in \mathscr{D}M_n \qquad \text{and} \qquad (\hat{G} : \hat{V}, \hat{Y}) \in \mathscr{D}N_n\,, \qquad \hat{Y} = \varphi(\hat{X}),$$

we conclude that the homeomorphism (16.22) is in fact a C^m-diff. Hence the presentations (16.21) are C^m-compatible and (16.18) is valid.

The proof that the implication (16.19) is valid is similar.

That the mapping of $\mathscr{D}M_n$ into $\mathscr{D}N_n$ implied by (16.18) is onto $\mathscr{D}N_n$, and that the mapping of $\mathscr{D}N_n$ into $\mathscr{D}M_n$ implied by (16.19) is onto $\mathscr{D}M_n$ follows from the fact that these two mappings are inverses.

Theorem 16.3b supplements Theorem 16.3a, and defines a C^m-manifold $\varphi(M_n)$:

Theorem 16.3b. *Let M_n be a C^m-manifold and φ a homeomorphism of $| M_n |$ onto a topological manifold Γ_n. The ensemble*

$$K = \{(\varphi \,\bar{\circ}\, F : U, \varphi(X)) \mid (F : U, X) \in \mathscr{D}M_n\} \tag{16.23}$$

of presentations of open subsets of Γ_n is a maximal set of pairwise C^m-compatible presentations of open subsets of Γ_n and covers Γ_n.

Moreover, φ is a C^m-diff of M_n onto a C^m-manifold, DENOTED BY *$\varphi(M_n)$, whose carrier is $\varphi(| M_n |)$ and for which $\mathscr{D}\varphi(M_n) = K$. Finally, φ is a C^m-diff of M_n onto no other C^m-manifold than $\varphi(M_n)$.*

Proof. Any two presentations

$$(\varphi \,\bar{\circ}\, F_1 : U_1\,, \varphi(X_1)) \qquad \text{and} \qquad (\varphi \,\bar{\circ}\, F_2 : U_2\,, \varphi(X_2))$$

in the set K are C^m-compatible, since the corresponding presentations

$$(F_1 : U_1\,, X_1) \in \mathscr{D}M_n \qquad \text{and} \qquad (F_2 : U_2\,, X_2) \in \mathscr{D}M_n$$

are C^m-compatible. It follows from Definition 13.1 of a C^m-manifold and from Theorem 13.1 that there exists a C^m-manifold $\hat{N}_n$ whose carrier is Γ_n and for which $\mathscr{D}\hat{N}_n \supset K$.

We shall verify the following:

(i) *The homeomorphism φ of $| M_n |$ onto Γ_n is a C^m-embedding of M_n onto $\hat{N}_n$.*

Verification of (i). Corresponding to a prescribed point $p_0 \in M_n$ there exist presentations

$$(F : U, X) \in \mathscr{D}M_n \qquad \text{and} \qquad (G : V, Y) \in \mathscr{D}\hat{N}_n \tag{16.24}$$

such that $p_0 \in X$ and

$$(G : V, Y) = (\varphi \bar{\circ} F : U, \varphi(X)). \tag{16.25}$$

The resultant mapping $G^{-1} \circ (\varphi \bar{\circ} F)$ reduces to the identity mapping of U onto U. It follows from Definition 16.2 that (i) is true.

We infer from Theorem 16.2b that φ is a C^m-diff of M_n onto $\hat{N}_n$ and from Theorem 16.3a that $\mathscr{D}\hat{N}_r = K$. Hence K is a maximal set of pairwise C^m-compatible presentations of open subsets of Γ_n.

Thus φ is a C^m-diff of M_n onto the C^m-manifold $\varphi(M_n)$ whose carrier is Γ_n and whose presentation set is K. That φ is a C^m-diff of M_n onto no other C^m-manifold follows from Theorem 16.3a.

This completes the proof of Theorem 16.3b.

Q_r-embedded Manifolds. The last four theorems have been concerned with mapping a differentiable manifold M_n onto a differentiable manifold of the same dimension. We now turn to C^m-embeddings φ (Definition 16.2) of a differentiable M_n into a differentiable manifold Q_r with $0 < n < r$. In this subsection we suppose that M_n and Q_r are of class at least C^μ and that $0 < m \leqslant \mu$.

Before going further it is essential to recall that two differentiable manifolds are regarded as *identical* if and only if they have the same carriers and the same set of presentations.

Embedding Terminology. A C^m-embedding φ of M_n into Q_r is understood in the sense of Definition 16.2. If φ is an arbitrary homeomorphism of $| M_n |$ onto a topological manifold $\varphi(| M_n |)$, $\varphi(M_n)$ has been defined as an n-manifold with carrier $\varphi(| M_n |)$ and set of presentations (16.23) (see Theorem 16.3b). If, in particular, φ is a homeomorphic mapping of $| M_n |$ into $| Q_r |$, φ is in general not an embedding of M_n into Q_r. If, however, φ *is* a C^m-embedding, we term $N_n = \varphi(M_n)$ a *Q_r-embedded* manifold.

The following theorem serves to simplify the problem of characterizing Q_r-embedded manifolds. In this theorem we understand that an *inclusion* mapping of a subset A of B into B maps each point $x \in A$ into $x \in B$. This is a conventional use of the term "inclusion mapping."

Theorem 16.4. *Let N_n and Q_r, $0 < n < r$, be manifolds of class at least C^μ. Suppose that $|N_n|$ is a subspace of $|Q_r|$ and that $0 < m \leqslant \mu$. Then N_n is a Q_r-embedded manifold if and only if the inclusion mapping I of $|N_n|$ into $|Q_r|$ is a C^m embedding, $N_n \to Q_r$.*

Proof. If I is an embedding $N_n \to Q_r$, then by definition the manifold $I(N_n) = N_n$, is a Q_r-embedded manifold.

If φ is a C^m-embedding $M_n \to Q_r$ such that $\varphi(M_n) = N_n$, it is a trivial consequence of Definition 16.2 that the inclusion mapping $I : N_n \to Q_r$ is a C^m-embedding [use the presentations (16.23) of N_n].

Thus Theorem 16.4 is true.

A Q_r-embedded C^m-Manifold $((\Gamma_n, Q_r, m))$. Let Γ_n be a topological n-manifold which is a subspace of Q_r, and m a positive integer. The principal objective of this subsection is the proof of a *uniqueness* theorem for a Q_r-embedded C^m-manifold whose carrier is Γ_n. If Γ_n is prescribed, such a manifold does not exist in general, but if it does exist, it is uniquely determined by Γ_n, Q_r, and m, as we shall see in Theorem 16.5, and will be denoted by $((\Gamma_n, Q_r, m))$.

Introduction to Theorem 16.5. Recall that a presentation $(H : U, Z) \in \mathscr{D}^0\Gamma_n$ (Definition 13.3) of an open subset Z of Γ_n and an integer $m > 0$ uniquely determine, in the sense of Definition 13.4, a C^m-manifold $\mathbf{Z}_n$ with carrier Z. Note also the following consequence of Definition 16.2: The inclusion map of $Z = |\mathbf{Z}_n|$ into $|Q_r|$ is a C^m-embedding $\mathbf{Z}_n \to Q_r$ if and only if each presentation $(G : V, Y) \in Q_r$ and each open restriction $(\hat{H} : \hat{U}, \hat{Z})$ of $(H : U, Z)$ for which $G^{-1} \circ \hat{H}$ is a well-defined composition yield a mapping

$$G^{-1} \circ \hat{H} : \hat{U} \to V \tag{16.26}$$

which is a C^m-embedding of $\hat{U}$ into V in the sense of Definition 5.1. In Theorem 16.5 we pass from this simple case to the general case.

Theorem 16.5. *For $0 < n < r$ let Q_r be a C^μ-manifold and Γ_n a topological n-manifold which is a subspace of $|Q_r|$. Suppose that $0 < m \leqslant \mu$ and that conditions (α) and (β) are defined as follows:*

(α) *Under condition* (α) *there shall exist a* Q_r*-embedded* C^m*-manifold* P_n *with carrier* Γ_n .

(β) *Under condition* (β) *there shall exist a covering of* Γ_n *by a countable subset* K *of presentations* $(H : U, Z) \in \mathscr{D}^0\Gamma_n$ *such that the corresponding prime* C^m*-manifolds* $\mathbf{Z}_n$ (Definition 13.4) *are* C^m*-embedded in* Q_r .

Then :

(h) *Conditions* (α) *and* (β) *are equivalent.*

(k) *A* Q_r*-embedded* C^m*-manifold with carrier* Γ_n , *if it exists, is uniquely determined by* (Γ_n, Q_r, m) *and will be denoted by* $((\Gamma_n, Q_r, m))$.

The uniqueness of P_n (if P_n exists) means its independence of the different C^m-embeddings $\varphi : M_n \to Q_r$ (if any exist) which yield a manifold $\varphi(M_n)$ with carrier Γ_n .

Proof that (α) *Implies* (β). If the manifold P_n of (α) exists, each presentation $(H : U, Z)$ in $\mathscr{D}P_n$ is a presentation in $\mathscr{D}^\circ\Gamma_n$. Since each open restriction of $(H : U, Z)$ is in $\mathscr{D}P_n$, the C^m-manifold $\mathbf{Z}_n$ defined by $(H : U, Z)$ in Definition 13.4 is a submanifold of P_n . Since P_n is by hypothesis Q_r-embedded, its submanifold $\mathbf{Z}_n$ is Q_r-embedded. (see *Remark* following Definition 16.2.) Condition (β) thus follows from (α).

Proof that (β) *Implies* (α). We begin this proof by verifying the following lemma:

Lemma 16.2. *Under the conditions of* (β) *the presentations in* K *are mutually* C^m*-compatible.*

Let there be given two presentations

$$(H' : U', Z') \in K \qquad \text{and} \qquad (H'' : U'', Z'') \in K \tag{16.27}$$

such that $Z' \cap Z'' \neq \varnothing$. It follows from Definition 5.4 of C^m-compatibility that H' and H'' are C^m-compatible if and only if the following statement (i) is true:

(i) Each point $p_0 \in Z' \cap Z''$ admits an open neighborhood Z° relative to $Z' \cap Z''$ such that the restrictions

$$(\hat{H}' : \hat{U}', Z^\circ) \qquad \text{and} \qquad (\hat{H}'' : \hat{U}'', Z^\circ) \tag{16.28}$$

of the presentations (16.27) are C^m-compatible.

We shall show that the presentations (16.27) are C^m-compatible by showing that the presentations (16.28) are C^m-compatible if Z° is a sufficiently small open neighborhood of $p_\circ$ in $Z' \cap Z''$.

Turning to Definition 16.2, let $(G : V, Y) \in \mathscr{D}Q_r$ be chosen so that $p_0 \in Y$, and suppose (as is possible) that $Z^\circ \subset Y$. By hypothesis of (β) the C^m-manifolds defined by the presentations (16.27), namely, $\mathbf{Z}_n'$ and $\mathbf{Z}_n''$, are C^m-embedded in Q_r, so that by Theorem 16.4 the inclusion mapping is an embedding of $\mathbf{Z}_n'$ and $\mathbf{Z}_n''$ in Q_r. Hence by Definition 16.2 the mappings

$$G^{-1} \circ \hat{H}' : \hat{U}' \to V \tag{16.29}$$

$$G^{-1} \circ \hat{H}'' : \hat{U}'' \to V \tag{16.30}$$

are C^m-embeddings in the sense of Definition 5.1. Since these two mappings are onto the *same* topological manifold, $G^{-1}(Z^\circ)$ in V, it follows from Lemma 5.1 that they are C^m-compatible. Hence the mapping

$$(G^{-1} \circ \hat{H}')^{-1} \circ (G^{-1} \circ \hat{H}'') = (\hat{H}')^{-1} \circ \hat{H}'' : \hat{U}'' \to \hat{U}' \tag{16.31}$$

is a C^m-diffeomorphism of $\hat{U}''$ onto $\hat{U}'$. Thus $\hat{H}'$ and $\hat{H}''$ are C^m-compatible. It follows from (i) that the presentations (16.27) are C^m-compatible.

Thus Lemma 16.2 is true.

The presentations K cover Γ_n by hypothesis of (β) and are C^m-compatible by Lemma 16.2. By Definition 13.1 Γ_n is thus the carrier of a C^m-manifold P_n such that $\mathscr{D}P_n$ contains the presentations in K. That P_n is C^m-embedded in Q_r by the inclusion mapping $|P_n| \to |Q_r|$ follows from the hypothesis of (β) that the submanifolds $\mathbf{Z}_n$ of P_n defined by the respective presentations $(H : U, Z) \in K$ are also C^m-embedded in Q_r. Thus (β) implies (α).

The equivalence of (α) and (β) is established.

Proof of (k); *the Uniqueness of* P_n. Let P_n' be an arbitrary n-manifold satisfying (α). We shall prove that $P_n = P_n'$.

The proof that (α) implies (β) shows that (β) holds if K is taken as the union of a countable subset of presentations $\mathscr{D}P_n$ covering Γ_n and a countable subset of presentations of $\mathscr{D}P_n'$ covering Γ_n. By

Lemma 16.2 the presentations in this set K are mutually C^m-compatible implying that $\mathscr{D}P_n = \mathscr{D}P'_n$ and hence $P_n = P'_n$.

Thus (k) is true and the proof of Theorem 16.5 is complete.

We record the following corollary of Theorem 16.5:

Theorem 16.6. *Let M_n and Q_r, $0 < n < r$, be manifolds of class at least C^μ, and for some positive $m \leqslant \mu$ let φ be a C^m-embedding $M_n \to Q_r$. If one sets $|\varphi(M_n)| = \Gamma_n$, then*

$$\boxed{\varphi(M_n) = ((\Gamma_n, Q_r, m)).} \tag{16.32}$$

Relation (16.32) is a consequence of the uniqueness of P_n of Theorem 16.5 given Γ_n, Q_r, and m. The C^m-manifold $\varphi(M_n)$ is defined in Theorem 16.3b, while $((\Gamma_n, Q_r, m))$ is defined in Theorem 16.5.

Corollary 16.2 will be needed:

Corollary 16.2. *Let $\varphi : P_r \to Q_r$ be a C^m-diffeomorphism of a C^m-manifold P_r onto a C^m-manifold Q_r and M_n, $0 < n < r$, a C^m-manifold such that $| M_n |$ is a subspace of $| P_r |$.*

If M_n is C^m-embedded in P_r, then $\varphi \mid | M_n |$ defines a C^m-embedding of M_n in Q_r as a manifold $\varphi(M_n)$.

Whether or not M_n is embedded in Q_r by $\varphi \mid | M_n |$, a C^m-manifold $\varphi(M_n)$ with carrier $\Gamma_n = \varphi(| M_n |)$ is well-defined in accord with Theorem 16.3*b*.

Since M_n is C^m-embedded in P_r, it is C^m-embedded in P_r by the inclusion mapping I of $| M_n |$ into $| P_r |$ (Theorem 16.4). By Definition 16.2 corresponding to a prescribed point $p_0 \in | M_n |$ there accordingly exist presentations

$$(F : U, X) \in \mathscr{D}M_n \qquad \text{and} \qquad (G : V, Y) \in \mathscr{D}P_r \tag{16.33}$$

such that $X \subset Y$ and the homeomorphism

$$G^{-1} \circ F : U \to V \tag{16.34}$$

of U into V is a C^m-embedding of U into V in the sense of Definition 5.1. The mapping

$$(\varphi \circ G)^{-1} \circ (\varphi \circ F) : U \to V \tag{16.35}$$

reduces to the C^m-embedding (16.34). Since

$$(\varphi \circ G : V, \varphi(Y)) \in \mathscr{D}Q_r\,, \tag{16.36}$$

in accord with Theorem 16.3b, mappings of the type (16.35) show that $\varphi \mid \mid M_n \mid$ is a C^m-embedding of M_n in Q_r, in accord with Definition 16.2.

Regular Arcs g on M_n. Let M_n be a C^m-manifold and I an interval of the t-axis. There are special mappings

$$g : t \to p(t) : I \to \mid M_n \mid$$

which are continuous and locally biunique and which are termed *regular* arcs on M_n. If g is a homeomorphism into $\mid M_n \mid$, the image γ of I in $\mid M_n \mid$ may be regarded as the *carrier* of the arc g. Let $(F : U, X) \in \mathscr{D}M_n$ be given. If g is a homeomorphism into X, a mapping $t \to u(t) : I \to U$ such that $p(t) = F(u(t))$ is called a *representation* of g in terms of F-coordinates u. In this case g is called *regular* if the mapping $t \to u(t)$ is of class C^1 and if $\| \dot{u}(t) \|$ never vanishes.

Whether g is a homeomorphism or not, g will be called *regular* if corresponding to each $t_0 \in I$ there exists a subinterval I_0 of I, open in I, which contains t_0 and is such that the mapping $t \to p(t) : I_0 \to \mid M_n \mid$ is a regular arc with carrier in the range of some presentation in $\mathscr{D}M_n$.

One sees that the "regularity" of a simple arc g on M_n is independent of the coordinate domains in which this regularity is tested.

We shall make use of a fundamental theorem due to Whitney [2]:

Theorem 16.7. *Whitney. An abstract differentiable manifold M_n of class C^m admits a C^m-embedding in a Euclidean space of dimension at most $2n + 1$.*

Whitney's result can be formulated as follows. Corresponding to an abstract C^m-manifold M_n there exists a C^m-diff φ of M_n onto a regular C^m-manifold N_n in a Euclidean space of dimension at most $2n + 1$.

In addition to Whitney's proof [2] of this theorem and of related theorems, there is a relatively short proof by de Rham appearing early in his book [1]. There is also an introductory treatment by

Munkres in [1], pp. 16–20. Because of the accessibility of these and other expositions, we shall not give a proof of Whitney's theorem.

This is perhaps the place to state a remarkable theorem due to Milnor [1]. The terminology is ours.

Theorem 16.8. *Milnor. There exist manifolds M_n and N_n of the same class C^m with identical carriers $| M_n | = | N_n |$ but such that M_n and N_n are not diffeomorphic.*

Milnor set up his first counterexample for the case $n = 7$. In this connection there are many unanswered questions.

Functions Corresponding under a Diff φ. We shall prove a theorem concerning real-valued functions *corresponding* under a diff $p \to \varphi(p)$ of M_n onto N_n.

Theorem 16.9. *Let $p \to \varphi(p) = q$ be a C^m-diff of a C^m-manifold M_n onto a C^m-manifold N_n. Let $p \to f(p)$ and $q \to g(q)$ be real-valued functions with values defined, respectively, for $p \in M_n$ and $q \in N_n$ and such that $f(p) = g(q)$ when $q = \varphi(p)$. Suppose that $0 < \mu \leqslant m$. Then f is of class C^μ on M_n if and only if g is of class C^μ on N_n. Moreover, critical points of f and g correspond under φ, with preservation of nondegeneracy and indices when $\mu > 1$.*

We recall conventions of § 13.

Given a presentation $(F : U, X) \in \mathscr{D}M_n$, $f \mid X$ is of class C^μ if and only if $f \circ F$ is of class C^μ on U. The critical points of $f \mid X$ are represented biuniquely under F by the critical points of $f \circ F$ on U. If $\mu > 1$, a critical point p_0 of $f \mid X$ is ND and has the index k if and only if the point $u_0 = F^{-1}(p_0)$ is a ND critical point of $f \circ F$ with index k.

Given a presentation $(F : U, X) \in \mathscr{D}M_n$, Theorem 16.3a affirms that $(\varphi \circ F : U, \varphi(X))$ is a presentation in $\mathscr{D}M_n$. If g is an *arbitrary* real-valued function on N_n, then the conventions of the preceding paragraph applied to N_n in place of M_n imply that the differentiability of $g \mid \varphi(X)$, its critical points and their nondegeneracy and indices are determined by the representation $g \,\bar{\circ}\, (\varphi \,\bar{\circ}\, F)$ of $g \mid \varphi(X)$ on U. In the *special* case in which $g(q) = f(p)$ under the diff $p \to \varphi(p) = q$,

$$g \,\bar{\circ}\, (\varphi \,\bar{\circ}\, F) = (f \circ \varphi^{-1}) \,\bar{\circ}\, (\varphi \,\bar{\circ}\, F) = f \circ F \qquad (\text{on } U)$$

and the theorem follows.

Theorem 16.10 is an extension of Theorem 6.4a. It is a consequence of Theorems 6.4a, 16.7, and 16.9.

Theorem 16.10. *Let M_n be an abstract manifold of class C^μ, $\mu > 1$, which admits a C^μ-diff onto a regular C^μ-manifold N_n in E_r, $0 < n < r$. There then exist r real-valued functions $\psi_1, \ldots, \psi_r$ of class C^μ on M_n with the following property*:

Corresponding to a prescribed real-valued function g of class C^m on M_n, $\mu \geqslant m > 1$, the function

$$q \to g(q) + \sum_{j=1}^{r} a_j \psi_j(q) = g^a(q), \qquad q \in M_n \tag{16.37}$$

is ND *for almost all choices of points $(a_1, \ldots, a_r) = a \in E_r$.*

Proof. One identifies N_n of this theorem with the regular manifold Σ of Theorem 6.4′. Let

$$p \to h(p) : \Sigma \to M_n$$

be a C^μ-diff of Σ onto M_n. Corresponding to the C^μ-functions $\varphi_1, \ldots, \varphi_r$ on Σ of Theorem 6.4′ let $\psi_1, \ldots, \psi_r$ be the respective C^μ-functions on M_n such that

$$\psi_j(q) = \varphi_j(p) \qquad \text{for} \quad q = h(p); \quad j = 1, \ldots, r.$$

Corresponding to the C^m-function $q \to g(q)$ prescribed on M_n, let $p \to f(p)$ be the C^m-function on Σ such that $g(q) = f(p)$, subject to the condition $q = h(p)$ for each $p \in \Sigma$. Then (16.37) and (6.25) show that

$$f^a(p) = g^a(q) \qquad \text{for} \quad q = h(p).$$

According to Theorem 16.9 f^a and g^a are both nondegenerate or both degenerate on Σ and M_n, respectively.

Theorem 16.10 follows from Theorem 6.4′.

EXERCISE 16.1. Let there be given a presentation

$$(F : U, X) \in \mathscr{D}M_n \qquad (U \text{ open in } E_n) \tag{16.38}$$

of a C^m-manifold M_n. Let $\mathbf{U}$ be the regular submanifold of E_n with carrier U (see § 5). Let $\mathbf{X}$ be the C^m-submanifold of M_n with carrier X. Use Definition 16.2 for a trivial verification that F C^m-embeds $\mathbf{U}$ in $\mathbf{X}$ onto $\mathbf{X}$. Then use Theorem 16.2b to verify the following lemma.

Lemma 16.3. *The presentation F of* (16.38) *can be regarded as a C^m-diff of the manifold* $\mathbf{U}$ *onto the manifold* $\mathbf{X}$.

We shall make use of this lemma.

EXERCISE 16.2. Prove the following lemma:

Lemma 16.4. *Let there be given positive integers $s \leqslant n \leqslant r$ and C^m-manifolds M_s and N_n in E_r, of which N_n is regular in E_r in the sense of Definition* 5.3. *If $| M_s | \subset | N_n |$, then M_s is C^m-embedded in N_n if and only if M_s is a regular C^m-manifold in E_r.*

Verify the validity of the following abstract of a proof: Given a point $x_0 \in | M_s |$, there exist presentations

$$(F : U, X) \in \mathscr{D}M_s \quad \text{and} \quad (G : V, Y) \in \mathscr{D}N_n, \quad x_0 \in X \subset Y$$

of which G is necessarily regular in E_r. If the coordinates of $x \in E_r$ are $x_1, \ldots, x_r$, we can suppose that for $x \in Y$, $x_i = G_i(v)$ for $i = 1, \ldots, r$ and suitable $v \in V$. Since $X \subset Y$, there exists a homeomorphism $u \to \varphi(u)$ of U onto $\varphi(U) \subset V$ such that for each i

$$F_i(u_1, \ldots, u_s) = G_i(\varphi_1(u), \ldots, \varphi_n(u)), \qquad u \in U. \tag{16.39}$$

Show that the homeomorphism φ is a C^m-diff of U onto $\varphi(U) \subset V$ if and only if F is a regular C^m-presentation of X in E_r.

Show also that φ is a C^m-diff if and only if the submanifold of M_s with carrier X is C^m-embedded in N_n by the inclusion. Use Theorem 16.4 to finish the proof.

§ 17

DYNAMICAL SYSTEMS ON A MANIFOLD M_n

As we have seen, a point q on a C^m-manifold M_n is represented by an infinity of points in E_n, namely, the antecedents of q under those presentations in $\mathscr{D}M_n$ whose coordinate domains contain q. A vector ξ "tangent" to M_n at q will be determined by an infinity of vectors in E_n, one at each point in E_n antecedent to q. It is an objective of this section to make clear in the classical sense how the vectors in E_n which represent ξ are defined and related.

A similar but somewhat more difficult problem is the proper definition of a "dynamical system" on M_n. Such systems, if defined on suitable open submanifolds of M_n, are essential in defining the homotopies necessary in proving the theorems of § 9.

The relevant relations of vectors in E_n one to the other are of two sorts, "dual" in a special sense, and termed *contravariant* and *covariant*. In this section we shall define the relation of contravariance and use it to define "dynamical systems" on M_n.

Notation. Given a point $u \in E_n$, n-tuples

$$(\eta^1(u),\ldots, \eta^n(u)) = \eta(u) \qquad \text{and} \qquad (c_1(u),\ldots, c_n(u)) = c(u) \tag{17.1}$$

are called *vectors* in E_n "at u". It is classical usage to employ superscripts in denoting the components of vectors involved in relations of contravariance and subscripts in denoting the components of vectors involved in relations of covariance. The symbols $\eta(u)$ and $c(u)$ as used to represent the vectors in (17.1) distinguish only by superscript or subscript the kind of vector they represent. However, note that η is a Greek letter and c a Roman letter. We make the convention that vectors whose components are to be distinguished by superscripts will be represented by *Greek* letters and vectors whose components are to be distinguished by subscripts will be represented by *Roman* letters.

The Vector Space $\mathbf{V}_u$ *over* R. For a fixed point $u \in E_n$ the set of vectors $(\eta(u))$ in E_n "at u" form a vector space $\mathbf{V}_u$ over the field of real numbers R. One adds vectors at u by adding their components. If $\rho \in R$, we understand that

$$\rho(\eta(u)) = (\rho\eta^1(u),..., \rho\eta^n(u)).$$

One sees that the vectors in E_n at u whose components are the rows of the n-square unit matrix I form a base for $\mathbf{V}_u$, so that $\mathbf{V}_u$ has the dimension n (see Birkhoff and MacLane [1], p. 192). A vector space $\mathbf{V}_u$ in E_n admits a natural isomorphism onto E_n in which a *vector* $\eta \in \mathbf{V}_u$ is mapped into the *point* in E_n with coordinates $\eta^1,..., \eta^n$.

We shall not always represent vectors in $\mathbf{V}_u$ in the form $\eta = (\eta^1,..., \eta^n)$, but at convenience shall represent a vector $g \in \mathbf{V}_u$ in the form $g = (g_1,..., g_n)$. The choice between these two ways of representing a vector in $\mathbf{V}_u$ will depend upon the potential use of the vector (see Theorem 18.2).

Let M_n be a C^∞-manifold and

$$(F : U, X) \in \mathscr{D}M_n \qquad \text{and} \qquad (G : V, Y) \in \mathscr{D}M_n \tag{17.2}$$

be overlapping presentations, that is, presentations such that $X \cap Y \neq \varnothing$. For $q \in X \cap Y$, points $u = F^{-1}(q)$ and $v = G^{-1}(q)$ will be said to be M_n-*related* by F and G. For such points a vector $\eta(u) \in \mathbf{V}_u$ will be symmetrically related by an isomorphism (denoted by $[F, G]_q$) of $\mathbf{V}_u$ onto $\mathbf{V}_v$ to a vector $\zeta(v) \in \mathbf{V}_v$. This isomorphism and its inverse $[G, F]_q$ will presently be defined.

EXAMPLE 17.1. Given the presentations (17.2) and $q \in X \cap Y$, let $\gamma : t \to p(t)$ be a simple regular arc in $X \cap Y$ meeting q when $t = t_0$. Let

$$\gamma_F : t \to u(t) \qquad \text{and} \qquad \gamma_G : t \to v(t) \tag{17.3}$$

be the arcs in U and V, respectively, antecedent to γ under F and G. Then γ_F meets $u_0 = F^{-1}(q)$ and γ_G meets $v_0 = G^{-1}(q)$ when $t = t_0$.

In the sense of a forthcoming definition, the vector $\dot{u}(t_0) \in \mathbf{V}_{u_0}$ and $\dot{v}(t_0) \in \mathbf{V}_{v_0}$ are *contravariantly related* by F and G.

We shall recall the "transition diffs" associated with F and G:

The Transition Diffs Defined by F and G. Given the overlapping presentations F and G of (17.2), set

$$F^{-1}(X \cap Y) = \hat{U} \qquad \text{and} \qquad G^{-1}(X \cap Y) = \hat{V}. \tag{17.4}$$

The C^∞-mapping

$$u \to \mathbf{v}(u) = (G^{-1} \circ F)(u) : \hat{U} \to \hat{V} \tag{17.5}$$

is the *transition diff* of $\hat{U}$ onto $\hat{V}$ defined by F and G. The inverse of this diff is the transition diff

$$v \to \mathbf{u}(v) = (F^{-1} \circ G)(v) : \hat{V} \to \hat{U} \tag{17.6}$$

of $\hat{V}$ onto $\hat{U}$. Points $u \in U$ and $v \in V$ are "M_n-related" by F and G if and only if $u \in \hat{U}$ and $v = \mathbf{v}(u)$, or equivalently, if $v \in \hat{V}$ and $u = \mathbf{u}(v)$.

Let $\mathbf{J}(\mathbf{v} : u)$ and $\mathbf{J}(\mathbf{u} : v)$ be, respectively, the Jacobian matrices of the above transition diffs $\mathbf{v}$ and $\mathbf{u}$ evaluated at u and v. When u and v are M_n-related by F and G one has the matrix relations

$$\mathbf{J}(\mathbf{v} : u)\, \mathbf{J}(\mathbf{u} : v) = I = \mathbf{J}(\mathbf{u} : v)\, \mathbf{J}(\mathbf{v} : u), \tag{17.7}$$

where I is the n-square unit matrix. For i, k on the range $1,\ldots, n$ let

$$J_i^k(\mathbf{v} : u) \qquad \text{and} \qquad J_i^k(\mathbf{u} : v) \tag{17.8}$$

be the elements in the kth row and ith column, respectively, of the matrices $\mathbf{J}(\mathbf{v} : u)$ and $\mathbf{J}(\mathbf{u} : v)$.

We introduce two fundamental isomorphisms:

The Isomorphisms $[F, G]_q$ *and* $[G, F]_q$. These isomorphisms are defined for each $q \in X \cap Y$. If $u = F^{-1}(q)$ and $v = G^{-1}(q)$, there exists an isomorphism (onto)

$$[F, G]_q : \eta \to \zeta : \mathbf{V}_u \to \mathbf{V}_v$$

of the form

$$\zeta^k = J_i^k(\mathbf{v} : u)\, \eta^i, \qquad k = 1,\ldots, n \tag{17.9}$$

with an inverse isomorphism

$$[G, F]_q : \zeta \to \eta : \mathbf{V}_v \to \mathbf{V}_u$$

of the form

$$\eta^k = J_i^k(\mathbf{u} : v)\, \zeta^i, \qquad k = 1,..., n. \tag{17.10}$$

Definition 17.1. **Contravariance.** Let F and G be the overlapping presentations given in (17.2). If $u \in U$ and $v \in V$ are M_n-related by F and G, a vector $\eta \in \mathbf{V}_u$ and a vector $\zeta \in \mathbf{V}_v$ are said to be *contravariantly related* if (17.9) and (17.10) hold.

EXAMPLE 17.1 (continued). Under the conditions of Example 17.1 the relation $F(u(t)) = G(v(t))$ is an identity, implying the identities

$$u(t) = \mathbf{u}(v(t)) \qquad \text{and} \qquad v(t) = \mathbf{v}(u(t)). \tag{17.11}$$

Since the points $u(t_0) = u_0$ and $v(t_0) = v_0$ are M_n-related by hypothesis, differentiation of the identities (17.11) shows that the vectors $\dot{u}(t_0) \in \mathbf{V}_{u_0}$ and $\dot{v}(t_0) \in \mathbf{V}_{v_0}$ are contravariantly related by F and G.

For us the most important application of the notion of vectors in E_n contravariantly related by presentations in $\mathscr{D}M_n$ is in the definition of a "dynamical system on M_n." Such systems are a major source of homotopies and isotopies on M_n.

Dynamical Systems on M_n. Before defining a dynamical system on M_n we introduce a convention belonging to Euclidean geometry:

Convention. Let W be a nonempty open subset of E_n with rectangular coordinates $w_1,..., w_n$. A C^m-mapping of W into E_n of the form

$$w \to \varphi(w) = (\varphi^1(w),..., \varphi^n(w)), \qquad w \in W \tag{17.12}$$

will be termed (at convenience) a C^m-*family* of *vectors* $(\varphi^1(w),..., \varphi^n(w)) = \varphi(w)$ in E_n over W. We shall term $\varphi(w)$ *the vector of the family "at w."* We understand that this vector is an element in the vector space $\mathbf{V}_w$.

Definition 17.2. *Contravariantly Related Vector Families in* E_n. Let $(F : U, X)$ and $(G : V, Y)$ be overlapping presentations given as in (17.2) and let

$$u \to \eta(u) = (\eta^1(u),..., \eta^n(u)), \qquad u \in U, \tag{17.13}$$

$$v \to \zeta(v) = (\zeta^1(v),..., \zeta^n(v)), \qquad v \in V, \tag{17.14}$$

be C^∞-mappings, respectively, of U and V into E_n. The vector families "defined" by these mappings will be termed *contravariantly related* by F and G if for each point $q \in X \cap Y$ and corresponding $u = F^{-1}(q)$ and $v = G^{-1}(q)$ the vectors $\eta(u) \in \mathbf{V}_u$ and $\zeta(v) \in \mathbf{V}_v$ are contravariantly related by F and G in the sense of Definition 17.1.

Trivial Contravariance. Let F and G be two presentations in $\mathscr{D}M_n$ which are *nonoverlapping*. Suppose vector families over U and V, respectively, are defined as in (17.13) and (17.14). There are no points $u \in U$ and $v \in V$ which are M_n-related by F and G. It is natural to say that the families $u \to \eta(u)$ and $v \to \zeta(v)$ are *trivially* contravariantly related by F and G.

Definition 17.3. *Dynamical Systems on* M_n. A dynamical system of class C^∞ on M_n satisfies the three following conditions. In these conditions the indexing subscript H, F, or G is a presentation in $\mathscr{D}M_n$.

Condition I. On the Euclidean domain of each presentation

$$(H : W_H, Z_H) \in \mathscr{D}M_n \qquad (W_H \text{ open in } E_n) \tag{17.15}$$

there is given a system of ordinary differential equations

$$dw^i/dt = \varphi_H{}^i(w^1,..., w^n), \qquad i = 1,..., n; \quad w \in W_H. \tag{17.16}$$

Condition II. For arbitrary $w \in W_H$ the vector $\varphi_H(w)$ is the vector "at w" of a "C^∞-family of vectors" over W_H vanishing at no point $w \in W_H$.

Condition III. If F and G are arbitrary presentations in $\mathscr{D}M_n$, the vector families φ_F and φ_G are contravariantly related by F and G. (Definition 17.2).

We shall indicate a sense in which Condition III is necessary if there is to be one and only one "trajectory" of a dynamical system on M_n meeting each point of M_n.

Notation. Given $H \in \mathscr{D}M_n$, a system of differential equations (17.16) in E_n satisfying Condition II will be denoted by d/φ_H. An ensemble of systems d/φ_H indexed by $H \in \mathscr{D}M_n$ will be denoted by $\{\mathbf{d}/\varphi_H\}$ and termed an ensemble $\{\mathbf{d}/\varphi_H\}$ over $\mathscr{D}M_n$. An ensemble $\{\mathbf{d}/\varphi_H\}$ over M_n upon whose members d/φ_H no Condition III is imposed might be called a *free ensemble* over $\mathscr{D}M_n$. An ensemble $\{\mathbf{d}/\varphi_H\}$ whose members d/φ_H, taken pairwise satisfy Condition III is called a *dynamical system* on M_n.

We shall term a system d/φ_H a *local* dynamical system to distinguish it from the ensemble $\{\mathbf{d}/\varphi_H\}$ of which it is a member.

Theorem 17.1 below justifies the imposition of Condition III in defining dynamical systems. In Theorem 17.1 we shall refer to trajectory-wise compatible local systems d/φ_F and d/φ_G of differential equations. We abbreviate "trajectory-wise compatible systems" as T-*comp systems*. We define this term as follows:

Definition 17.4. T-*Comp Systems* d/φ_F *and* d/φ_G. Let F and G be overlapping presentations as given in (17.2). Given $q \in X \cap Y$, set $u_0 = F^{-1}(q)$ and $v_0 = G^{-1}(q)$. Given t_0, let $t \to u(t)$ and $t \to v(t)$ be solutions, respectively, of d/φ_F and d/φ_G whose graphs meet, respectively, (t_0, u_0) and (t_0, v_0). Then d/φ_F and d/φ_G are said to be T-*comp at* q if for some $e > 0$

$$F(u(t)) = G(v(t)), \qquad |t - t_0| < e. \tag{17.17}$$

More generally, d/φ_F and d/φ_G are termed T-*comp* if T-comp at each point $q \in X \cap Y$.

If F and G are nonoverlapping, d/φ_F and d/φ_G are considered T-comp in a trivial sense.

Theorem 17.1. *Let F and G be arbitrary presentations in $\mathscr{D}M_n$. A necessary and sufficient condition that local systems d/φ_F and d/φ_G be trajectorywise compatible is that the vector families φ_F and φ_G be contravariantly related by F and G* (Definition 17.2).

The Condition is Necessary. This is trivial when F and G are nonoverlapping.

Suppose then that F and G are overlapping and given as in (17.2). By hypothesis d/φ_F and d/φ_G are "T-comp" at a prescribed point $q \in X \cap Y$, so that, in the notation of Definition 17.4, (17.17) holds. It follows, as in Example 17.1, that the vectors $\dot{u}(t_0) \in \mathbf{V}_{u_0}$ and $\dot{v}(t_0) \in \mathbf{V}_{v_0}$ are contravariantly related by F and G. But $\dot{u}(t_0) = \varphi_F(u_0)$ and $\dot{v}(t_0) = \varphi_G(v_0)$. Since q was prescribed in $X \cap Y$, we can infer that the vector families φ_F and φ_G are contravariantly related by F and G in the sense of Definition 17.2.

The Condition is Sufficient. Suppose that F and G are again given as in (17.2). Let solutions $t \to u(t)$ and $t \to v(t)$ be defined as in Definition 17.4. Assuming that the vector families φ_F and φ_G are contravariantly related, we must prove that (17.17) holds for some $e > 0$.

Let e be so small that the solutions $t \to u(t)$ and $t \to v(t)$ are defined for $| t - t_0 | < e$. In terms of the diff $u \to \mathbf{v}(u)$ of (17.5), set $\bar{v}(t) = \mathbf{v}(u(t))$ for $| t - t_0 | < e$. Then by virtue of (17.5)

$$G(\bar{v}(t)) = F(u(t)), \qquad | t - t_0 | < e. \tag{17.18}$$

To show that (17.17) holds it is sufficient to show that $\bar{v}(t) = v(t)$ when $| t - t_0 | < e$. Since $\bar{v}(t_0) = \mathbf{v}(u_0) = v_0$ by hypothesis, it is sufficient to prove the following:

(α) *The mapping $t \to \bar{v}(t)$ is a solution of the local system d/φ_G.*

Since $\bar{v}(t) = \mathbf{v}(u(t))$ for $| t - t_0 | < e$, we infer that

$$\dot{\bar{v}}^k(t) = J_i^k(\mathbf{v} : u(t))\, \dot{u}^i(t) = J_i^k(\mathbf{v} : u(t))\, \varphi_F{}^i(u(t)) = \varphi_G{}^k(\bar{v}(t)), \qquad k = 1,..., n$$

where the last equality is valid because the vector families φ_F and φ_G are by hypothesis contravariantly related by F and G. Thus $t \to \bar{v}(t)$ is a solution of d/φ_G, as affirmed in (α).

It follows that (17.17) holds, so that the condition of the theorem is sufficient.

This completes the proof of Theorem 17.1.

Definition 17.5. *Solutions of a Dynamical System.* A simple regular arc $\gamma : t \to p(t)$ on M_n whose t-domain is an open interval of the t-axis will be called a *solution* of a dynamical system $\{\mathbf{d}/\varphi_H\}$ on M_n if the following is true: Corresponding to a prescribed point q of γ there exists a solution $t \to u(t)$ of some local system $d/\varphi_F \in \{\mathbf{d}/\varphi_H\}$ whose F-image is a subarc of γ meeting q.

We shall characterize the special dynamical systems most useful in defining homotopies in the critical point theory.

f-Transverse Dynamical Systems on M_n^f. Let f be a real-valued ND C^∞-function on M_n. Let M_n^f denote the submanifold of M_n from which the critical points of f have been deleted. We suppose M_n^f given the differentiable structure induced by that of M_n. That is, the presentations in $\mathscr{D}M_n^f$ shall be the subset of the presentations in $\mathscr{D}M_n$ whose coordinate domains are open subspaces of $| M_n^f |$.

A simple regular arc $t \to p(t)$, $a < t < b$, on M_n^f will be termed *f-transverse* if the t-derivative of $f(p(t))$ is positive for all $t \in (a, b)$. A dynamical system $\{\mathbf{d}/\varphi_H\}$ on M_n^f will be termed *f-transverse* if each simple arc $t \to p(t)$ on M_n^f which is a solution of the dynamical system is f-transverse.

Given a ND C^∞-function f on M_n, we seek an f-transverse dynamical system on M_n^f. By converting M_n into a Riemannian manifold of class C^∞ as in § 19 this objective will be reached. The local system $d/\varphi_H \in \{\mathbf{d}/\varphi_H\}$ will be uniquely determined by $f \circ H$ and the Riemannian form to be associated with the presentation H.

The Second Reduction Theorem for f. The classical conversion of M_n into a Riemannian manifold in § 19 will be followed in § 22 by a nonclassical modification of the resulting Riemannian metric of M_n as introduced by Morse [9] Lemma 6.1.

A presentation $(F : U, X)$ in $\mathscr{D}M_n$ such that the Euclidean length of a regular arc $g \in U$ equals the Riemannian length of $F(g)$ on M_n will be called an *isometric* presentation. Such presentations will exist if the Riemannian metric on M_n is specially conditioned as in § 22. However, Lemma 6.1 of Morse [9] can be formulated as follows:

Theorem 17.2. *The Second Reduction Theorem for f. Given a C^∞-real-valued* ND *function f on a C^∞-manifold M_n, M_n can be converted into a Riemannian manifold in such a manner that the following is true:*

Corresponding to each critical point q of f of index k and critical value c there exists in $\mathscr{D}M_n$ an isometric presentation $(F : U, X)$ *of a neighborhood X of q on M_n such that $F(0) = q$ and*

$$(f \circ F)(u) - c = -u_1^2 - \cdots - u_k^2 + u_{k+1}^2 + \cdots + u_n^2, \qquad u \in U. \tag{17.19}$$

The existence of these isometric presentations will simplify problems of homology, homotopy, and "surgery" related to the critical points of f and their addition and elimination.

Program. f-Transverse dynamical systems have been defined in this section. Once M_n has been assigned a standard Riemannian structure in § 19 f-transverse dynamical systems can be proved to exist on M_n^f, that is, on M_n with the critical points of f deleted. But the standard Riemannian structure, namely the first Riemannian structure assigned to M_n in § 19, must be further modified in § 22 to satisfy the Second Reduction Theorem. Theorem 22.2 serves this purpose.

§ 18

TANGENT AND COTANGENT VECTORS OF M_n

Let M_n be a C^∞-manifold. We shall presently define a vector tangent to M_n at a point $q \in M_n$. To motivate this definition we shall examine tangents to a regular manifold in a Euclidean space E_r.

Vectors Tangent to a Regular manifold M_n in E_r. Suppose presentations F and $G \in \mathscr{D}M_n$ are given as in (17.2). For $u \in U$ and $v \in V$, $F(u)$ and $G(v)$ are points

$$(F^1(u),\ldots, F^r(u)) \qquad \text{and} \qquad (G^1(v),\ldots, G^r(v))$$

in E_r. Suppose that $q \in X \cap Y$. Set $u = F^{-1}(q)$. A non-null vector [cf. (17.1)] $\eta \in \mathbf{V}_u$ may be associated with that non-null vector ξ_η in E_r tangent to M_n at q, whose μth component in E_r is

$$\xi_\eta{}^\mu = \frac{\partial F^\mu}{\partial u^i}(u)\eta^i, \qquad \mu = 1,\ldots, r. \tag{18.1}$$

Set $v = G^{-1}(q)$. A non-null vector $\zeta \in \mathbf{V}_v$ may be similarly associated with the vector ξ_ζ in E_r tangent to M_n at q with components

$$\xi_\zeta{}^\mu = \frac{\partial G^\mu}{\partial v^i}(v)\zeta^i, \qquad \mu = 1,\ldots, r. \tag{18.2}$$

The following lemma serves as a partial justification of Definition 18.1 of a vector tangent to M_n at q:

Lemma 18.1. *Suppose that M_n is a regular manifold in E_r, that F and G are the overlapping presentations in $\mathscr{D}M_n$ given in* (17.2), *and that $q \in X \cap Y$. Set $u_0 = F^{-1}(q)$, $v_0 = G^{-1}(q)$.*

A necessary and sufficient condition that the vectors ξ_η and ξ_ζ tangent to M_n at q and defined respectively by (18.1) *and* (18.2) *be equal is that the vectors $\eta \in \mathbf{V}_{u_0}$ and $\zeta \in \mathbf{V}_{v_0}$ be contravariantly related by F and G.*

Prior to the proof proper recall the following: If $\gamma : t \to p(t)$ is a simple regular arc in $X \cap Y$ meeting q when $t = t_0$ and if $t \to u(t)$ and $t \to v(t)$ are the arcs in U and V antecedent to γ under F and G, respectively, then $F(u(t)) = G(v(t))$, and hence

$$\frac{\partial F^\mu}{\partial u^i}(u(t))\,\dot{u}^i(t) = \frac{\partial G^\mu}{\partial v^i}(v(t))\,\dot{v}^i(t), \qquad \mu = 1,\ldots, r. \tag{18.3}$$

Note that $u_0 = u(t_0)$ and $v_0 = v(t_0)$. According to Example 17.1 the vectors $\dot{u}(t_0) \in \mathbf{V}_{u_0}$ and $\dot{v}(t_0) \in \mathbf{V}_{v_0}$ are contravariantly related by F and G.

The Condition of the Lemma is Sufficient. We are assuming that the vectors $\eta \in \mathbf{V}_{u_0}$ and $\zeta \in \mathbf{V}_{v_0}$ are contravariantly related by F and G and wish to prove $\xi_\eta = \xi_\zeta$.

The above arc γ can be chosen in many ways so that $\eta = \dot{u}(t_0)$. If γ is so chosen, $\zeta = \dot{v}(t_0)$, since $\dot{v}(t_0)$ is contravariantly related to $\dot{u}(t_0)$ and ζ is contravariantly related to η. Hence (18.3) holds with $\dot{u}(t_0)$ and $\dot{v}(t_0)$ replaced, respectively, by η and ζ. It follows that $\xi_\eta = \xi_\zeta$.

The Condition of the Lemma is Necessary. By hypothesis

$$\xi_\eta{}^\mu = \frac{\partial F^\mu}{\partial u^i}(u_0)\eta^i = \frac{\partial G^\mu}{\partial v^i}(v_0)\zeta^i = \xi_\zeta{}^\mu, \qquad \mu = 1,\ldots, r. \tag{18.4}$$

We wish to prove that η and ζ are contravariantly related.

Choose γ as previously so that $\dot{u}(t_0) = \eta$. We infer from (18.3) that

$$\frac{\partial F^\mu}{\partial u^i}(u_0)\eta^i = \frac{\partial G^\mu}{\partial v^i}(v_0)\,\dot{v}^i(t_0), \qquad \mu = 1,\ldots, r, \tag{18.5}$$

and from (18.4) and (18.5) that $\zeta = \dot{v}(t_0)$. Thus $\eta \in \mathbf{V}_{u_0}$ and $\zeta \in \mathbf{V}_{v_0}$ are contravariantly related by F and G, since $\dot{u}(t_0) \in \mathbf{V}_{u_0}$ and $\dot{v}(t_0) \in \mathbf{V}_{v_0}$ are so related.

This establishes Lemma 18.1.

Vectors Tangent to M_n at q. The definition of a vector tangent to M_n at q calls for a notational introduction.

The Subset $(\mathscr{D}M_n)_q$ of $\mathscr{D}M_n$. Given $q \in M_n$, let $(\mathscr{D}M_n)_q$ denote the subset of the presentations $H \in \mathscr{D}M_n$ whose coordinate domains contain q.

The Product Space Π_q. Corresponding to a presentation $H \in (\mathscr{D}M_n)_q$, for $w = H^{-1}(q)$ denote $\mathbf{V}_w$ by $\mathbf{V}_H{}^q$. We introduce the infinite formal set product

$$\Pi_q = \underset{H \in (\mathscr{D}M)_q}{\text{Set product}} \ \mathbf{V}_H{}^q. \tag{18.6}$$

An element $\mathbf{z} \in \Pi_q$ is an ensemble of vectors one, z_H, from each "factor space" $\mathbf{V}_H{}^q$ of Π_q. The vector z_H will be represented as convenient in one of the two forms: $((z_H)^1,..., (z_H)^n)$, and $((z_H)_1, ..., (z_H)_n)$. The vector z_H will be termed the "factor" of $\mathbf{z} \in \Pi_q$ with presentation index H. The word "factor" used in this connection has no connotation beyond the one just assigned to it. For us this terminology is permanent. The set product Π_q can be converted into a vector space over R as follows.

If $\mathbf{z}'$ and $\mathbf{z}''$ are in Π_q, the sum $\mathbf{z} = \mathbf{z}' + \mathbf{z}''$ is defined by the condition that

$$z_H = z'_H + z''_H, \qquad H \in (\mathscr{D}M_n)_q. \tag{18.7}$$

If $\rho \in R$ and $\mathbf{z} \in \Pi_q$, $\rho\mathbf{z}$ is defined by the condition

$$(\rho\mathbf{z})_H = \rho z_H, \qquad H \in (\mathscr{D}M_n)_q. \tag{18.8}$$

Definition 18.1. *A vector tangent to M_n at q is a vector $\mathbf{z} \in \Pi_q$ with the following property*: Whenever F and G are presentations in $(\mathscr{D}M_n)_q$, the "factors" z_F and z_G of $\mathbf{z}$ are vectors in $\mathbf{V}_F{}^q$ and $\mathbf{V}_G{}^q$, respectively, which are contravariantly related by F and G.

The Vector Space T_q Tangent to M_n at q. The vectors $\mathbf{z} \in \Pi_q$ which are "tangent" to M_n at q, form a subset T_q of Π_q which is a "vector subspace" of Π_q over R, since the subset T_q of Π_q is "closed" in Π_q under the operations of addition and of multiplication by $\rho \in R$, as defined above over Π_q. This is a consequence of the properties of the isomorphisms $[F, G]_q$ defined in §17 for each pair of presentations F and G in $(\mathscr{D}M_n)_q$.

We add the following theorem:

Theorem 18.1. *Let q be an arbitrary point in M_n and H a presentation in $(\mathscr{D}M_n)_q$. Then the vector space T_q tangent to M_n at q is isomorphic to the vector space $\mathbf{V}_H{}^q$ under the linear homomorphism φ_q which maps each vector $\mathbf{z}$ in T_q into its "factor" z_H in $\mathbf{V}_H{}^q$.*

Proof. The mapping φ_q is obviously biunique and onto. As a linear homomorphism, it is then an isomorphism (see Birkhoff and Mac Lane [1] p. 224).

Covariance. Let F and G be arbitrary overlapping presentations in $\mathscr{D}M_n$ as given in (17.2) and $u \in U$ and $v \in V$ be such that

$$F(u) = G(v) = q \in M_n . \tag{18.9}$$

The relation of covariance, as it will be defined by F and G between two vectors

$$h = (h_1, \ldots, h_n) \in \mathbf{V}_u \qquad \text{and} \qquad g = (g_1, \ldots, g_n) \in \mathbf{V}_v , \tag{18.10}$$

is regarded as "dual" to the relation of contravariance as defined by F and G between two vectors $\eta \in \mathbf{V}_u$ and $\zeta = \mathbf{V}_v$ (Definition 17.1). In defining this relation, the matrix $\mathbf{J}(\mathbf{u} : v)$ of the transformation (17.10) corresponds formally to the matrix $\mathbf{J}^{\text{tr}}(\mathbf{u} : v)$ of the transformation

$$g_k = J_k{}^j(\mathbf{u} : v) h_j , \qquad k = 1, \ldots, n. \tag{18.11}$$

Vectors $h \in \mathbf{V}_u$ and $g \in \mathbf{V}_v$ which are related as in (18.11) are said to be *convariantly related* by F and G.

Similarly, the matrix $\mathbf{J}(\mathbf{v} : u)$ of the transformation (17.9) corresponds to the matrix $\mathbf{J}^{\text{tr}}(\mathbf{v} : u)$ of the transformation

$$h_k = J_k{}^j(\mathbf{v} : u)\, g_j , \qquad k = 1, \ldots, n, \tag{18.12}$$

inverse to the transformation (18.11).

Covariantly Related Gradients. Corresponding to a prescribed $f \in C^\infty(M_n)$ (see §13) and to an arbitrary presentation $H \in \mathscr{D}M_n$, $f \circ H$ is the *representation* of f on the E_n-domain of H. With gradients understood in the classical sense, the vector $(\text{grad}(f \circ H))(w)$, evaluated at a point w in the E_n-domain of H, will be regarded as a vector in $\mathbf{V}_w$.

We refer to the presentations F and G of (17.2) and to points $u \in U$, $v \in V$ such that $F(u) = G(v)$, and introduce the vectors

$$(\text{grad}(f \circ F))(u) = (h_1, \ldots, h_n) \in \mathbf{V}_u \tag{18.13}$$

and

$$(\text{grad}(f \circ G))(v) = (g_1, \ldots, g_n) \in \mathbf{V}_v. \tag{18.14}$$

In the notation of (17.5) and (17.6) the identities

$$(f \circ F)(u) = (f \circ G)(\mathbf{v}(u)), \qquad u \in \hat{U}, \tag{18.15}$$

and

$$(f \circ G)(v) = (f \circ F)(\mathbf{u}(v)), \qquad v \in \hat{V}, \tag{18.16}$$

are valid. On differentiating the members of (18.15) with respect to u^k and the members of (18.16) with respect to v^k, one finds that

$$h_k = J_k{}^j(\mathbf{v} : u) g_j, \qquad k = 1, \ldots, n, \tag{18.17}$$

and

$$g_k = J_k{}^j(\mathbf{u} : v) h_j, \qquad k = 1, \ldots, n. \tag{18.18}$$

We have thus proved the following:

Lemma 18.2. *If F and G are the overlapping presentations given in (17.2) and if $u \in U$ and $v \in V$ are such that $F(u) = G(v)$, then the gradients*

$$(\text{grad}(f \circ F))(u) \in \mathbf{V}_u \qquad \text{and} \qquad (\text{grad}(f \circ G))(v) \in \mathbf{V}_v$$

are covariantly related by F and G.

The Cotangent Vector Space T_q^* at $q \in M_n$. We shall define a vector space T_q^* over R associated with a prescribed point $q \in M_n$

and formed by using the notion of covariance as T_q was formed by using the dual notion of contravariance. Use will be made of the product space Π_q introduced in (18.6) with its "factors" $\mathbf{V}_H{}^q$ indexed by the presentations $H \in (\mathscr{D}M_n)_q$.

Definition 18.2. *A Cotangent Vector at $q \in M_n$.* Such a vector is an element $\mathbf{z} \in \Pi_q$ with the following property: Whenever F and G are presentations in $(\mathscr{D}M_n)_q$, the "factors" z_F and z_G of $\mathbf{z}$ in $\mathbf{V}_F{}^q$ and $\mathbf{V}_G{}^q$, respectively, are *covariantly* related by F and G.

The Cotangent Vector Space T_q^*. The set of all cotangent vectors $\mathbf{z}$ in Π_q form a vector subspace T_q^* of Π_q over R termed the "dual" of T_q. By a proof similar to that of Theorem 18.1 one shows that if $H \in (\mathscr{D}M_n)_q$, the vector space T_q^* is isomorphic to the vector space $\mathbf{V}_H{}^q$, "factoring" Π_q, under an isomorphism in which a cotangent in T_q^* goes into its "factor" in $\mathbf{V}_H{}^q$.

The set of cotangent vectors $(\mathbf{grad}\, f)(q)$ now to be defined contains all cotangent vectors in T_q^*.

Definition 18.3. *The Cotangent Vector* $(\mathbf{grad}\, f)(q)$. If $q \in M_n$ and $f \in C^\infty(M_n)$, the cotangent vector at $q \in M$ whose "factor" with index $H \in (\mathscr{D}M_n)_q$ is the vector

$$(\text{grad}(f \circ H))(w) \in \mathbf{V}_H{}^q = \mathbf{V}_w, \qquad w = H^{-1}(q),$$

will be denoted by $(\mathbf{grad}\, f)(q)$.

If a cotangent vector $\mathbf{z}$ is prescribed at $q \in M_n$, it is clear that for infinitely many choices of $f \in C^\infty(M)$, $\mathbf{z} = (\mathbf{grad}\, f)(q)$.

We shall prove a fundamentatal theorem underlying the notion of duality of contravariance and covariance:

Theorem 18.2. *Given* $\mathbf{y}$ *and* $\mathbf{z}$ *in* Π_q, *let the sum with respect to* k

$$S_H = (y_H)^k (z_H)_k, \qquad k = 1,..., n, \tag{18.19}$$

be formed for each $H \in (\mathscr{D}M_n)_q$. *Then the following hold*:

(i) *If* $\mathbf{y} \in T_q$ *and* $\mathbf{z} \in T_q^*$, S_H *is independent of* H, *thus depending only on* q.

(ii) *If for fixed* $\mathbf{z} \in \Pi_q$, S_H *is independent of* H *for each choice of* $\mathbf{y} \in T_q$, *then* $\mathbf{z} \in T_q^*$.

(iii) *If for fixed* $\mathbf{y} \in \Pi_q$, S_H *is independent of* H *for each choice of* $\mathbf{z} \in T_q^*$, *then* $\mathbf{y} \in T_q$.

Proof of (i). Let F and G be overlapping presentations, given as in (17.2), with $q \in X \cap Y$. Set $u = F^{-1}(q)$ and $v = G^{-1}(q)$. Then the factors

$$y_F, y_G \qquad \text{and} \qquad z_F, z_G \tag{18.20}$$

of $\mathbf{y}$ and $\mathbf{z}$ are, respectively, vectors

$$\eta \in \mathbf{V}_u, \quad \zeta \in \mathbf{V}_v \qquad \text{and} \qquad h \in \mathbf{V}_u, \quad g \in \mathbf{V}_v. \tag{18.21}$$

By virtue of (17.9) and (18.11)

$$\zeta^k g_k = (J_i^{\ k}(\mathbf{v} : u)\eta^i)(J_k^{\ j}(\mathbf{u} : v)h_j) = \delta_i^{\ j}\eta^i h_j = \eta^i h_i,$$

thereby establishing (i).

Proof of (ii). When the vectors (18.20) are the respective vectors (18.21), it is given in (ii) that the relation

$$\eta^i h_i = \zeta^k g_k \tag{18.22}$$

holds for fixed vectors $h \in \mathbf{V}_u$ and $g \in \mathbf{V}_v$, and for arbitrary vectors $\eta \in \mathbf{V}_u$ and $\zeta \in \mathbf{V}_v$ such that in accord with (17.9)

$$\zeta^k = J_i^{\ k}(\mathbf{v} : u)\eta^i, \qquad k = 1,\ldots, n. \tag{18.23}$$

It follows that

$$\eta^i[J_i^{\ k}(\mathbf{v} : u)g_k - h_i] = 0 \tag{18.24}$$

for an arbitary n-tuple $\eta^1,\ldots, \eta^n$. We infer that

$$h_i = J_i^{\ k}(\mathbf{v} : u)g_k, \qquad i = 1,\ldots, n. \tag{18.25}$$

According to (18.12) h and g are covariantly related by F and G. Now (ii) follows, and the proof (iii) is similar.

§ 19

M_n AS RIEMANNIAN MANIFOLD

Let M_n be a C^∞-manifold. With each presentation

$$(H : W_H, Z_H) \in \mathscr{D}M_n, \qquad W_H \subset E_n, \tag{19.1}$$

and each point $w = (w^1,..., w^n) \in W_H$ let there be associated a positive-definite, symmetric, quadratic form

$$Q_H{}^w(\alpha) = a_{ij}(H : w)\, \alpha^i \alpha^j, \qquad w \in W_H, \tag{19.2}$$

in the n variables $\alpha^1,..., \alpha^n$ subject to the condition that each mapping $w \to a_{ij}(H : w)$ of W_H into R be of class C^∞. Then Q_H is a family of quadratic forms over W_H, or as we shall say more briefly, a *Q-family* over the E_n-domain W_H of H.

Given the overlapping presentations F and G as in (17.2), let the forms

$$Q_F{}^u(\eta) = b_{ij}(u)\, \eta^i \eta^j, \qquad u \in U, \tag{19.3}$$

and

$$Q_G{}^v(\zeta) = c_{ij}(v)\, \zeta^i \zeta^j, \qquad v \in V, \tag{19.4}$$

define the Q-families Q_F and Q_G indexed by F and G.

We give a basic definition:

Definition 19.1. *M_n-Compatibility of Q_F and Q_G.* When F and G are arbitrary overlapping presentations in $\mathscr{D}M_n$ given as in (17.2), the corresponding Q-families Q_F and Q_G will be said to be *M_n-compatible* if for arbitrary $q \in X \cap Y$ and for $u = F^{-1}(q)$ and $v = G^{-1}(q)$

$$b_{ij}(u)\, \eta^i \eta^j = c_{ij}(v)\, \zeta^i \zeta^j \tag{19.5}$$

for vectors $\eta \in \mathbf{V}_u$ and $\zeta \in \mathbf{V}_v$ contravariantly related by F and G.

A Convention. When F and G are in $\mathscr{D}M_n$, but not overlapping, the Q-families Q_F and Q_G will be regarded as M_n*-compatible* in a trivial sense.

In any case the relation of M_n-compatibility of Q-families Q_F and Q_G is symmetric, since the relation of contravariance between vectors $\eta \in \mathbf{V}_u$ and $\zeta \in \mathbf{V}_v$ defined by F and G is symmetric.

When F and G are the overlapping presentations of (17.2) a necessary and sufficient condition that Q_F and Q_G be M_n-compatible is that when $F(u) = G(v) \in X \cap Y$ the matrix equality

$$\| c_{ij}(v) \| = \mathbf{J}^{\text{tr}}(\mathbf{u} : v) \cdot \| b_{ij}(u) \| \cdot \mathbf{J}(\mathbf{u} : v) \tag{19.6}$$

hold. This is a consequence of the classical law on the transformation of a quadratic form, such as $Q_F{}^u$, when subjected to a linear transformation, here the transformation (17.10) (Bôcher [1], p. 129).

Since the relation of M_n-compatibility of Q_F and Q_G is symmetric, the relation (19.6) is equivalent to the matrix relation

$$\| b_{ij}(u) \| = \mathbf{J}^{\text{tr}}(\mathbf{v} : u) \cdot \| c_{ij}(v) \| \cdot \mathbf{J}(\mathbf{v} : u), \tag{19.7}$$

which is also directly derivable from (19.6).

A Riemannian manifold M_n of class C^∞ will now be defined. The restriction to manifolds of class C^∞ is not necessary but is made for simplicity.

Definition 19.2. *Riemannian Manifolds and Forms.* Let M_n be a C^∞-manifold to each of whose presentations H there has been assigned a Q-family Q_H. If this assignment is such that for arbitrary presentations F and G in $\mathscr{D}M_n$, Q_F and Q_G are M_n-compatible, M_n will be said to admit a C^∞*-Riemannian structure* defined by these Q-families. In such a structure Q_H will be termed the *Riemannian form* indexed below by H. With such a structure M_n will be termed a *Riemannian manifold.*

On a Riemannian manifold the length of a regular arc $\gamma : t \to p(t)$ $a \leqslant t \leqslant b$ admits a classical definition. We shall limit ourselves to the case in which $| M_n |$ is connected.

Length on M_n. Suppose that M_n is a Riemannian manifold of class C^∞. Suppose that the above regular arc γ on M_n is included in the coordinate domain X of a presentation $(F : U, X) \in \mathscr{D}M_n$. On U, γ is represented by a regular arc $\gamma_F : t \to u(t)$, $a \leqslant t \leqslant b$. If (19.3) gives the form $Q_F{}^u(\eta)$, we shall assign γ the length

$$L(\gamma) = \int_a^b (b_{ij}(u(t))\, \dot{u}^i(t)\, \dot{u}^j(t))^{1/2}\, dt. \tag{19.8}$$

Returning to the overlapping presentations F and G of (17.2), suppose that the carrier $|\gamma| \subset X \cap Y$. Then γ is represented by a regular arc $\gamma_G : t \to v(t)$, $a \leqslant t \leqslant b$, in V. We shall see that $L(\gamma)$, as defined by (19.8), is also equal to the integral

$$\int_a^b (c_{ij}(v(t))\, \dot{v}^i(t)\, \dot{v}^j(t))^{1/2}\, dt. \tag{19.9}$$

The equality of the integrals (19.8) and (19.9) is a consequence of the identity

$$b_{ij}(u(t))\, \dot{u}^i(t)\, \dot{u}^j(t) = c_{ij}(v(t))\, \dot{v}^i(t)\, \dot{v}^j(t), \qquad a \leqslant t \leqslant b. \tag{19.10}$$

Relation (19.10) is valid because $F(u(t)) = G(v(t))$, and hence the vectors $\dot{u}(t) \in \mathbf{V}_{u(t)}$ and $\dot{v}(t) \in \mathbf{V}_{v(t)}$ are contravariantly related by F and G, as affirmed in Example 17.1.

A regular arc $\gamma : t \to p(t)$, $a \leqslant t \leqslant b$, on M_n not restricted to a single coordinate domain is a finite sequence of regular arcs $\gamma_1, \gamma_2, \ldots, \gamma_r$ whose t-domains are successive subintervals of a partition of $[a, b]$ so chosen that each carrier $|\gamma_i|$ has a closure included in some coordinate domain of M_n. The length of γ can be defined as the sum

$$L(\gamma) = L(\gamma_1) + \cdots + L(\gamma_r),$$

where $L(\gamma_i)$ is evaluated as above on any coordinate domain which includes $\mathrm{Cl}\,|\gamma_i|$.

One sees that $L(\gamma)$ so defined is independent of the partitions of $[a, b]$ which are admissible for this purpose. In fact, if P_1 and P_2 are two such partitions of $[a, b]$, there is a third partition of $[a, b]$ which is also a partition both of P_1 and of P_2. If $L_1(\gamma)$, $L_2(\gamma)$, and $L_3(\gamma)$ are the corresponding lengths assigned to γ, one sees that

$$L_1(\gamma) = L_3(\gamma) = L_2(\gamma).$$

A Metric for M_n. Given any two points p and q of M_n, let $d(p, q)$ denote the inferior limit of lengths of regular arcs on M_n joining p to q on M_n. A distance $d(p, q)$ so defined satisfies the usual three axioms on a metric space and gives back to $| M_n |$ its original topology. No variational theory is reguired to prove these elementary facts.

The Existence of Riemannian Structures.

We shall prove that each C^∞-manifold admits a Riemannian structure. A first step in proving this is to verify Theorem 19.1:

Theorem 19.1. (i) *If M_n is a regular C^∞-manifold in a Euclidean space E_r, with the presentations $H \in \mathscr{D}M_n$, there can then be associated pairwise M_n-compatible Q-families Q_H of quadratic forms.*

(ii) *These forms can be chosen so that the resultant Riemannian length of a regular arc $t \to p(t)$, $a \leqslant t \leqslant b$, on M_n is its ordinary length in E_r.*

Definition of the Q-Families Q_H. The general definition is sufficiently indicated by defining Q_F and Q_G when F and G are presentations given as in (17.2). Given $u \in U$ and independent variables $du^1,\ldots, du^n$, we introduce the symmetric, positive-definite, quadratic form

$$Q_F{}^u(du) = b_{ij}(u)\, du^i\, du^j = \sum_{\mu=1}^{r} (dx^\mu)^2 = \sum_{\mu=1}^{r} \left(\frac{\partial F^\mu}{\partial u^k}(u)\, du^k\right)^2, \qquad (19.11)$$

where dx^μ is the linear form in $du^1,\ldots, du^r$ given by the final parentheses in (19.11). Given $v \in V$ and independent variables $dv^1,\ldots, dv^n$, we similarly introduce the quadratic form

$$Q_G{}^u(dv) = c_{ij}(v)\, dv^i\, dv^j = \sum_{\mu=1}^{r} (dx^\mu)^2 = \sum_{\mu=1}^{r} \left(\frac{\partial G^\mu}{\partial v^k}(v)\, dv^k\right)^2. \qquad (19.12)$$

Proof of the M_n-Compatibility of Q_F and Q_G. If $u \to \mathbf{v}(u)$ is the transition diff (17.5) defined by F and G, one has the r identities

$$G^\mu(\mathbf{v}(u)) = F^\mu(u), \qquad \mu = 1,\ldots, r;\ u \in \hat{U};$$

implying that

$$\frac{\partial G^\mu}{\partial v^k}(v)\, dv^k = \frac{\partial F^\mu}{\partial u^k}(u)\, du^k, \qquad \mu = 1,\ldots, r, \qquad (19.13)$$

subject to the conditions

$$v = \mathbf{v}(u), \quad u \in \hat{U}, \quad dv^k = \frac{\partial \mathbf{v}^k}{\partial u^i}(u)\, du^i, \quad k = 1,..., n. \tag{19.14}$$

Thus (19.13) holds subject to the condition that $F(u) = G(v) \in X \cap Y$ and that the vectors $du \in \mathbf{V}_u$ and $dv \in \mathbf{V}_v$ be contravariantly related by F and G, as are the vectors in (17.9).

Subject to these same conditions, the right-hand sums in (19.11) and (19.12) are equal, proving that Q_F and Q_G are M_n-compatible in accord with Definition 19.1.

Thus (i) of Theorem 19.1 is true. That (ii) is true follows from the middle equality in the definitions (19.11) and (19.12).

This completes the proof of Theorem 19.1.

Theorem 19.1 leads to the following:

Theorem 19.2. *Each C^∞-manifold N_n admits a C^∞-Riemannian structure.*

According to Theorem 16.7 there exists a C^∞-diff $\psi : N_n \to M_n$ of N_n onto a regular C^∞-manifold M_n in some Euclidean space E_r. Let φ be the C^∞-diff which is the inverse of ψ. According to Theorem 16.3a there is a 1–1 correspondence of the ensemble of presentations $\mathscr{D}M_n$ with the ensemble $\mathscr{D}N_n$, in which $H \in \mathscr{D}M_n$ corresponds to $\varphi \circ H$ in $\mathscr{D}N_n$. According to Theorem 19.1 there exists an ensemble $\{Q_H\}$, indexed by $H \in \mathscr{D}M_n$, of pairwise M_n-compatible Riemannian forms Q_H, where Q_H is "over" the E_n-domain of H. To prove Theorem 19.2 it is accordingly sufficient to prove (α):

(α) *If to each presentation $H \in \mathscr{D}M_n$ and corresponding presentation $\varphi \circ H \in \mathscr{D}N_n$ one assigns a common Q-family over the common E_n-domain of H and $\varphi \circ H$, the resulting ensemble of Q-families is N_n-compatible if M_n-compatible.*

Proof of (α). Let F and G be arbitrary overlapping presentations in $\mathscr{D}M_n$ given as in (17.2). To F and G in $\mathscr{D}M_n$ correspond the presentations

$$(\varphi \circ F : U, \varphi(X)) \quad \text{and} \quad (\varphi \circ G : V, \varphi(Y)) \tag{19.15}$$

in $\mathscr{D}N_n$. If $u \in U$ and $v \in V$ are such that $F(u) = G(v)$, then

$$(\varphi \circ F)(u) = (\varphi \circ G)(v). \tag{19.16}$$

The transition diff (17.5) defined by F and G over $\hat{U}$ is the transition diff of $\varphi \bar{\circ} F$ and $\varphi \bar{\circ} G$. Hence vectors $\eta \in \mathbf{V}_u$ and $\zeta \in \mathbf{V}_v$ which are contravariantly related by the linear isomorphism (17.9) defined by F and G are contravariantly related by the same linear isomorphism regarded as defined by $\varphi \bar{\circ} F$ and $\varphi \bar{\circ} G$.

Suppose the quadratic form on the left of (19.5) has been assigned to F and to $\varphi \bar{\circ} F$, and the form on the right of (19.5) has been assigned to G and to $\varphi \bar{\circ} G$. If (19.5) holds when $\eta \in \mathbf{V}_u$ and $\zeta \in \mathbf{V}_v$ are contravariantly related by F and G, then (19.5) holds when $\eta \in \mathbf{V}_u$ and $\zeta \in \mathbf{V}_v$ are contravariantly related by $\varphi \bar{\circ} F$ and $\varphi \bar{\circ} G$.

Statement (α) is accordingly true and Theorem 19.2 follows.

The condition that Q_F and Q_G of Definition 19.1 be M_n-compatible can be usefully and equivalently stated in terms of the bilinear forms defined by Q_F and Q_G.

Lemma 19.1. *The equality* (19.5) *is valid subject to the conditions of Definition* 19.1 *if and only if*

$$b_{ij}(u)\, \eta^i \hat{\eta}^j = c_{ij}(v)\, \zeta^i \hat{\zeta}^j \tag{19.17}$$

for arbitrary vectors $\eta \in \mathbf{V}_u$ *and* $\zeta \in \mathbf{V}_v$ *contravariantly related by* F *and* G *and arbitrary vectors* $\hat{\eta} \in \mathbf{V}_u$ *and* $\hat{\zeta} \in \mathbf{V}_v$ *similarly contravariantly related by* F *and* G.

This is because the matrix equalities (19.6) and (19.7) are necessary and sufficient conditions that (19.17) hold when η is the image of ζ and $\hat{\eta}$ the image of $\hat{\zeta}$ under the linear transformation with matrix $\mathbf{J}(\mathbf{u} : v)$.

Riemannian Forms and Coforms. We began §19 by associating with each presentation $(H : W_H, Z_H) \in \mathscr{D}M_n$ a family Q_H of positive-definite, symmetric, quadratic forms

$$Q_H{}^w(\alpha) = a_{ij}(H : w)\, \alpha^i \alpha^j, \qquad w \in W_H, \tag{19.18}$$

in the n variables $(\alpha^1, \ldots, \alpha^n)$. With H we now associate a *dual* family Q^H of quadratic forms, termed *coforms*,

$$Q_w{}^H(e) = a^{ij}(H : w)\, e_i e_j, \qquad w \in W_H, \tag{19.19}$$

in the n variables $e_1, \ldots, e_n$, where the matrix $\| a^{ij}(H : w) \|$ is the inverse of the matrix $\| a_{ij}(H : w) \|$. Each form $Q_w{}^H(e)$ is symmetric. It is positive-definite by virtue of Corollary 3.1.

We shall be concerned with pairs of families Q^F and Q^G, indexed above by presentations F and G given as in (17.2), with Q-families Q_F and Q_G represented, respectively, as in (19.3) and (19.4). The dual families Q^F and Q^G may be given the representations

$$Q_u{}^F(h) = b^{ij}(u)\, h_i h_j\,, \qquad u \in U, \tag{19.20}$$

and

$$Q_v{}^G(g) = c^{ij}(v)\, g_i g_j\,, \qquad v \in V. \tag{19.21}$$

The following definition is dual to the Definition 19.1.

Definition 19.3. *M_n-Compatibility of Q^F and Q^G.* When F and G are arbitrary overlapping presentations in $\mathscr{D}M_n$ given as in (17.2) the families Q^F and Q^G, as given respectively by (19.20) and (19.21), will be termed *M_n-compatible* if for arbitrary $q \in X \cap Y$ and for $u = F^{-1}(q)$ and $v = G^{-1}(q)$

$$b^{ij}(u)\, h_i h_j = c^{ij}(v)\, g_i\, g_j \tag{19.22}$$

for $h \in \mathbf{V}_u$ and $g \in \mathbf{V}_v$ covariantly related by F and G.

We have found the conditions (19.6) and (19.7) necessary and sufficient that the Q-families Q_F and Q_G be M_n-compatible in the sense of Definition 19.1. Referring to the linear transformations (18.17) and (18.18) defining the relation of covariance between vectors $h \in \mathbf{V}_u$ and $g \in \mathbf{V}_v$, we see that the matrix equalities

$$\| c^{ij}(v) \| = \mathbf{J}(\mathbf{v} : u) \cdot \| b^{ij}(u) \| \cdot \mathbf{J}^{\mathrm{tr}}(\mathbf{v} : u) \tag{19.23}$$

and

$$\| b^{ij}(u) \| = \mathbf{J}(\mathbf{u} : v) \cdot \| c^{ij}(v) \| \cdot \mathbf{J}^{\mathrm{tr}}(\mathbf{u} : v) \tag{19.24}$$

are necessary and sufficient that the cofamilies Q^F and Q^G of Definition 19.3 be M_n-compatible.

A Convention. One must supplement Definition 19.3 by the convention that when F and G are presentations in $\mathscr{D}M_n$ which are not overlapping then Q^F and Q^G are to be regarded as M_n-compatible in a trival sense.

The condition that Q^F and Q^G of Definition 19.3 be M_n-compatible can be usefully and equivalently expressed in terms of the bilinear forms defined by Q^F and Q^G. The lemma is a dual of Lemma 19.1.

Lemma 19.2. *The equality* (19.22) *is valid subject to the conditions of Definition* 19.3 *if and only if*

$$b^{ij}(u)\, h_i \hat{h}_j = c^{ij}(v)\, g_i \hat{g}_j \tag{19.25}$$

for arbitrary vectors $h \in \mathbf{V}_u$ *and* $g \in \mathbf{V}_v$ *covariantly related by* F *and* G *and arbitrary vectors* $\hat{h} \in \mathbf{V}_u$ *and* $\hat{g} \in \mathbf{V}_v$ *similarly covariantly related by* F *and* G.

The proof is similar to that of Lemma 19.1.

Recall the matrix law that the inverse of the transpose of a nonsingular matrix is the transpose of the inverse. Using this law, one finds that the matrix relation (19.23) is derivable from the matrix relation (19.6) on equating the inverses of two members of (19.6). One similarly derives (19.24) from (19.7), and, if one pleases, (19.6) from (19.23) and (19.7) from (19.24).

We draw the following conclusion:

Theorem 19.3. *When* F *and* G *are in* $\mathscr{D}M_n$ *a necessary and sufficient condition that* Q*-families* Q_F *and* Q_G *be* M_n*-compatible is that the dual* Q*-families* Q^F *and* Q^G *be* M_n*-compatible.*

Riemannian Forms and Coforms. When the ensemble $\{Q_H\}$ of Q-families, indexed by $H \in \mathscr{D}M_n$, is such that its Q-families are pairwise M_n-compatible, we have said that M_n admits a "Riemannian structure" in which Q_H is the Riemannian form over the E_n-domain of H. Under these conditions the Q-families of the dual ensemble $\{Q^H\}$ are pairwise M_n-compatible in the dual sense, and we term Q^H the *Riemannian coform* over the E_n-domain of H.

We have seen that C^∞-manifold always admits a Riemannian structure. From this point on we shall suppose that M_n is provided with such a structure and that Q_H and Q^H are the corresponding form and coform with presentation index H.

The Fundamental Tensors. Corresponding to an arbitrary presentation $H \in \mathscr{D}M_n$ the matrices

$$\| a_{ij}(H : w) \| \quad \text{and} \quad \| a^{ij}(H : w) \|, \qquad w \in W_H, \tag{19.26}$$

of the Riemannian form Q_H and coform Q^H respectively represent at w the fundamental *covariant* and *contravariant tensors* of second order at the point $q = H(w)$ of the Riemannian manifold M_n. In the context of tensor algebra the distinction between the fundamental covariant and contravariant tensors arises from the difference between the transformations (19.6) and the transformation (19.23) of the respective local representations.

"Conjugacy" of Tangent Vectors and Cotangent Vectors. Given $q \in M_n$, with the aid of the "fundamental" tensors of the Riemannian structure of M_n one can define a unique isomorphism of the tangent vector space T_q onto the cotangent vector space T_q^*. We shall call this isomorphism the *conjugacy isomorphism.* A tangent vector $\mathbf{y} \in T_q$ and a cotangent vector $\mathbf{z} \in T_q^*$ which correspond under this isomorphism will be called *conjugates* one of the other, as will their "factors" in Π_q with the same presentation index H [see (18.6) for definition of Π_q].

Three lemmas are required to define and characterize the conjugacy isomorphism.

Lemma 19.3. *Given a tangent vector* $\mathbf{y} \in T_q$, *the vector* $\mathbf{z} \in \Pi_q$ *(termed the conjugate of* $\mathbf{y}$*) whose "factor"* z_H, *with index* $H \in (\mathscr{D}M_n)_q$, *has the n components*

$$(z_H)_i = a_{ij}(H : w)(y_H)^j, \qquad i = 1,\ldots, n, \quad w = H^{-1}(q), \tag{19.27}$$

in $\mathbf{V}_w$ *is a cotangent vector in* T_q^*.

Lemma 19.4. *Given a cotangent vector* $\mathbf{z} \in T_q^*$, *the vector* $\mathbf{y} \in \Pi_q$ *(termed the conjugate of* $\mathbf{z}$*) whose "factor"* y_H, *with index* $H \in (\mathscr{D}M_n)_q$ *in* Π_q *has the n components*

$$(y_H)^k = a^{ki}(H : w)(z_H)_i, \qquad k = 1,\ldots, n, \quad w = H^{-1}(q), \tag{19.28}$$

in $\mathbf{V}_w$ *is a tangent vector in* T_q.

Lemma 19.5 indicates precisely what it means for the conjugate of the conjugate of a tangent vector or a cotangent vector at $q \in M_n$ to be the original vector.

Lemma 19.5. *If* $\mathbf{y} \to \Gamma_q(\mathbf{y})$ *maps* $\mathbf{y} \in T_q$ *into its conjugate in* T_q^* *and* $\mathbf{z} \to \Lambda_q(\mathbf{z})$ *maps* $\mathbf{z} \in T_q^*$ *into its conjugate in* T_q, *then* Γ_q *and* Λ_q *are onto and inverses one of the other.*

Proof of Lemma 19.3. Given $\mathbf{y} \in T_q$, if $\mathbf{z} \in \Pi_q$ is defined by (19.27), then for arbitrary tangent vector $\hat{\mathbf{y}} \in T_q$ the sum $(z_H)_i(\hat{y}_H)^i$ is independent in value of $H \in (\mathscr{D}M_n)_q$, in accord with Lemma 19.1. It follows from Theorem 18.2 (ii) that $\mathbf{z}$ is in T_q^*.

Proof of Lemma 19.4. This proof is similar, using Lemma 19.2 and Theorem 18.2(iii).

Proof of Lemma 19.5. Given $\mathbf{y} \in T_q$, set $\mathbf{z} = \Gamma_q(\mathbf{y})$. We wish to show that $\Lambda_q(\mathbf{z}) = \mathbf{y}$. Given $H \in (\mathscr{D}M_n)_q$, set $w = H^{-1}(q)$.

By definition of Λ_q and Γ_q the "factor" of $\Lambda_q(\mathbf{z})$ with presentation index H has a kth component in $\mathbf{V}_w$,

$$a^{ki}(H : w)(z_H)_i = a^{ki}(H : w)\, a_{ij}(H : w)(y_H)^j = \delta_j{}^k(y_H)^j = (y_H)^k,$$

showing that $(\Lambda_q \circ \Gamma_q)(\mathbf{y}) = \mathbf{y}$ for $\mathbf{y} \in T_q$.

One shows similarly that $(\Gamma_q \circ \Lambda_q)(\mathbf{z}) = \mathbf{z}$ for $\mathbf{z} \in T_q^*$, and the lemma follows.

We summarize and complete these results as follows:

Theorem 19.4. *The mapping of* T_q *onto* T_q^* *in which each tangent vector* $\mathbf{y} \in T_q$ *goes into its "conjugate" cotangent vector in* T_q^* *is an isomorphism of* T_q *onto* T_q^* *under whose inverse each cotangent vector in* T_q^* *goes into its "conjugate" tangent vector in* T_q.

The Existence of f-Transverse Dynamical Systems in M_n. The fundamental theorem is Theorem 19.6.

Introduction to Theorem 19.5. Let a function $f \in C^\infty(M_n)$ be "ordinary" on M_n, that is, without critical points on M_n. Corresponding to each $H \in \mathscr{D}M_n$ Theorem 19.5 makes use of the representation $f \bar{\circ} H$ of f on the E_n-domain W_H of H.

According to Definition 17.3 of a dynamical system $\{\mathbf{d}/\varphi_H\}$ on M_n such a system is determined by an ensemble $\{\varphi_H\}$, indexed by $H \in \mathscr{D}M_n$, of C^∞-families φ_H of nonvanishing vectors $\varphi_H(w)$ defined for w on the respective E_n-domains W_H and subject to Condition III of Definition 17.3, that the families φ_H be pairwise contravariantly related by their indexing presentations.

Theorem 19.5. *Given an ordinary function $f \in C^\infty(M_n)$, the system of differential equations $\{\mathbf{d}/\varphi_H\}$, indexed by $H \in \mathscr{D}M_n$, in which for each $H \in \mathscr{D}M_n$*

$$\varphi_H{}^k(w) = a^{ki}(H : w)\,(\partial(f \circ H)(w)/\partial w^i), \qquad k = 1,\ldots, n, \ w \in W_H, \quad (19.29)$$

is an f-transverse dynamical system on M_n.

We shall show that the three conditions on a dynamical system in Definition 17.3 are satisfied.

Condition I is trivially satisfied. Condition II is satisfied if the vectors $\varphi_H(w)$ never vanish. This is the case because grad$(f \circ H)(w)$ never vanishes by hypothesis and the matrix $\| a^{ki}(H : w)\|$ is non-singular.

Condition III is satisfied if for each $q \in M_n$ the vector $\mathbf{y} \in \Pi_q$ whose factors are the vectors $\varphi_H(w)$ indexed by $H \in (\mathscr{D}M_n)_q$ is a tangent vector in T_q. This is the case since $\mathbf{y}$ as defined by (19.29) is the conjugate of the cotangent vector $(\mathbf{grad}\, f)(q)$ (see Definition 18.3).

The dynamical system $\{\mathbf{d}/\varphi_H\}$ thereby defined is *f-transverse*, as we now verify.

Each nontrivial solution of the dynamical system on M_n is a regular arc, since the vectors $\varphi_H(w)$ never vanish. If $t \to p(t)$ is a solution on M_n which has a representative $t \to w(t)$ in W_H, then

$$f(p(t)) = (f \circ H)(w(t)).$$

The t-derivative of $f(p(t))$ is thus the dot product $h(t) \cdot \dot{w}(t)$ of the gradient, say $h(t)$, of $f \circ H$ evaluated at $w(t)$ and the vector $\dot{w}(t) = \varphi_H(w(t))$. If $\dot{w}(t)$ is represented by the right member of (19.29), this dot product reduces to

$$a^{ij}(H : w(t))\, h_i(t)\, h_j(t)$$

and is positive, since each Riemannian coform is positive-definite.

This establishes the theorem.

Notation. The Submanifold $M_n{}^f$ of M_n. Given a ND function f on M_n, let $M_n{}^f$ denote the open submanifold of M_n from which the critical points of f have been deleted.

A final *canonical* form for an f-transverse dynamical system is given by Theorem 19.6. The characteristic condition (19.31) is motivated by the desire to make $df/dt = 1$ along the image in M_n of solutions of the corresponding local dynamical systems (19.30).

Theorem 19.6. *If f is a* ND *function in $C^\infty(M_n)$, there exists on $M_n{}^f$ an f-transverse dynamical system $\{\mathbf{d}/\psi_H\}$, termed canonical, in which the local system d/ψ_H*

$$dw^k/dt = \psi_H{}^k(w), \qquad w \in \text{domain } H, \quad k = 1,\ldots, n, \tag{19.30}$$

indexed by $H \in \mathscr{D}M_n{}^f$, satisfies the condition

$$\psi_H{}^k(w)\,\frac{\partial(f \circ H)}{\partial w^k}(w) = 1, \qquad w \in \text{domain } H. \tag{19.31}$$

Definition of $\psi_H{}^k(w)$. With $\varphi_H{}^k(w)$ defined as in (19.29) we set

$$\psi_H{}^k(w) = \rho_H(w)\,\varphi_H{}^k(w), \qquad w \in \text{domain } H, \qquad k = 1,\ldots, n, \tag{19.32}$$

where $\rho_H(w)$ is an "invariant" factor equal to the reciprocal of

$$\varphi_H{}^k(w)\,\frac{\partial(f \circ H)}{\partial w^k}(w), \qquad w \in \text{domain } H. \tag{19.33}$$

The sum (19.33) is "invariant" in the sense that its value depends only on the point $p = H(w) \in M_n$ and not on the particular presentation $H \in \mathscr{D}M_n{}^f$ of p. This "invariance" follows from (i) of Theorem 18.2.

Moreover, the sum (19.33) equals

$$a^{ki}(H : w)\,\frac{\partial(f \circ H)}{\partial w^i}(w)\,\frac{\partial(f \circ H)}{\partial w^k}(w), \qquad w \in \text{domain } H,$$

and so is positive. Because $\rho_H(w)$ is invariant the vector $\psi_H(w)$, like $\varphi_H(w)$, is transformed contravariantly. One verifies (19.31).

Thus Theorem 19.6 is true.

f-Trajectories on $M_n{}^f$. If $t \to w_1(t)$ is a solution of the local dynamical system d/ψ_H given by (19.30), then

$$d(f \circ H)(w_1(t))/dt = 1 \tag{19.34}$$

by virtue of (19.31). On such a solution $(f \circ H)(w_1(t)) = t + \text{const.}$ Given such a solution, there exists another solution $t \to w(t)$ with the same carrier and such that

$$(f \circ H)(w(t)) = t. \tag{19.35}$$

Abbreviating "parameterized" by *par*, we say that $t \to w(t)$ is an *$(f \circ H)$-par solution of d/ψ_H* when (19.35) holds.

Given an $(f \circ H)$-par solution $t \to w(t)$ of the *local* system d/ψ_H, the corresponding "solution" $t \to p(t) = H(w(t))$ on M_n^f of the system $\{\mathbf{d}/\psi_H\}$ has the property that $f(p(t)) = t$, and will be termed an *arc* of an *f-trajectory* on M_n^f of $\{\mathbf{d}/\psi_H\}$. Any simple regular arc of M_n^f which is the union of open "arcs of f-trajectories" of M_n^f will be called an *f-trajectory* of M_n^f.

The f-transverse dynamical system $\{\mathbf{d}/\psi_H\}$, conditioned as in Theorem 19.6, has the local form

$$\frac{dw^k}{dt} = \frac{a^{ki}(H:w)\dfrac{\partial(f \circ H)}{\partial w^i}(w)}{a^{ji}(H:w)\dfrac{\partial(f \circ H)}{\partial w^i}(w)\dfrac{\partial(f \circ H)}{\partial w^j}(w)}, \qquad k = 1, 2, \ldots, n, \tag{19.36}$$

where w is a point in the domain of H. The differential equations are classically interpreted as differential equations of the orthogonal trajectories of the f-level manifolds on M_n^f.

A Simplified Determination of Riemannian Structures. The following theorem is useful when it is necessary, as in §22, to modify a given Riemannian structure:

Theorem 19.7. *Let M_n be a C^∞-manifold and K a subset of presentations in $\mathscr{D}M_n$ which cover $|M_n|$. With each presentation $H \in K$ let there be associated a form Q_H as in* (19.2) *in such a manner that when F and G are overlapping presentations in K, Q_F and Q_G are M_n-compatible in the sense of Definition* 19.1.

There then exists a unique C^∞-Riemannian structure on M_n whose Riemannian forms Q_H include those assigned to presentations $H \in K$.

Theorem 19.7 will follow once we have verified Lemma 19.6.

Lemma 19.6. *In* $\mathscr{D}M_n$ *let there be given presentations* $(F : U, X)$, $(G : V, X)$, *and* $(H : W, X)$ *with a common range* X, *of which* F *is in* K *and has been admissibly assigned a form* Q_F *with matrix* $\| b_{ij}(u)\|$. *If* G *and* H *are assigned forms* Q_G *and* Q_H *with matrices respectively* $\| c_{ij}(v)\|$ *and* $\| e_{ij}(w)\|$ *such that* Q_G *and* Q_H *are* M_n*-compatible with* Q_F, *then* Q_G *and* Q_H *are* M_n*-compatible.*

Proof of Lemma 19.6. With the ordered pairs (H, F), (F, G), and (H, G) there are associated, respectively, the transition diffs

$$w \to \mathbf{u}(w), \qquad u \to \mathbf{v}(u), \qquad \text{and} \qquad w \to \mathbf{v}(w), \tag{19.36'}$$

as in §17, with respective Jacobian matrices $\mathbf{J}(\mathbf{u} : w)$, $\mathbf{J}(\mathbf{v} : u)$, and $\mathbf{J}(\mathbf{v} : w)$.

By hypothesis Q_H and Q_G have been so defined that subject to (19.36′)

$$\| e_{ij}(w)\| = \mathbf{J}^{\text{tr}}(\mathbf{u} : w) \, \| b_{ij}(u)\| \, \mathbf{J}(\mathbf{u} : w) \tag{19.37}$$

[by (19.6)] and

$$\| b_{ij}(u)\| = \mathbf{J}^{\text{tr}}(\mathbf{v} : u) \, \| c_{ij}(v)\| \, \mathbf{J}(\mathbf{v} : u) \tag{19.38}$$

[by (19.7)]. It follows from (19.37) and (19.38) and the relation $\mathbf{J}(\mathbf{v} : w) = \mathbf{J}(\mathbf{v} : u)\, \mathbf{J}(\mathbf{u} : w)$ that subject to (19.36′)

$$\| e_{ij}(w)\| = \mathbf{J}^{\text{tr}}(\mathbf{v} : w) \, \| c_{ij}(v)\| \, \mathbf{J}(\mathbf{v} : w),$$

so that Q_H and Q_G are M_n-compatible by (19.7).

This establishes Lemma 19.6.

Proof of Theorem 19.7. Let K' be a subset of presentations in $\mathscr{D}M_n$ which includes K of Theorem 19.7 and contains each presentation which is a restriction F of some presentation $G \in K$. To such a presentation $F \in K'$ let a Q-family Q_F be assigned by restricting the parameter of the family Q_G to the domain of F. The families Q_F and Q_G are clearly M_n-compatible. It follows from Lemma 19.6 that for each $F' \in K'$ the families $Q_{F'}$ are pairwise compatible.

If H is given in $\mathscr{D}M_n$ but is not in K', the range of H can be covered by the ranges of a countable set $F_1, F_2, \ldots$ of presentations in K'. A matrix $\| e_{ij}(w)\|$ of the required family Q_H will be uniquely and consistently defined on the ranges of the respective presentations $F_1, F_2, \ldots$ in accord with Lemma 19.6, by imposing the condition that Q_H be M_n-compatible with the families $Q_{F_1}, Q_{F_2}, \ldots$. The resultant families Q_H, $H \in \mathscr{D}M_n$, will be M_n-compatible by virtue of Lemma 19.6.

Thus Theorem 19.7 is true.

§ 20

ff-PRESENTATIONS IN $\mathscr{D}M_n{}^f$

Having defined f-trajectories of the canonical f-transverse dynamical system $\{\mathbf{d}/\psi_H\}$, we shall set up fields of such trajectories. Such fields are required in all *homotopy* considerations dependent on f. Our initial study is local. We begin by defining certain special presentations in $\mathscr{D}M_n{}^f$ termed *f-presentations.*

As in §19, there is given a C^∞-manifold M_n and a ND $f \in C^\infty(M_n)$. Again $M_n{}^f$ is the submanifold of M_n obtained by deleting from M_n the critical points of f.

Topological Manifolds f^c. Corresponding to each value c of f, set

$$f^c = \{p \in M_n \mid f(p) = c\} \tag{20.1}$$

(the *c-level subset* of M_n) and let the set f^c be topologized by $| M_n |$. If c is an *ordinary* value of f, f^c is a topological $(n - 1)$-manifold, called an *f-level* manifold. In any case $f^c \cap M_n{}^f$ is empty or a topological $(n - 1)$-manifold.

Special Coordinates in E_n. Let the Euclidean space E_n of coordinates $x^1,\ldots, x^n$ be represented as a product $E_{n-1} \times R$ of a coordinate subplane E_{n-1} of coordinates $y^1,\ldots, y^{n-1}$ and an axis R of coordinate τ. We are thus setting

$$(x^1,\ldots, x^n) = (y^1,\ldots, y^{n-1}, \tau). \tag{20.2}$$

Let (α, β) be an open interval of R, and U a nonempty open subset of the space E_{n-1} of points y.

Definition 20.1. *f-Presentations* $\mathscr{F}$. A presentation

$$(\mathscr{F} : U \times (\alpha, \beta), X) \in \mathscr{D}M_n{}^f \qquad (U \text{ open in } E_{n-1}) \tag{20.3}$$

will be called an *f-presentation* if

$$(f \bar{\circ} \mathscr{F})(y, \tau) = \tau, \qquad y \in U; \quad \tau \in (\alpha, \beta). \tag{20.4}$$

We shall denote f-presentations by script letters $\mathscr{F}$, $\mathscr{G}$, etc. The use of f-presentations depends upon the validity of Lemma 20.1.

Lemma 20.1. *The subset of presentations in $\mathscr{D}M_n^f$ which are f-presentations covers $| M_n^f |$.*

Proof of Lemma 20.1. Let p_0 be a point prescribed in M_n^f. To prove Lemma 20.1, it is sufficient to show that there exists an f-presentation $\mathscr{F} \in \mathscr{D}M_n^f$ of form (20.3) such that $p_0 \in X$.

There exists a presentation $(G : V, Y) \in \mathscr{D}M_n^f$ such that $p_0 \in Y$. Since f is ordinary on M_n^f, $\text{grad}(f \bar{\circ} G)(v) \neq 0$ for $v \in V$. We shall suppose the coordinates of points $v \in V$ so numbered that the nth component of $\text{grad}(f \bar{\circ} G)(v_0)$ fails to vanish when $v_0 = G^{-1}(p_0)$. For simplicity we shall suppose that v_0 is the origin in the Euclidean n-plane of V.

We introduce a C^∞-mapping $V \to E_{n-1} \times R$ of the form

$$y^i = v^i, \qquad \tau = (f \bar{\circ} G)(v), \qquad i = 1, \ldots, n-1. \tag{20.5}$$

Under (20.5) v goes into the point (y, τ); v_0 in particular goes into a point (y_0, τ_0), and

$$\frac{D(y^1, \ldots, y^{n-1}, \tau)}{D(v^1, \ldots, v^{n-1}, v^n)} (v_0) \neq 0 \tag{20.6}$$

because of the condition on $\text{grad}(f \bar{\circ} G)$. We can accordingly suppose that V and the presentation $(G : V, Y)$ have been so chosen that the mapping (20.5) has an inverse λ which is a C^∞-diff, $(y, \tau) \to v$, of the form

$$(y, \tau) \to \lambda(y, \tau) : U \times (\alpha, \beta) \to V \qquad (\text{onto } V)$$

with a domain $U \times (\alpha, \beta)$ which is the product of an open neighborhood U of the origin y_0 in E_{n-1} and an open interval (α, β) containing τ_0. The image points are points $\lambda(y, \tau) = v \in V$.

If one sets $\mathscr{F} = G \circ \lambda$ on the domain $U \times (\alpha, \beta)$, then

$$\mathscr{F}(U \times (\alpha, \beta)) = G(\lambda(U \times (\alpha, \beta)) = G(V).$$

The presentation

$$(\mathscr{F} : U \times (\alpha, \beta), G(V)) \in \mathscr{D}^0 \mid M_n{}^f \mid$$

(for $\mathscr{D}^0$ see Definition 13.3) is C^∞-compatible with G because G and $\mathscr{F}$ have the same range and are related by the transition diff $\lambda = G^{-1} \circ \mathscr{F}$. One show readily (cf. Example 13.1) that $\mathscr{F}$ is C^∞-compatible with each presentation in $\mathscr{D}M_n{}^f$. It follows that $\mathscr{F} \in \mathscr{D}M_n{}^f$.

We now verify that $\mathscr{F}$ is an "f-presentation" in the sense of Definition 20.1. If (y, τ) is prescribed in $U \times (\alpha, \beta)$, then by definition of $\mathscr{F}$ as $G \circ \lambda$ and of λ,

$$(f \bar{\circ} \mathscr{F})(y, \tau) = (f \bar{\circ} G)(\lambda(y, \tau)) = \tau,$$

in accord with (20.5).

We have thus shown the existence of an f-presentation $\mathscr{F} \in \mathscr{D}M_n{}^f$ whose range $G(V)$ contains the point p_0 prescribed in $M_n{}^f$.

This establishes Lemma 20.1.

Definition 20.2. *Partial Presentations* $\mathscr{F}^c$. Corresponding to each f-presentation

$$(\mathscr{F} : U \times (\alpha, \beta), X) \in \mathscr{D}M_n{}^f \tag{20.7a}$$

and ordinary value c of f fixed in (α, β) we introduce the partial mapping

$$\mathscr{F}^c : y \to \mathscr{F}(y, c) : U \to X \cap f^c \qquad (\text{onto } X \cap f^c).$$

This mapping defines a presentation

$$(\mathscr{F}^c : U, X^c) \in \mathscr{D}^0 f^c, \qquad X^c = X \cap f^c, \tag{20.7b}$$

termed a *partial presentation induced by* $\mathscr{F}$.

M_n-Embedded Manifolds $\mathbf{f}^c$. We shall prove the following theorem:

Theorem 20.1. *Corresponding to each ordinary value c of a* ND *$f \in C^\infty(M_n)$ there exists a unique M_n-embedded C^∞-manifold $\mathbf{f}^c$ of dimension $n - 1$ whose carrier is f^c. The set $\mathscr{D}\mathbf{f}^c$ for such a manifold* *contains the partial presentation* $\mathscr{F}^c$ *of each f-presentation* $\mathscr{F}$ *whose range meets* f^c.

We begin the proof of Theorem 20.1 by establishing (i):

(i) *Corresponding to any two f-presentations* $(\mathscr{F}_1 : U_1 \times (\alpha_1, \beta_1), X_1)$ and $(\mathscr{F}_2 : U_2 \times (\alpha_2, \beta_2), X_2)$ in $\mathscr{D}M_n^f$ *whose ranges meet* f^c, $\mathscr{F}_1^c$ *and* $\mathscr{F}_2^c$ *are* C^∞*-compatible.*

According to the definition of C^∞-compatibility it is sufficient to prove (i) in the special case in which $X_1 = X_2 = X \neq \varnothing$ and $(\alpha_1, \beta_1) = (\alpha_2, \beta_2) = (\alpha, \beta)$. In this case there exists, by hypothesis of C^∞-compatibility, a C^∞-diff (onto) of form

$$\mathscr{F}_2^{-1} \circ \mathscr{F}_1 : U_1 \times (\alpha, \beta) \to U_2 \times (\alpha, \beta). \tag{20.8}$$

Because of the condition (20.4) on *f*-presentations the restriction of the diff (20.8) to $U_1 \times \{c\}$ yields the C^∞-diff

$$(\mathscr{F}_2^c)^{-1} \circ \mathscr{F}_1^c : U_1 \to U_2 \qquad (\text{onto } U_2),$$

thereby establishing (i).

Since the C^∞-presentations $\mathscr{F}^c \in \mathscr{D}^0 f^c$ admitted in Theorem 20.1 cover f^c there exists a C^∞*-manifold* $\mathbf{f}^c$ whose carrier is f^c and whose set $\mathscr{D}\mathbf{f}^c$ contains the presentations $\mathscr{F}^c$ admitted in Theorem 20.1. We continue with a proof of (ii):

(ii) *The inclusion mapping* φ *of* f^c *into* $|M_n|$ *defines a* C^∞*-embedding of* f^c *in* M_n.

To prove (ii) we apply the test of Definition 16.2 on an "embedding." Corresponding to a point p_0 prescribed in f^c there exists by Lemma 20.1 an *f*-presentation

$$(\mathscr{F} : U \times (\alpha, \beta), X) \in \mathscr{D}M_n^f \tag{20.9}$$

such that $p_0 \in X$. By definition of $\mathbf{f}^c$ the partial presentation $(\mathscr{F}^c : U, X \cap f^c)$ is in $\mathscr{D}\mathbf{f}^c$. The test of Definition 16.2, associated with p_0 and the inclusion φ, is satisfied since the mapping

$$\mathscr{F}^{-1} \bar{\circ} \mathscr{F}^c : U \to U \times (\alpha, \beta) \tag{20.10}$$

induces the identity mapping of $U \times \{c\}$ onto $U \times \{c\}$.

Hence (ii) is true.

That $\mathbf{f}^c$ is an M_n-embedded C^∞-manifold now follows from (ii). By Theorem 16.5(k) $\mathbf{f}^c$ is unique among M_n-embedded C^∞-manifolds with carrier f^c.

A System d/ψ_H *when* $H = \mathscr{F}$. When $\mathscr{F}$ is an f-presentation in $\mathscr{D}M_n{}^f$ the f-transverse dynamical system of Theorem 19.6 takes a special form in terms of $\mathscr{F}$-coordinates (y, τ). By hypothesis $(f \circ \mathscr{F})(y, \tau) = \tau$ and in terms of the coordinates $y^1, \ldots, y^{n-1}$, τ condition (19.31) of Theorem 19.6 implies that $\psi^n_{\mathscr{F}}(y, \tau) = 1$. Hence the nth equation in the local system $d/\psi_{\mathscr{F}}$ takes the form

$$d\tau/dt = \psi^n_{\mathscr{F}}(y, \tau) = 1, \qquad (y, \tau) \in U \times (\alpha, \beta). \tag{20.11}$$

***ff*-Presentations in $\mathscr{D}M_n{}^f$.** Theorem 20.2 is the principal theorem of §20. It requires the following definition:

Definition 20.3. *ff-Presentations. An f-presentation* $\mathscr{F} \in \mathscr{D}M_n{}^f$, *of form* (20.9), *will be said to be an ff-presentation if each partial mapping*

$$t \to \mathscr{F}(u, t) : (\alpha, \beta) \to X \tag{20.12}$$

is an f-trajectory of the system $\{\mathbf{d}/\psi_H\}$.

Theorem 20.2. *Given a* ND $f \in C^\infty(M_n)$ *and point* $p_0 \in M_n{}^f$, *there exists an ff-presentation* $\mathscr{F}$ *in* $\mathscr{D}M_n{}^f$ *of form* (20.9) *whose range* X *is an open neighborhood of* p_0 *in* $M_n{}^f$.

Notation for Lemma 20.2. Theorem 20.2 will follow from Lemma 20.2 below. In Lemma 20.2 there is given an f-presentation

$$(\mathscr{G} : V \times (a, b), W) \in \mathscr{D}M_n{}^f \qquad (V \text{ open in } E_{n-1}) \tag{20.13}$$

such that $p_0 \in W$ and $a < f(p_0) < b$. We shall refer to points $u = (u^1, \ldots, u^{n-1}) \in V$ and values $\tau \in (a, b)$. Let (u_0, τ_0) be the $\mathscr{G}$-coordinates of p_0.

Lemma 20.2. *If* (α, β) *is a sufficiently small open subinterval of* (a, b) *such that* $\alpha < \tau_0 < \beta$, *and if* U *is a sufficiently small open neighborhood in* V *of* u_0, *there exists a* C^∞*-diff*

$$(u, t) \to \Theta(u, t) : U \times (\alpha, \beta) \to N \tag{20.14}$$

onto a neighborhood N *in* $V \times (\alpha, \beta)$ *of* (u_0, τ_0) *such that the following is true: For each* $u \in U$ *and for the given* τ_0 *the partial mapping*

$$t \to \Theta(u, t) : (\alpha, \beta) \to E_n \tag{20.15}$$

gives a t-parameterized trajectory

$$y^1 = \Theta^1(u, t), \ldots, y^{n-1} = \Theta^{n-1}(u, t), \qquad \tau = \Theta^n(u, t) = t, \tag{20.16}$$

of the local dynamical system $d/\psi_{\mathscr{G}}$ *which is in* N, *satisfies the initial conditions*

$$\Theta(u, \tau_0) = (u, \tau_0), \qquad u \in U, \tag{20.17}$$

and is identical with any trajectory in N *which meets it at the same time* t.

Proof of Lemma 20.2. Except for the affirmation that the mapping Θ is a C^∞-diff Lemma 20.2 is a classical theorem on ordinary differential equations (see Valiron [1], Vol. 2, pp. 308–313).

That Θ is a diff for U and (α, β) sufficiently restricted follows on noting that the Jacobian

$$\frac{D(\Theta^1, \ldots, \Theta^{n-1}, \Theta^n)}{D(u^1, \ldots, u^{n-1}, t)}(u_0, \tau_0) = \frac{\partial \Theta^n}{\partial t}(u_0, \tau_0) = 1$$

by virtue of the initial conditions

$$\Theta^1(u, \tau_0) = u^1, \ldots, \Theta^{n-1}(u, \tau_0) = u^{n-1}$$

of (20.17) and the identity $\Theta^n(u, t) = t$, valid for $t \in (\alpha, \beta)$. This identity is a consequence of the differential equation (20.11) and the initial condition $\Theta^n(u, \tau_0) = \tau_0$.

Proof of Theorem 20.2. Let p_0 be the point prescribed in $M_n{}^f$. In the notation of Lemma 20.2 the presentation

$$\mathscr{F} = (\mathscr{G} \,\bar{\circ}\, \Theta : U \times (\alpha, \beta), \mathscr{G}(N)) \in \mathscr{D}^0 \mid M_n{}^f \mid, \tag{20.18}$$

is well-defined and $p_0 = \mathscr{G}(u_0, \tau_0) \in \mathscr{G}(N)$. We shall show that $\mathscr{F}$ satisfies Theorem 20.2 by proving the following: (A_1) $\mathscr{F}$ is in $\mathscr{D}M_n{}^f$; (A_2) $\mathscr{F}$ is an f-presentation; (A_3) $\mathscr{F}$ is an ff-presentation.

Verification of (A_1). We shall compare $\mathscr{F}$ with the presentation

$$(\mathscr{G} \mid N : N, \mathscr{G}(N)) \in \mathscr{D}M_n{}^f,$$

noting first that $\mathscr{G}$ is defined in N, since $N \subset V \times (\alpha, \beta)$ by virtue of Lemma 20.2. Moreover, $\mathscr{F}$ and $\mathscr{G} \mid N$ have the same range $\mathscr{G}(N)$ and are C^∞-compatible, since

$$(\mathscr{G} \mid N)^{-1} \circ (\mathscr{G} \bar{\circ} \Theta) = \Theta \qquad \text{on} \quad U \times (\alpha, \beta).$$

Statement (A_1) follows readily.

Verification of (A_2). For $(u, \tau) \in U \times (\alpha, \beta)$

$$(f \bar{\circ} (\mathscr{G} \bar{\circ} \Theta))(u, \tau) = (f \bar{\circ} \mathscr{G})(\Theta(u, \tau)) = \Theta^n(u, \tau) = \tau. \tag{20.19}$$

The second equality in (20.19) is valid because $\mathscr{G}$ is an f-presentation, and the third equality is valid in accord with (20.16) of Lemma 20.2.

Verification of (A_3). For each $u \in U$ the partial mappings

$$t \to \mathscr{G}(\Theta(u, t)) = \mathscr{F}(u, t), \qquad t \in (\alpha, \beta), \tag{20.20}$$

are solutions on $M_n{}^f$ of the dynamical system $\{\mathbf{d}/\psi_H\}$, since the partial mappings $t \to \Theta(u, t)$ are solutions of d/ψ_H when $H = \mathscr{G}$, in accord with Lemma 20.2. The solutions (20.20) are f-parameterized by virtue of (20.19).

This completes the proof of Theorem 20.2.

We note the following consequence of the unique determination of the solutions of a sufficiently restricted local dynamical system by each of their points, as implied by Lemma 20.2:

Two f-trajectories on $M_n{}^f$ of the system $\{\mathbf{d}/\psi_H\}$ which intersect in a point are overlapping.

EXERCISE 20.1. Establish the following lemma:

Lemma 20.3. *If ff-presentations $\mathscr{F}$ and $\mathscr{G}$ in $\mathscr{D}M_n{}^f$ have the same range $X \subset M_n{}^f$ and Euclidean domains*

$$U \times (\alpha, \beta), \qquad V \times (\alpha, \beta) \qquad \text{in} \quad E_n \tag{20.21}$$

with $\mathscr{F}$-coordinates (u, t) and $\mathscr{G}$-coordinates (v, t), respectively, the transition diff λ [cf. (5.6)] *of the first of the domains* (20.21) *onto the second is represented by the identity mapping $t = \tau$ of (α, β) onto (α, β) and a C^∞-diff $u \to \mu(u) = v$ of U onto V.*

EXERCISE 20.2. Given a ND $f \in C^\infty(M_n)$, an *ff*-presentation

$$(\mathscr{F} : U \times (\alpha, \beta), X) \in \mathscr{D}M_n^f, \qquad U \text{ open in } E_n \tag{20.22}$$

and values c and e in (α, β), prove the following lemma:

Lemma 20.4. *There exists a C^∞-diff of $\mathbf{f}^c \cap X$ onto $\mathbf{f}^e \cap X$ in which points in $f^c \cap X$ and $f^e \cap X$ correspond which are on the same f-trajectory.*

Suggestion. Apply Lemma 16.3 to the partial presentations $(\mathscr{F}^c : U, X^c) \in \mathscr{D}\mathbf{f}^c$, $X^c = X \cap f^c$, and $(\mathscr{F}^e : U, X^e) \in \mathscr{D}\mathbf{f}^e$, $X^e = X \cap f^e$.

§ 21

f-TRAJECTORIES ON M_n^f

The objective of this section is to prove Theorem 21.1. Theorem 21.1 is one of two principal aids in proving the fundamental homotopy theorem, Theorem 23.2.

Introduction to Theorem 21.1. As in §20 let there be given a C^∞-manifold M_n and a ND $f \in C^\infty(M_n)$.

The Manifold $\mathbf{f}_{(a,b)}$. Let (a, b) be an open interval of ordinary values of f. Set

$$f_{(a,b)} = \{p \in M_n \mid a < f(p) < b\}. \tag{21.1}$$

We suppose $f_{(a,b)}$ topologized by $|M_n|$. Let $\mathbf{f}_{(a,b)}$ be the C^∞-submanifold of M_n whose carrier is $f_{(a,b)}$. Among the presentations in $\mathscr{D}\mathbf{f}_{(a,b)}$ are those ff-presentation $\mathscr{F}$ of §20 whose ranges are subsets of $f_{(a,b)}$. The manifold $\mathbf{f}_{(a,b)}$ is covered by such presentations by virtue of Theorem 20.2.

We define a basic condition on M_n and f:

Definition 21.1. *Bounded f-Compactness of* M_n. Given a ND $f \in C^\infty(M_n)$, we say that M_n is *boundedly f-compact* if whenever $[a, b]$ is a finite closed interval of values of f on M_n, the subset

$$f_{[a,b]} = \{p \in M_n \mid a \leqslant f(p) \leqslant b\} \tag{21.2}$$

of M_n is compact.

Let c be a value of f. If M_n is boundedly f-compact, the level set f^c of $|M_n|$ is compact, as one readily proves. If M_n is boundedly

f-compact and if an arbitrary ND $g \in C^\infty(M_n)$ is given, then M_n is not necessarily boundedly g-compact. If, however, M_n is compact, then for each ND $g \in C^\infty(M_n)$ M_n is boundedly g-compact. The hypothesis of compactness for M_n is too restrictive for many problems in global analysis and differential topology. Bounded f-compactness will be a principal hypothesis in Theorem 21.1.

The Product Manifold $\mathbf{f}^c \times (a, b)$. Let c be an ordinary value of f in the interval (a, b). According to Theorem 20.1 f^c is the carrier of a C^∞-manifold $\mathbf{f}^c$ admitting a C^∞-embedding in M_n and having a differentiable structure thereby uniquely determined by M_n. Let (a, b) represent the one dimensional regular C^∞-manifold whose carrier is the interval (a, b). In Theorem 21.1 we shall refer to the product manifold $\mathbf{f}^c \times (a, b)$ with the product differentiable structure defined in §13.

The principal theorem follows.

Theorem 21.1. *Let (a, b) be an open interval, possibly infinite, of ordinary values of f, and let c be prescribed in (a, b). If M_n is boundedly f-compact, there exists a homeomorphism*[†]

$$(q, t) \to \Gamma(q, t) : \mathbf{f}^c \times (a, b) \to \mathbf{f}_{(a,b)} \tag{21.3}$$

onto the manifold $\mathbf{f}_{(a,b)}$ such that for each point $q \in f^c$ the partial mapping

$$t \to \Gamma(q, t) : (a, b) \to \mathbf{f}_{(a,b)} \tag{21.4}$$

is an f-trajectory on $\mathbf{f}_{(a,b)}$ of the canonical dynamical system $\{\mathbf{d}/\psi_H\}$.

Note. Theorem 21.1, altered by affirming that the homeomorphism (21.3) is a C^∞-diff, is true. This extension of Theorem 21.1 is not needed and accordingly will not be proved.

Before coming to the proof proper of Theorem 21.1 we add the following remarks on f-trajectories:

f-Trajectories on $\mathbf{f}_{(a,b)}$. The symbols ξ, ξ_1, etc. will denote f-trajectories on $\mathbf{f}_{(a,b)}$ of the canonical dynamical system $\{\mathbf{d}/\psi_H\}$.

[†] $f^c \times (a, b)$ onto $f_{(a,b)}$ (Definition 16.1).

On these trajectories the range of f is an open subinterval of (a, b). As noted at the end of §20, two f-trajectories ξ_1 and ξ_2 which intersect in a point are overlapping. The f-trajectories are f-parameterized, and consequently are simple curves. Because they are simple curves, no difficulty will arise if ξ denotes both a mapping $t \to p(t)$ into $\mathbf{f}_{(a,b)}$ and the carrier of the image of this mapping.

Theorem 21.1 will follow from statements I–IV:

I. *Corresponding to a prescribed point $q \in f^c$ there exists a unique f-trajectory $t \to \eta_q(t)$ on $\mathbf{f}_{(a,b)}$ which meets q and admits an extension for $a < t < b$.*

Verification of I. Let a point $q \in f^c$ be given. According to Theorem 20.2 there is an *ff*-presentation $\mathscr{F}$ in $\mathscr{D}\mathbf{f}_{(a,b)}$ whose range contains q. There then exists an f-trajectory ξ^0 which meets q. Of the f-trajectories ξ on $\mathbf{f}_{(a,b)}$ which contain ξ^0 as a subarc there is a unique f-trajectory ξ_q whose f-parameter t has a maximal open domain $(h, k) \subset (a, b)$. We shall show that $(h, k) = (a, b)$.

Assuming that $k < b$, we shall arrive at a contradiction. If $k < b$, $f_{[c,k]}$ is by hypothesis a compact subset of $f_{(a,b)}$ and Cl ξ_q contains a point $p_0 \in f^k$. By virtue of Theorem 20.2 there exists an *ff*-presentation $(\mathscr{F} : U \times (\alpha, \beta), X) \in \mathscr{D}\mathbf{f}_{(a,b)}$ with $p_0 \in X$ and $h < \alpha < k < \beta$. There is clearly a point $p_1 \in \xi_q \cap X$. Since $\mathscr{F}$ is an *ff*-presentation, there is a point $u \in U$ such that the partial mapping $t \to \mathscr{F}(u, t) : (\alpha, \beta) \to \mathbf{f}_{(a,b)}$ is an f-trajectory ξ' meeting p_1. The f-trajectories ξ_q and ξ' both meet p_1 and thus overlap. If ξ_q is continued by ξ', an f-trajectory ξ'' is formed on which the range of f is (h, β). Since $\beta > k$, the assumption that $k < b$ is false.

Hence $k = b$. One proves similarly that $h = a$, thereby completing the proof of I.

The Point $\Gamma(q, t)$ Defined. For each point $q \in f^c$ and value of t on the interval (a, b) let $\Gamma(q, t)$ be the point $\eta_q(t)$ on the f-trajectory ξ_q meeting q.

We continue with a proof of II.

II. *The mapping*

$$(q, t) \to \Gamma(q, t) : f^c \times (a, b) \to f_{(a,b)} \tag{21.5}$$

is onto $f_{(a,b)}$ and biunique.

Let p be prescribed in $f_{(a,b)}$ with $f(p) = \gamma \neq c$. On replacing the level set f^c in the formulation of I by the level set f^γ we are led to the following conclusion: Of the f-trajectories on $f_{(a,b)}$ which meet p there is one, say ξ, on which the range of values of f is (a, b). Since $a < c < b$, the f-trajectory ξ must meet f^c, so that the mapping Γ is *onto* $f_{(a,b)}$.

The mapping Γ is *biunique*, since f-trajectories meeting different points of f^c do not intersect. These f-trajectories thus form a "field."

III. *If e is a value in the interval (a, b), the partial mapping*

$$q \to \Gamma(q, e) : f^c \to f^e \tag{21.6}$$

is a homeomorphism Γ_c^e of f^c onto f^e.

The mapping Γ_c^e is biunique and onto f^e by virtue of I and II. Since f^c and f^e are compact, to prove that Γ_c^e is a homeomorphism, it is sufficient to verify (i):

(i) *The mapping Γ_c^e is continuous at each point $p_1 \in f^c$.*

Set $p_e = \Gamma(p_1, e)$. For simplicity we suppose that $e > c$, so that $p_e \neq p_1$. Let ξ be the closed arc of the f-trajectory joining p_1 to p_e. If $e - c$ is so small that there exists an *ff*-presentation $\mathscr{F}$ whose open range in $\mathbf{f}_{(a,b)}$ includes ξ, statement (i) follows readily from the properties of $\mathscr{F}$ as an *ff*-presentation.

In the general case ξ is the union of successive closed subarcs

$$\xi_1, \xi_2, \ldots, \xi_{r+1} \tag{21.7}$$

bearing values of f on closed intervals separated by values

$$c = c_1 < c_2 < \cdots < c_{r+1} = e$$

of f so chosen that the arc ξ_i, $i = 1, \ldots, r$, is included in the range of an *ff*-presentation $\mathscr{F}_i$. Let

$$p_1, p_2, \ldots, p_{r+1}, \qquad p_{r+1} = p_e, \tag{21.8}$$

be the successive endpoints of the arcs (21.7). Mappings

$$\Gamma_{c_i}^{c_{i+1}} : f^{c_i} \to f^{c_{i+1}}, \qquad i = 1, \ldots, r, \tag{21.9}$$

defined as was Γ_c^e are continuous on sufficiently small open neighborhoods N_i in f^{c_i} of p_i.

If N_1 is sufficiently small, and if one sets

$$N_{i+1} = \Gamma_{c_i}^{c_{i+1}}(N_i), \qquad i = 1,..., r,$$

then the restriction

$$\Gamma_c^{\ e} \mid N_1 = (\Gamma_{c_r}^{c_{r+1}} \mid N_r) \circ \cdots \circ (\Gamma_{c_1}^{c_2} \mid N_1), \tag{21.10}$$

and so is continuous.

This establishes (i) and hence III.

We conclude the proof of Theorem 21.1 by verifying IV.

IV. *The mapping* Γ *of* (21.3) *is a homeomorphism of* $f^c \times (a, b)$ *onto* $f_{(a,b)}$.

If $[h, k]$ is a closed subinterval of (a, b), Γ maps the compact subset $f^c \times [h, k]$ of $f^c \times (a, b)$ onto the compact subset $f_{[h,k]}$ of $f_{(a,b)}$. It follows that IV will be established if we show that Γ is continuous at each point $(p, e) \in f^c \times (a, b)$.

Notation for Proof. Set $p_e = \Gamma(p, e)$. We wish to show that corresponding to a prescribed open neighborhood W of p_e in $f_{(a,b)}$ there exists an open neighborhood Z of (p, e) in $f^c \times (a, b)$ so small that

$$\Gamma(Z) \subset W. \tag{21.11}$$

To that end, let

$$(\mathscr{F} : U \times (\alpha, \beta), X) \in \mathscr{D}\mathbf{f}_{(a,b)} \tag{21.12}$$

be an *ff*-presentation such that

$$e \in (\alpha, \beta) \subset (a, b), \qquad p_e \in X \subset W.$$

Such a presentation exists. Let N be so small an open neighborhood in f^c of $p \in f^c$ that

$$\Gamma(N, e) \subset f^e \cap X. \tag{21.13}$$

N exists by virtue of III.

Let Z be the open neighborhood $N \times (\alpha, \beta)$ of (p, e) in $f^c \times (a, b)$. Each point $q \in \Gamma(Z)$ is on an arc of an f-trajectory bearing the interval of f-values (α, β) and meeting $f^e \cap X$ in accord with (21.13). The point q is accordingly in the range X of the presentation (21.12). Thus

$$\Gamma(Z) \subset X \subset W. \tag{21.14}$$

The continuity of Γ at (p, e) follows, implying the truth of IV. This completes the proof of Theorem 21.1.

EXERCISE 21.1. *Show that the homeomorphism* Γ_c^e *of* (21.6) *is a* C^∞*-diff of* $\mathbf{f}^c$ *onto* $\mathbf{f}^e$.

Suggestion. Parallel the proof of III(i), making use of Lemmas 20.3 and 20.4 to show that the homeomorphisms in the composition (21.10) are C^∞-diffs.

§ 22

f-PREFERRED RIEMANNIAN STRUCTURES $\mathbf{S}^f$

Let M_n be a C^∞-manifold with a C^∞-Riemannian structure denoted by $\mathbf{S}$. Given a ND $f \in C^\infty(M_n)$, the determination of f-trajectories of the dynamical system of Theorem 19.6 near a critical point q_0 of f can be very difficult unless this dynamical system is induced by a structure $\mathbf{S}$ specially chosen near q_0. In this section we shall show how to modify $\mathbf{S}$ suitably in a prescribed neighborhood X of q_0 while leaving M_n, $\mathscr{D}M_n$, and f unchanged and changing $\mathbf{S}$ at most on X. The principal difficulty arises from the requirement that the new forms Q_F assigned to the presentations $F \in \mathscr{D}M_n$ be M_n-compatible in the sense of Definition 19.1.

Theorem 22.1 is the principal theorem.

Introduction to Theorem 22.1. The simplest differential form Q_F which can be assigned to a presentation $(F : U, X) \in \mathscr{D}M_n$ is one whose matrix $\| g_{ij}(u) \|$ of coefficients is the unit matrix. We term such a form *canonically Euclidean.* Distances and angles defined on U by such a form are Euclidean.

Two C^∞-Riemannian structures $\mathbf{S}$ and $\mathbf{S}^*$ defined on M_n are understood to be *identical* on any open subset Z of $| M_n |$ if corresponding to each presentation $F \in \mathscr{D}M^n$ whose range is included in Z the forms Q_F assigned to F in $\mathbf{S}$ and $\mathbf{S}^*$ are identical.

The following theorem was proved with a somewhat different formulation by Morse [9] under Lemma 6.1:

Theorem 22.1. *Let a point q_0 be prescribed in M_n, together with a presentation $(F : U, X) \in \mathscr{D}M_n$, such that $q_0 \in X$. Corresponding to a C^∞-Riemannian structure $\mathbf{S}$ given on M_n there exists a C^∞-Riemannian structure $\mathbf{S}^*$ on M_n with the following two properties*:

(i) *On* $|M_n| - \bar{X}$ *the two Riemannian structures are identical.*

(ii) *There is a restriction* $(F_0 : U_0, X_0)$ *of* $(F : U, X)$ *such that* $q_0 \in X_0$, *that* U_0 *is a relatively compact subset of* U, *and that the Riemannian form* $Q^*_{F_0}$ *assigned to* F_0 *in* $\mathbf{S}^*$ *is canonically euclidean.*

We begin the proof of Theorem 22.1 by establishing Lemma 22.1.

Notation for Lemma 22.1. *Let* $\| c_{ij}(u) \|$ be the matrix of coefficients in the given C^∞-Riemannian form Q_F. For simplicity suppose that $F^{-1}(q_0)$ is the origin in $U \subset E_n$. For each positive r set $D_r = \{u \in E_n \mid \| u \| < r\}$. Let σ be so small a positive constant that

$$\mathrm{Cl}\, D_{4\sigma} \subset U. \tag{22.1}$$

Lemma 22.1. *There exists a positive-definite, symmetric, quadratic form* Q^*_F *with variable coefficients* $g_{hk}(u)$ *of class* C^∞ *on* U *such that*

$$\| g_{hk}(u) \| = \| c_{ij}(u) \|, \qquad u \in U - D_{4\sigma}, \tag{22.2}$$

and

$$\| g_{hk}(u) \| = \| \delta_{hk} \|, \qquad u \in D_\sigma. \tag{22.3}$$

An Auxiliary Mapping η. In proving Lemma 22.1 we shall make use of a C^∞-mapping $t \to \eta(t)$ of the real axis of t onto $[0, 1]$ such that

$$\eta(t) = 0, \qquad 0 \leqslant |t| \leqslant 1, \tag{22.4}$$

$$\eta(t) = 1, \qquad |t| \geqslant 2, \tag{22.5}$$

$$0 < \eta(t) < 1, \qquad 1 < |t| < 2. \tag{22.6}$$

Such a mapping exists, as is readily seen.

Definition of $g_{hk}(u)$. With h and k on the range $1, \ldots, n$ set

$$g_{hk}(u) = \eta\left(\frac{\| u \|}{\sigma}\right) c_{hk}(u) + \left(1 - \eta\left(\frac{\| u \|}{\sigma}\right)\right) \delta_{hk}, \qquad u \in D_{4\sigma}. \tag{22.7}$$

It follows from (22.7) and (22.4) that

$$g_{hk}(u) = \delta_{hk}, \qquad \| u \| \leqslant \sigma. \tag{22.8}$$

A consequence of (22.8) and (22.7) is that $g_{hk}(u)$, as defined over $D_{4\sigma}$, is of class C^∞.

It follows from (22.7) and (22.5) that

$$g_{hk}(u) = c_{hk}(u), \qquad 2\sigma \leqslant \| u \| \leqslant 4\sigma. \tag{22.9}$$

Taking account of (22.9), we can consistently extend $g_{hk}(u)$ so as to be of class C^∞ over U by requiring that

$$g_{hk}(u) = c_{hk}(u), \qquad u \in U - D_{2\sigma}.$$

The matrix $\| g_{hk}(u)\|$ thereby defined over U satisfies the lemma provided we show that $\| g_{hk}(u)\|$ is the matrix of a positive-definite quadratic form. Since the matrices $\| c_{hk}(u)\|$ and $\| \delta_{hk} \|$ are both positive-definite over U, we have merely to show that $\| g_{hk}(u)\|$ is positive-definite for $\sigma < \| u \| < 2\sigma$.

Positive-Definiteness of $\| g_{hk}(u)\|$. By virtue of (22.6)

$$0 < \eta(\| u \|/\sigma) < 1, \qquad \sigma < \| u \| < 2\sigma,$$

so that for these values of $\| u \|$ the values $\lambda(u) = \eta(\| u \|/\sigma)$ and $\mu(u) = 1 - \eta(\| u \|/\sigma)$ are positive. According to (22.7)

$$\| g_{hk}(u)\| = \lambda(u) \| c_{hk}(u)\| + \mu(u) \| \delta_{hk} \|, \tag{22.10}$$

so that the matrix $\| g_{hk}(u)\|$ is positive-definite when $\sigma < \| u \| < 2\sigma$. This establishes Lemma 22.1.

Completion of Proof of Theorem 22.1. Referring to $(F : U, X) \in \mathscr{D}M_n$ of Theorem 22.1 and to $D_\sigma \subset U$ of Lemma 22.1, we shall show that the restriction

$$(F_0 : U_0 , X_0) \in \mathscr{D}M_n , \qquad U_0 = D_\sigma , \tag{22.11}$$

of F is a presentation such that Theorem 22.1 will be satisfied by F_0 and a suitable choice of $\mathbf{S}^*$.

Set $X_1 = F(D_{4\sigma})$. We shall apply Theorem 19.7, letting K be the subset of presentations in $\mathscr{D}M_n$ which contains the presentation F, as given in Theorem 22.1, together with those presentations in $\mathscr{D}M_n$ whose ranges do not meet X_1. The presentations in K clearly cover $| M_n |$, and are M_n-compatible since $K \subset \mathscr{D}M_n$.

To the presentations $H \in K$ we shall assign M_n-compatible forms Q_H^*. When $H = F$ we assign to H the form Q_F^* of Lemma 22.1. To each other presentation $H \in K$ we assign the form Q_H^* assigned to H in the given Riemannian structure $\mathbf{S}$.

For $H \in K$ these forms Q_H^* are M_n-compatible, since this compatibility is implied by the M_n-compatibility of the forms Q_H of $\mathbf{S}$ for $H \in K$. It follows from Theorem 19.7 that there exists a C^∞ Riemannian structure $\mathbf{S}^*$ whose forms include the above forms Q_H^*, $H \in K$. In particular, $\mathbf{S}^*$ admits the forms Q_F^* of Lemma 22.1 and hence a canonically Euclidean form assigned to the restriction F_0 of F.

This establishes Theorem 22.1.

The Reduction Theorem 4.1 and Theorem 22.1 imply the following fundamental theorem on modifying a Riemannian structure on M_n :

Theorem 22.2. *Let q be a critical point of f of index k, N an open neighborhood of q, and $\mathbf{S}$ a C^∞-Riemannian structure given on M_n.*

There then exists a presentation

$$(F : D_\sigma, X) \in \mathscr{D}M_n, \qquad q = F(\mathbf{0}) \tag{22.12}$$

for which

$$(f \circ F)(u) = -u_1{}^2 - \cdots - u_k{}^2 + u_{k+1}^2 + \cdots + u_n{}^2 + f(q), \qquad \|u\| < \sigma, \tag{22.13}$$

and $\bar{X} \subset N$, and with which there can be associated a modified C^∞-Riemannian structure $\mathbf{S}'$ on M_n which differs from $\mathbf{S}$ at most on N and in which the Riemannian form Q_F is canonically Euclidean.

A presentation F associated as in Theorem 22.2 with a critical point q and a Riemannian structure $\mathbf{S}'$ will be termed *preferred* relative to f.

Corollary 22.1. *Given a* ND $f \in C^\infty(M_n)$, *it is possible to associate presentations F_μ chosen from $\mathscr{D}M_n$ biuniquely with the respective critical points of f and to define a C^∞-Riemannian structure $\mathbf{S}^f$ on M_n in such a manner that relative to f the presentations F_μ are of "preferred" type and have disjoint range closures.*

Definition 22.1. *f-Preferred Riemannian Structures $\mathbf{S}^f$.* Let a ND $f \in C^\infty(M_n)$ be given. A C^∞-Riemannian structure on M_n associated as in Corollary 22.1 with a set of "preferred" presentations of the respective critical points of f will be called an *f-preferred Riemannian structure* $\mathbf{S}^f$.

In the remainder of this section and in §23 we shall assume that a ND $f \in C^\infty(M_n)$ is given and that M_n has an f-preferred Riemannian structure $\mathbf{S}^f$.

f-Trajectories near a Critical Point. We are concerned with f-trajectories on M_n of a canonical f-transverse dynamical system $\{\mathbf{d}/\psi_H\}$ induced by an f-preferred Riemannian structure $\mathbf{S}^f$. We shall examine the local representation in F-coordinates of f-trajectories in a system d/ψ_F in which F is a preferred presentation of a neighborhood of a critical point q.

We are assuming that F has the form (22.12), that (22.13) holds, and that the Riemannian form Q_F is canonically Euclidean. The differential equations of the local system d/ψ_F have the form (19.36), or here

$$\begin{aligned} \frac{du^i}{dt} &= \frac{-u^i}{2\,\|u\|^2}, \qquad i = 1,\ldots,k, \qquad\qquad 0 < \|u\| < \sigma, \\ \frac{du^j}{dt} &= \frac{u^j}{2\,\|u\|^2}, \qquad j = k+1,\ldots,n, \quad 0 < \|u\| < \sigma, \end{aligned} \tag{22.14}$$

in accord with Theorems 19.5 and 19.6 and the nature of Q_F as canonically Euclidean.

For simplicity in characterizing the $(f \circ F)$-parameterized solutions of (22.14) we shall here suppose that $f(q) = 0$. When $f(q) = 0$, $(f \circ F)(u)$ is a quadratic function whose values will be denoted by $P_k(u)$. When $0 < k < n$ we shall make use of the cone

$$P_k^0 = \{u \in E_n \mid P_k(u) = 0\}.$$

When $0 < k < n$ we introduce the coordinate planes

$$\begin{aligned} \pi_k &= \{u \in E_n \mid u_{k+1} = \cdots = u_n = 0\} \\ \pi_{n-k} &= \{u \in E_n \mid u_1 = \cdots = u_k = 0\}. \end{aligned} \tag{22.15}$$

The *solution arcs* of (22.14) can be characterized as follows:

The Case $k = 0$. In this case each solution arc is radial and $(f \circ F)(u) = \|u\|^2$.

The Case $k = n$. The solution arcs are again radial, with $(f \circ F)(u) = -\| u \|^2$.

The Case $0 < k < n$. Each solution arc which meets π_k or π_{n-k} is radial, remaining in π_k and π_{n-k}, respectively.

All other solution arcs in D_σ *are arcs of equilateral hyperbolas explicitly characterized as follows*:

Let two sensed lines meeting the origin be prescribed in π_k and π_{n-k}, respectively, and termed x- and y-axes. These lines are mutually orthogonal. Relative to the (x, y)-plane π thereby defined let x and y be rectangular coordinates chosen so as to define a metric on π consistent with the metric on π induced by E_n.

Each solution arc of (22.14) which meets π remains in π and is a solution arc of the differential equations

$$\frac{dx}{dt} = \frac{-x}{2(x^2 + y^2)}, \qquad \frac{dy}{dt} = \frac{y}{2(x^2 + y^2)} \tag{22.16}$$

on the domain $0 < x^2 + y^2 < \sigma^2$, as one readily sees. If not radial, such a solution arc, if extended without limit in π, is a branch of an equilateral hyperbola with the x- and y-axes as asymptotes. There is one and only one such hyperbolic arc meeting each point of π not on the x- or y-axes. These arcs are orthogonal to the level arcs of $-x^2 + y^2$ and are to be parameterized by the values of $-x^2 + y^2$ on these arcs.

We shall explicitly record certain important properties of solution arcs of (22.14) in D_σ.

A Special Field of Solution Arcs. When $0 < k < n$ there is a hyperbolic solution arc h_u meeting each point $u \in P_k^0 \cap D_\sigma$ not the origin. The point u bisects h_u and is the nearest point to the origin on h_u. The solution arc h_u varies continuously with a point $u \in P_k^0 \cap D_\sigma$ not the origin.

Entrance Properties of Solution Arcs in D_σ. Let Bd D_σ denote the boundary of D_σ. If $k < n$, corresponding to each point $u \in$ Bd D_σ at which $P_k(u) > 0$ there is an extended $(f \circ F)$-parameterized solution arc which enters $\bar{D}_\sigma$ at u, with t decreasing, and if not radial leaves $\bar{D}_\sigma$ at a point $v \in$ Bd D_σ at which $P_k(v) < 0$.

Descending and Ascending Bowls. Suppose that $f \in C^{\infty}(M_n)$ is ND and that M_n is provided with an f-preferred Riemannian structure $\mathbf{S}^f$. With each critical point z of f of index $k > 0$ we shall associate a *descending* k-bowl $B_-(z, k)$ and with each critical point z of index $k < n$ an *ascending* $(n - k)$-bowl $B_+(z, n - k)$ defined as follows:

Definition 22.2. *If $k > 0$, $B_-(z, k)$ shall be the union of z and the maximally extended f-trajectories which have z as an upper limiting endpoint.*

If $k < n$, $B_+(z, n - k)$ shall be the union of z and the maximally extended f-trajectories which have z as lower limiting endpoint.

The limit points to which reference is made in these definitions are of points $p(t)$ on f-trajectories as the f-parameter t increases or decreases to a limiting value. The bowls are to be topologized by $|M_n|$. We term z the *pole* of $B_-(z, k)$ and of $B_+(z, n - k)$.

For the purposes of this book two introductory lemmas on bowls will suffice and will proved. All results on bowls in this book presuppose the existence of an f-preferred Riemannian structure $\mathbf{S}^f$. In the first of the following lemmas it is not assumed that M_n is *boundedly f-compact*. However, this assumption is necessary in Lemma 22.3.

In the exercises at the end of the section no assumption that M_n is boundedly f-compact is needed.

Lemma 22.2. (i) *An ascending $(n - k)$-bowl is a topological $(n - k)$-manifold.*

(ii) *A descending k-bowl is a topological k-manifold.*

Proof of (i). By the "manifold condition" on $B_+ = B_+(z, n - k)$ at a point q of B_+ is meant the condition that there exist a neighborhood of q relative to B_+ which is a topological $(n - k)$-ball.

That this manifold condition is satisfied at the pole z of B_+ is seen as follows. Let z be identified with the critical point q for which $(F : D_\sigma, X)$ of (22.12) is a preferred presentation of a neighborhood X of q. As we have just seen, the f-trajectories which tend to z as a lower limiting endpoint meet X in the topological $(n - k)$-ball $F(\pi_{n-k} \cap D_\sigma)$.

Lemma 22.2(i) is trivial if $n - k = 1$. Set $r = n - k - 1$ and suppose that $r > 0$.

With each point p on an f-trajectory ξ in $B_+ = B_+(z, n - k)$ we associate a condition K_p defined as follows but not *a priori* satisfied:

The condition K_p. *Set* $f(p) = \eta$. *Under condition* K_p *there shall exist on* f^η *an arbitrarily small neighborhood* N_p *of* p *such that* $N_p \cap B_+$ *is a topological* r-*ball* $A_p{}^r$, $r = n - k - 1$.

The trajectory ξ is an open arc with the critical point z as lower limiting endpoint but not a point of ξ. The condition K_p is satisfied by all points p on ξ sufficiently near z; it is obviously satisfied by all points $p \in \xi$ in the set $F(D_\sigma) = X$ of Theorem 22.1 associated with the critical point $q = z$.

We continue with a proof of (α):

(α) *If* ξ *is an* f-*trajectory in* B_+, *the condition* K_p *is satisfied at each point of* ξ.

Let $t \to p(t)$ be a representation of ξ in terms of the f-parameter t. Were (α) false, there would exist values a and b among the values of f on ξ such that the condition $K_{p(t)}$ is satisfied when $a < t < b$ but that the condition $K_{p(b)}$ is not satisfied. Let

$$(\mathscr{G} : V \times (\alpha, \beta), Y) \in \mathscr{D}M_n{}^f, \qquad a < \alpha < b < \beta,$$

be an ff-presentation such that $p(b) \in Y$.

Choose c so that $\alpha < c < b$. The condition $K_{p(c)}$ is satisfied by hypothesis by an open neighborhood $N_{p(c)}$ of $p(c)$ in f^c and by an open r-ball $A^r_{p(c)} \subset N_{p(c)}$. By hypothesis $N_{p(c)}$ can be supposed arbitrarily small. Since the presentation $\mathscr{G}$ can be restricted by restricting V as a neighborhood of $\mathscr{G}^{-1}(p(b))$ without altering (α, β), we can suppose that $N_{p(c)} = \mathscr{G}(V, c)$.

The condition $K_{p(b)}$ can then be satisfied by taking $N_{p(b)}$ as $\mathscr{G}(V, b)$ and $A^r_{p(b)}$ as the homeomorph of $A^r_{p(c)}$ under the mapping in which points in $A^r_{p(b)}$ and $A^r_{p(c)}$ correspond which are on the same f-trajectory.

This establishes (α).

Verification of Lemma 22.2(i). Let $p \neq z$ be an arbitrary point in $B_+ - z$ and set $f(p) = \eta$ as in condition K_p. By (α) the condition K_p is satisfied. Let the topological r-ball $A_p{}^r$ of condition K_p be so small a neighborhood of p on f^η that it is included in the range of an ff-presentation in $\mathscr{D}M_n{}^f$. If $e > 0$ is sufficiently small, the *union* of the subarcs of f-trajectories that meet the above r-ball A_p^r and on which the range of f is $(\eta - e, \eta + e)$ will be a topological $(n - k)$-ball on B_+ containing p.

Note. The process of proving (α) is called a *continuation process.* Lemma 22.2(i) follows. The proof of Lemma 22.2(ii) is similar.

The following lemma is easily verified:

Lemma 22.3. *Given an ascending bowl* $B_+ = B_+(z, n - k)$, *set* $f(z)$ a. = *Corresponding to an interval* $(a, c]$ *of ordinary values of* f *set*

$$\overset{c}{\underset{a}{B_+}} = \{p \in B_+ \mid a \leqslant f(p) \leqslant c\}. \tag{22.17}$$

If M_n *is boundedly f-compact,* $\overset{c}{\underset{a}{B_+}}$ *is the homeomorph of a closed* $(n - k)$*-ball* $\mathscr{B}$ *of radius* $c - a$ *under a mapping* Λ *in which* z *corresponds to the center of* $\mathscr{B}$ *and each f-trajectory in* $\overset{c}{\underset{a}{B_+}}$ *is mapped linearly with respect to its f-parameter* t *onto an open radial arc of* $\mathscr{B}$.

One shows first that a mapping Λ of $\overset{c}{\underset{a}{B_+}}$ onto an $(n - k)$-ball $\mathscr{B}$ exists and is biunique, as in the proofs of I and II under Theorem 21.1, and then shows that Λ is a homeomorphism as in the proof of Lemma 22.2.

Definition 22.3. *Traces of Bowls.* Under the conditions of Lemma 22.3 the intersection $B_+ \cap f^c$ will be called the *trace* of the *bowl* B_+ *on* f^c. This trace is a topological sphere of dimension $(n - k) - 1$.

Similarly, if a descending bowl $B_- = B_-(z, k)$ is given, set $f(z) = b$. If $[c, b)$ is an interval of ordinary values of f and if M_n is boundedly f-compact, the intersection $B_- \cap f^c$ is a topological $(k - 1)$-sphere and will be called the *trace* of the bowl B_- on f^c.

Modification of Critical Values of f. The following lemma is needed in proving the homotopy theorem of §23 (a study in depth of such modifications with the aid of bowls is given by Morse [16]):

Lemma 22.4. *Given a* ND $f \in C^\infty(M_n)$, *let* z *be a critical point of* f *and* N *a compact neighborhood of* z *which contains no critical point of* f *other than* z. *Corresponding to any sufficiently small open neighborhood* $N_0 \subset N$ *of* z *and sufficiently small positive constant* e *there exists a* ND $g \in C^\infty(M_n)$ *such that* z *is the only critical point of* g *in* N *and*

$$g(p) = f(p), \qquad p \in M_n - N, \tag{22.18}$$

and

$$g(p) = f(p) \pm e, \qquad p \in N_0 . \tag{22.19}$$

As a consequence the critical points of f *and* g *are identical and have identical indices, and if* M_n *is boundedly* f*-compact,* M_n *is boundedly* g*-compact* (see Morse [9] §2).

Notation for Proof. Let r be a positive constant. Let $t \to h_r(t)$ be a C^∞-mapping of R onto $[0, 1]$ such that $h_r(t) = 1$ when $|t| \leqslant r$ and $h_r(t) = 0$ when $|t| \geqslant 2r$.

Let $(F : U, X) \in \mathscr{D}M_n$ be such that $z \in X \subset N$ and $F^{-1}(z)$ is the origin in U. Set

$$\omega(u) = (f \circ F)(u), \qquad u \in U. \tag{22.20}$$

Then ω is ND on U with the origin its only critical point. Let D_{2r} be an open, origin-centered n-ball in E_n of radius $2r$. Suppose that $\bar{D}_{2r} \subset U$.

A modification of ω. Set

$$\theta(u) = \omega(u) \pm e h_r(\| u \|), \qquad u \in U, \tag{22.21}$$

and note that

$$\theta(u) = \omega(u), \qquad u \in U - \bar{D}_{2r} , \tag{22.22}$$

and

$$\theta(u) = \omega(u) \pm e, \qquad u \in D_r . \tag{22.23}$$

The origin is a ND critical point of θ. Any other critical point of θ must be in the closed set $\bar{D}_{2r} - D_r$ on which ω has no critical point. Hence if $e > 0$ is sufficiently small, θ has no critical point other than the origin. We suppose e so conditioned.

To prove the lemma, it is sufficient to show that for *some* sufficiently small open neighborhood $N_0 \subset N$ of z the conclusion of Lemma 22.4 is true. We shall prove the lemma for $N_0 = F(D_r)$.

Definition of g. The function g is overdefined by the conditions

$$(g \circ F)(u) = \theta(u), \qquad u \in U, \tag{22.24}$$

$$g(p) = f(p), \qquad p \in M_n - F(\bar{D}_{2r}). \tag{22.25}$$

This overdefinition is consistent, since (22.24), (22.22), and (22.20) imply that

$$(g \circ F)(u) = (f \circ F)(u), \qquad u \in U - \bar{D}_{2r}. \tag{22.26}$$

Moreover, the sets $F(U)$ and $M_n - F(\bar{D}_{2r})$ are open and have M_n as union. It follows from (22.24) and (22.25) that g is of class C^∞ on M_n.

In view of the definitions of g in (22.24) and N_0 as $F(D_r)$, (22.23) implies (22.19). Since $N \supset F(\bar{D}_{2r})$, (22.25) implies (22.18). Since θ has no critical point in U other than the origin and since (22.25) holds, z is the only critical point of g in N.

That the critical points of f and g are identical and have the same indices is now clear. Since f and g have different values at most on the compact set N, we infer that M_n is boundedly f-compact if and only if boundedly g-compact.

This establishes the lemma for $N_0 = F(D_r)$.

Exercises on Bowls

Open Radial Sets Defined. An open subset Z of E_n which contains the origin will be termed *radial* if whenever a point x is in Z, the straight arc joining x to the origin is in Z. The boundary of such a set is not in general a topological manifold or even bounded. Prove the following:

EXERCISE 22.1. An open radial subset Z of E_n is the homeomorph of an open n-ball.

EXERCISE 22.2. A bowl $B_+(z, n - k)$ on M_n is the homeomorph of an open radial subset Z of E_{n-k} under a mapping in which the pole z corresponds to the origin in Z and each f-trajectory ξ in B_+ is mapped linearly with respect to its f-parameter t onto an open radial arc in Z.

EXERCISE 22.3. $B_+(z, n-k)$ is the carrier of an M_n-embedded C^∞-manifold N_{n-k}.

EXERCISE 22.4. The manifold N_{n-k} is the C^∞-diffeomorph of an open radial subset of E_{n-k}.

EXERCISE 22.5. An open radial subset Z of E_n is a real analytic diffeomorph of an open n-ball (see Morse [11] Part I, Bowls).

These exercises imply the following fundamental theorem (see Morse [11]). No assumption that M_n is boundedly f-compact is needed.

Theorem 22.3. (i) *An ascending $(n-k)$-bowl is the carrier of an M_n-embedded C^∞-manifold and is the C^∞-diffeomorph of an open $(n-k)$-ball.*

(ii) *A descending k-bowl is the carrier of an M_n-embedded C^∞-manifold and is the C^∞-diffeomorph of an open k-ball.*

§ 23

THE BASIC HOMOTOPY THEOREM

Hypotheses and Notation. There is given a ND $f \in C^{\infty}(M_n)$ and a closed interval $[a, b]$ of values of f of which a and b alone are critical. Taking account of Lemma 22.4, we suppose for the present that there is just one critical point z_a on f^a and just one critical point z_b on f^b. Suppose that $k_a < n$ and $k_b > 0$ are, respectively, the indices of z_a and z_b. Let c be a value fixed in (a, b) and let $T^c_{z_a}$ and $T^c_{z_b}$ be the "traces" on f^c, respectively, of the bowl ascending from z_a and the bowl descending from z_b. The *bowl traces* $T^c_{z_a}$ and $T^c_{z_b}$ on f^c play a vital role (Definition 22.3.).

We shall regard the product space $f^c \times [a, b]$ as a subspace of the product space $f^c \times R$, terming $f^c \times \{a\}$ the *lower boundary* of $f^c \times [a, b]$ and $f^c \times \{b\}$ the *upper boundary* of $f^c \times [a, b]$.

Geometric Homotopy. The type of homotopy developed in this section will be termed *geometric* to distinguish it from the algebraic homotopy involved in the chain homotopies of §27.

An Extension of Theorem 21.1. Our basic homotopy theorem is a corollary of Theorem 23.1. Theorem 23.1 characterizes the family of f-trajectories on $f_{[a,b]}$. In this family the parameter q of an f-trajectory ξ_q will be taken as the point q of intersection of ξ with f^c. Theorem 23.1 is an extension of Theorem 21.1 in the following sense: The values a, b define an open interval (a, b) of ordinary values of f to which Theorem 21.1 applies. The mapping Ω of $f^c \times [a, b]$ onto $f_{[a,b]}$, which Theorem 23.1 characterizes, is an extension of the homeomorphism Γ of $f^c \times (a, b)$ onto $f_{(a,b)}$, which Theorem 21.1 affirms to exist. Theorem 23.1 is concerned with this extension, in particular with its continuity and its biuniqueness (insofar as it is biunique). We refer to Definition 21.1 of bounded f-compactness.

The Mapping Ω of Theorem 23.1. Suppose that M_n is *boundedly f-compact.*

Under the conditions of the first paragraph of this section, a mapping

$$(q, t) \to \Omega(q, t) : f^c \times [a, b] \to f_{[a,b]}$$

onto $f_{[a,b]}$ is uniquely defined by requiring that for each $q \in f^c$ the partial mapping into $f_{[a,b]} = |\mathbf{f}_{[a,b]}|$,

$$t \to \Omega(q, t) : [a, b] \to f_{[a,b]}, \tag{23.1}$$

shall be an f-trajectory ξ_q, which in particular shall be closed by z_a at $t = a$ when $q \in T^c_{z_a}$ and shall be closed by z_b at $t = b$ when $q \in T^c_{z_b}$.

It is clear that Ω maps $f^c \times [a, b]$ onto $f_{[a,b]}$. It is not clear *a priori* that Ω is continuous.

Theorem 23.1. *Ω is a mapping of $f^c \times [a, b]$ onto $f_{[a,b]}$ with the following properties:*

(i) *The restriction $\Omega | (f^c \times (a, b))$ is the homeomorphism Γ onto $\mathbf{f}_{(a,b)}$ of Theorem* 21.1.

(ii) *The antecedents of z_a and z_b under Ω are, respectively, the subsets $T^c_{z_a} \times \{a\}$ and $T^c_{z_b} \times \{b\}$ of the lower and upper boundaries of $f^c \times [a, b]$.*

(iii) *The restriction of Ω^{-1} to $f_{[a,b]} - z_a - z_b$ is a homeomorphism onto*

$$f^c \times [a, b] - (T^c_{z_a} \times \{a\}) - (T^c_{z_b} \times \{b\}). \tag{23.2}$$

(iv) *Ω is continuous.*

Verification of (i). (i) follows immediately from the definition (21.5) of Γ and the above definition of Ω.

Verification of (ii). (ii) follows from the definition of Ω and of the traces $T^c_{z_a}$ and $T^c_{z_b}$ on f^c, in view of Lemma 22.3.

Verification of (iii). Let

$$(F_a : D_{\sigma_a}, X_a) \in \mathscr{D}M_n \qquad \text{and} \qquad (F_b : D_{\sigma_b}, X_b) \in \mathscr{D}M_n \tag{23.3}$$

be preferred presentations of neighborhoods, respectively, of z_a and z_b of form (22.12) with $\bar{X}_a \cap \bar{X}_b = \varnothing$. Since X_a and X_b can be taken as arbitrarily small neighborhoods of z_a and z_b, (iii) will follow if we verify (α):

(α) *The restriction of* Ω^{-1} *to* $f_{[a,b]} - X_a - X_b$ *is a homeomorphism into the set* (23.2).

Proof of (α). To prove (α) we shall modify f slightly near z_a and z_b in accord with Lemma 22.4.

Let $N_a \subset X_a$ and $N_b \subset X_b$ be compact neighborhoods of z_a and z_b in $M_n - f^c$. By virtue of Lemma 22.4 there exists a ND function $g \in C^\infty(M_n)$ which differs from f at most on $N_a \cup N_b$ and for which $[a, b]$ is an interval of *ordinary* values of g and M_n is boundedly g-compact. Note that $f^c = g^c$.

There accordingly exists an open interval (α, β) of ordinary values of g such that $(\alpha, \beta) \supset [a, b]$. Let g-trajectories be defined in terms of the g-transverse dynamical system determined by g and the f-preferred Riemannian metric $\mathbf{S}^f$. By virtue of Theorem 21.1 there exists a homeomorphism

$$(q, t) \to \Psi(q, t) : \mathbf{g}^c \times (\alpha, \beta) \to \mathbf{g}_{(\alpha,\beta)}, \qquad g^c = f^c, \tag{23.4}$$

onto $\mathbf{g}_{(\alpha,\beta)}$ such that for each $q \in f^c$ the partial mapping

$$t \to \Psi(q, t) : (\alpha, \beta) \to \mathbf{g}_{(\alpha,\beta)} \tag{23.5}$$

is a g-trajectory on $\mathbf{g}_{(\alpha,\beta)}$. The identity of g and f on $M_n - N_a - N_b$ implies the set equality

$$f_{[a,b]} - \bar{X}_a - \bar{X}_b = g_{[a,b]} - \bar{X}_a - \bar{X}_b = W, \tag{23.6}$$

introducing the subset W of $| M_n |$. The set W is open relative to $f_{[a,b]}$ and $g_{[a,b]}$.

It follows that if ξ'_q and ξ''_q are, respectively, maximally extended f- and g-trajectories meeting a common point q of $f^c = g^c$, then

$$(\xi'_q \cap W) = (\xi''_q \cap W) \tag{23.7}$$

and these subarcs have identical parameterizations. The validity of (23.7) depends on the fact that $(\xi'_q \cap W)$ is a *connected* subarc of W, with a closure terminating at any intersection with Bd W (see "Entrance Properties of Solution arcs in D_σ" in §22).

Statement (α) is true if read with g replacing f. It follows from the equality (23.7) that (α) holds as stated for f, and (iii) follows.

Verification of (iv). Because (ii) and (iii) are true, to establish the continuity of Ω it is sufficient to establish (iv)a and (iv)b, unconditioned by the parentheses (which are added because their conditions are needed later in establishing the homotopy theorem):

(iv)a. *Corresponding to a prescribed open neighborhood N_a of z_a in $f_{[a,c]}$ the set $\omega_a = \Omega^{-1}(N_a)$ is an open neighborhood of $T^c_{z_a} \times \{a\}$ in $f^c \times [a, c]$ (arbitrarily small if N_a is sufficiently small).*

(iv)b. *Corresponding to a prescribed open neighborhood N_b of z_b in $f_{[c,b]}$ the set $\omega_b = \Omega^{-1}(N_b)$ is an open neighborhood of $T^c_{z_b} \times \{b\}$ in $f^c \times [c, b]$ (arbitrarily small if N_b is sufficiently small).*

Proof of (iv)a. Set $Y = f^c \times [a, c] - (T^c_{z_a} \times \{a\})$.

Note that $Z = f_{[a,c]} - N_a$ is a compact subset of $f_{[a,c]} - z_a$. It follows from (iii) that the subset $\Omega^{-1}(Z)$ of Y is compact. Hence the complement $\omega_a = \Omega^{-1}(N_a)$ of $\Omega^{-1}(Z)$ in $f^c \times [a, c]$ is *open*.

To verify the parenthetical supplement of (iv)a, let ω_a be a prescribed open neighborhood of $T^c_{z_a} \times \{a\}$ in $f^c \times [a, c]$. Then Comp. ω_a in $f^c \times [a, c]$ is compact. By virtue of (iii) Ω (Comp. ω_a) is compact and hence has an open complement N_a in $f_{[a,c]}$. Since $N_a = \Omega(\omega_a)$ and $z_a \in N_a$, (iv)a follows as supplemented.

The proof of (iv)b is similar.

Thus (iv) is true and the proof of Theorem 23.1 is complete.

Retracting Deformations. We shall establish the homotopy theorem in the general setting of retracting deformations (see Borsuk [1]). Several definitions are needed (see Crowell and Fox [1], pp. 54–60).

Definition 23.1. *A Retracting Deformation* **D**. Let Z be a topological space (Hausdorff) and B a nonempty subspace of Z. Let τ be a variable, termed the *time*, with domain [0, 1]. A continuous mapping

$$(p, \tau) \to \mathbf{D}(p, \tau) : Z \times [0, 1] \to Z \tag{23.8}$$

will be called a *deformation* $\mathbf{D}$ *retracting* Z *on itself onto* B if the following conditions are satisfied

$$
\begin{aligned}
&(\mathrm{A}_1): &\quad \mathbf{D}(p, 0) = p &\quad p \in Z, \\
&(\mathrm{A}_2): &\quad \mathbf{D}(p, 1) \in B &\quad p \in Z, \\
&(\mathrm{A}_3): &\quad \mathbf{D}(p, \tau) = p &\quad (p, \tau) \in B \times [0, 1].
\end{aligned}
\tag{23.9}
$$

For each $p \in Z$ the partial mapping

$$\tau \to \mathbf{D}(p, \tau) = \mathbf{D}^p(\tau) : [0, 1] \to Z, \tag{23.10}$$

introducing $\mathbf{D}^p$, is called the *deformation arc* of p. For each $\tau \in [0, 1]$ the partial mapping

$$p \to \mathbf{D}(p, \tau) = \mathbf{D}_\tau(p) : Z \to Z, \tag{23.11}$$

introducing $\mathbf{D}_\tau$, is called the τ-*mapping* of Z into Z. A retracting deformation $\mathbf{D}$ each of whose τ-mappings $\mathbf{D}_\tau$ is a homeomorphism will be said to be *isotopic*.

Condition (A_1) requires that the "initial" mapping $\mathbf{D}_0$ of Z into Z be the identity. Condition (A_2) requires that the "terminal" mapping $\mathbf{D}_1$ be into B. Condition (A_3) requires that the deformation arc of a point $p \in B$ have p as carrier.

Retracting deformations may have some, none, or all of the following three properties:

Property P_1. Each deformation arc of $\mathbf{D}$ is either simple or has a point carrier on B.

Property P_2. If the carriers of two deformation arcs intersect other than in a point of B, one of these carriers is included in the other.

Property P_3. Property P_3 is a property of $\mathbf{D}$ relative to a real-valued continuous function f defined on Z. Deformations with this property are such that if p_1 and p_2 are points in Z at the same f-level, then

$$f(\mathbf{D}^{p_1}(\tau)) = f(\mathbf{D}^{p_2}(\tau)), \qquad 0 \leqslant \tau \leqslant 1. \tag{23.12}$$

We shall study deformations $\mathbf{D}$ retracting a subset of $f_{[a,b]}$ onto f^a assuming that (a, b) is an interval of ordinary values of f and that

a and b are critical values. One could similarly study deformations retracting subsets of $f_{[a,b]}$ onto f^b. Our homotopy theorem affirms the existence of an "$\downarrow f$-linear" deformation $\mathbf{D}$ characterized as follows:

Definition 23.2. A deformation $\mathbf{D}$ retracting a subset Z of M_n on which $f(p) \geqslant a$ onto f^a will be called *$\downarrow f$-linear* if conditions (u) and (v) are satisfied:

Condition (u). For each point $p \in Z$ such that $f(p) > a$ the carrier of a deformation arc $\tau \to \mathbf{D}^p(\tau)$ is either an f-trajectory joining p to an ordinary point of f^a, or an f-trajectory joining p to a limiting critical point of f on f^a.

Condition (v). On each deformation arc $\tau \to \mathbf{D}^p(\tau)$ of a point $p \in Z$, f is a nonincreasing *linear* function of τ, constant if $p \in f^a$.

An *$\uparrow f$-linear* deformation retracting a subset Z of M_n on which $f(p) \leqslant a$ onto f^a is similarly defined.

The $\downarrow$ Homotopy Theorem. The hypotheses of the $\downarrow$ homotopy theorem are as follows: There is given a ND $f \in C^\infty(M_n)$ and a closed interval $[a, b]$ of values of f of which a and b alone are critical values, taken on, respectively, by critical points z_a and z_b with indices $k_a < n$ and $k_b > 0$. We assume that M_n is boundedly f-compact.

Theorem 23.2. *Under these conditions there exists a unique $\downarrow f$-linear deformation $\mathbf{D}$ retracting the set $Z = f_{[a,b]} - z_b$ on itself onto f^a in such a manner that just one f-trajectory is retracted onto each point of f^a, excepting the critical point z_a, onto which the intersection with $f_{[a,b]} - z_b$ of the bowl ascending from z_a is retracted.*

Note. The affirmation of uniqueness of an $\downarrow f$-linear deformation $\mathbf{D}$ presupposes a unique "preferred" Riemannian structure $\mathbf{S}^f$ on M_n.

In proving Theorem 23.2 use will be made of the projections pr_1 and pr_2 of $f^c \times [a, b]$ onto f^c and $[a, b]$. If $w = (q, t) \in f^c \times [a, b]$, $\mathrm{pr}_1 w$ and $\mathrm{pr}_2 w$ are defined by setting $\mathrm{pr}_1 w = q$ and $\mathrm{pr}_2 w = t$. Referring to the mapping Ω of Theorem 23.1, we shall verify the following lemma.

Lemma 23.1 (λ) *A mapping* $\mathbf{D}$ *satisfying the homotopy theorem must have as values the points in* Z

$$\mathbf{D}(p, \tau) = \Omega(\text{pr}_1 \Omega^{-1}(p), (1 - \tau)f(p) + \tau a), \quad (p, \tau) \in Z \times [0, 1]. \tag{23.13}$$

(μ) *Conversely, a mapping* $\mathbf{D}$ *with values given by* (23.13) *defines a deformation* $\mathbf{D}$ *which satisfies the first homotopy theorem.*

Proof of (λ). Suppose that $\mathbf{D}$ is an $\downarrow f$-linear deformation satisfying Theorem 23.2. If a point $p \in Z$ is ordinary, that is, if $p \neq z_a$, a point $\Omega^{-1}(p) \in f^c \times [a, b]$ is uniquely determined with projections

$$\text{pr}_1 \Omega^{-1}(p) \in f^c \quad \text{and} \quad \text{pr}_2 \Omega^{-1}(p) = f(p) \in [a, b]. \tag{23.14}$$

The antecedent under Ω in $f^c \times [a, b]$ of the deformation subarc $\tau \to \mathbf{D}^p(\tau)$ on which $0 \leqslant \tau < 1$ must then be an arc

$$\tau \to \Omega^{-1}(\mathbf{D}^p(\tau)) = (\text{pr}_1 \Omega^{-1}(p), (1 - \tau)f(p) + \tau a) \in f^c \times [a, b], \quad 0 \leqslant \tau < 1,$$

in accord with the conditions of Definition 23.2 on an $\downarrow f$-linear deformation $\mathbf{D}$. The formula (23.13) follows when $p \in Z$ is ordinary.

When $\mathbf{D}$ satisfies Theorem 23.2, (23.13) remains valid when $p = z_a$. In this case we note that

$$\Omega(T^c_{z_a} \times \{a\}) = z_a, \qquad \Omega^{-1}(z_a) = T^c_{z_a} \times \{a\}. \tag{23.15}$$

Since $f(z_a) = a$, both sides of (23.13) reduce to z_a when $p = z_a$.

Proof of (μ). Conversely, a mapping $\mathbf{D}$ of $Z \times [0, 1]$ into Z whose values $\mathbf{D}(p, \tau)$ are given by (23.13) defines a deformation satisfying Theorem 23.2 provided the mapping $\mathbf{D}$ so defined is continuous. This continuity follows at once from (iii) of Theorem 23.1, except at pairs (p, τ) at which $p = z_a$. To establish the continuity of $\mathbf{D}$ at pairs (z_a, τ), one notes that $\mathbf{D}(z_a, \tau) = z_a$ and confirms statments 1–4:

1. The real-valued function

$$(p, \tau) \to (1 - \tau)f(p) + \tau a, \qquad (p, \tau) \in Z \times [0, 1], \tag{23.16}$$

is continuous and takes on the value a when $(p, \tau) = (z_a, \tau)$.

2. The set-valued function $p \to \text{pr}_1 \Omega^{-1}(p)$ takes on the set value $T^c_{z_a} \subset f^c$ when $p = z_a$.

3. According to (iv)a of the proof of Theorem 23.1 the following is true: Corresponding to a prescribed open neighborhood N_a of z_a in $f_{[a,c]}$ the set $\mathrm{pr}_1\, \Omega^{-1}(N_a)$ is an open neighborhood of $T^c_{z_a}$ in f^c arbitrarily small if N_a is sufficiently small.

4. The mapping Ω is continuous and (23.15) holds.

It follows that a neighborhood $\hat{N}$ in $Z \times [0, 1]$ of a pair (z_a, τ) is mapped by **D**, as defined by (23.13), into a neighborhood of z_a in Z, arbitrarily small if $\hat{N}$ is sufficiently small.

Thus (μ) is true and the proof of Lemma 23.1 is complete.

Theorem 23.2 follows from Lemma 23.1 and Theorem 23.1.

Note. An $\downarrow f$*-linear deformation* **D** *retracting* $f_{[a,b]} - z_b$ *onto* f^a *has properties* P_1 *and* P_2 *and property* P_3 *relative to* f.

If for an arbitrary value α of f one sets

$$f_\alpha = \{p \in M_n \mid f(p) \leqslant \alpha\}, \tag{23.17}$$

one obtains a basic corollary:

Corollary 23.1. *Under the hypotheses of homotopy Theorem* 23.2 *there exists a unique deformation* **D**′ *(termed* $\downarrow f$*-linear) retracting* $f_b - z_b$ *onto* f_a *, under which* $f_{[a,b]} - z_b$ *is deformed as in Theorem* 23.2 *and points in* f_a *remain fixed.*

It is clear that in the homotopy theorem and its corollary the roles of a and b can be interchanged. If one sets

$$f_{\alpha^+} = \{p \in M_n \mid f(p) \geqslant \alpha\}, \tag{23.18}$$

one is lead to a deformation (termed $\uparrow f$*-linear*) retracting $f_{a^+} - z_a$ onto f_{b^+} and similar to the $\downarrow f$-linear deformation of Corollary 23.1.

The Sequence of Critical Values. We are assuming that there is given a ND $f \in C^\infty(M_n)$ and that M_n is boundedly f-compact. The case in which M_n is compact is included. There are two other conditions on M_n less restrictive than compactness but more restrictive than bounded f-compactness. These conditions are defined as follows: If for each value α of f, f_α is compact, M_n is termed *f-compact below*. If for each value α of f, f_{α^+} is compact, M_n is termed *f-compact above*. If M_n is f-compact below and above, M_n is compact.

We shall suppose for the present that each critical value a of f is taken on at *just one critical point* z_a (cf. Lemma 22.4).

If M_n is *compact*, the critical values of f form a finite sequence $a_0 < a_1 < \cdots < a_s$ with at least two critical values. If M_n is *f-compact below*, the critical values form a sequence

$$a_0 < a_1 < a_2 < \cdots \tag{23.19}$$

which may be finite but has at least one critical value. If M_n is *f-compact above*, a similar descending sequence of critical values exists. If f is *boundedly f-compact*, the critical values (if any exist) form a sequence

$$\cdots < a_{-2} < a_{-1} < a_0 < a_1 < a_2 < \cdots, \tag{23.20}$$

which may terminate on the right or on the left. The sequence (23.20) may be empty or have just one critical value.

A Supplement to Corollary 23.1. In the sequence (23.20) of critical values of f there can be a maximum (or minimum) critical value a which is not a maximum (or minimum) value of f on M_n. Consider the case in which a is a maximum critical value. In this case M_n can be retracted down onto f_a, but not by means of Corollary 23.1.

Hypotheses of Theorem 23.3. To cover the above case, suppose that $[a, b)$ is an interval of values of f of which a alone is critical and is taken on by a single critical point z_a. The number b may or *may not be* a value of f. As in Theorem 21.1, b may be equal to $+\infty$. One assumes that M_n is boundedly f-compact, and for simplicity that index $z_a < n$.

Theorem 23.3 supplements Theorem 23.2:

Theorem 23.3. *Under the hypotheses of the preceding paragraph there exists a unique $\downarrow$f-linear deformation $\hat{\mathbf{D}}$ retracting $f_{[a,b)}$ onto f^a in such a manner that there is just one f-trajectory in $f_{[a,b)}$ retracted onto each point of f^a, excepting the point z_a, onto which the intersection with $f_{[a,b)}$ of the bowl ascending from z_a is retracted.*

To prove Theorem 23.3, one replaces the mapping Ω of Theorem 23.1 by a mapping

$$(q, t) \to \hat{\Omega}(q, t) : f^c \times [a, b) \to f_{[a,b)}, \qquad c \in (a, b), \tag{23.21}$$

onto $f_{[a,b)}$ uniquely determined by requiring that for each $q \in f^c$ the partial mapping

$$t \to \hat{\Omega}(q, t) : [a, b) \to f_{[a,b)} \tag{23.22}$$

shall be an f-trajectory ξ_q which in particular shall be closed by z_a at $t = a$ when $q \in T^c_{z_a}$.

The properties of Ω were enumerated in Theorem 23.1. A similar enumeration of characteristics of $\hat{\Omega}$ follows.

Lemma 23.2. *The above mapping $\hat{\Omega}$ of $f^a \times [a, b)$ onto $f_{[a,b)}$ has the following properties:*

(i) *The restriction $\hat{\Omega} \mid (f^c \times (a, b))$ is the homeomorphism Γ onto $f_{(a,b)}$ of Theorem* 21.1.

(ii) *The antecedent of z_a under $\hat{\Omega}$ is the subset $T^c_{z_a} \times \{a\}$ of $f^c \times \{a\}$.*

(iii) *The restriction of $\hat{\Omega}^{-1}$ to $f_{[a,b)} - z_a$ is a homeomorphism onto*

$$f^c \times [a, b) - (T^c_{z_a} \times \{a\}). \tag{23.23}$$

(iv) *The mapping $\hat{\Omega}$ is continuous.*

The proof is similar to that of Theorem 23.1.

Proof of Theorem 23.3. As in the proof of Theorem 23.2, a mapping $\hat{\mathbf{D}}$ satisfying Theorem 23.3 must have values given by the formula (23.13) with $\hat{\Omega}$ in place of Ω and

$$(p, \tau) \in f_{[a,b)} \times [0, 1].$$

Conversely, a mapping $\hat{\mathbf{D}}$ so defined will satisfy the conditions on $\hat{\mathbf{D}}$ of Theorem 23.3, as one shows by a similar proof.

Theorem 23.3 follows readily.

Theorem 23.3 was stated and proved as a supplement to Theorem 23.2 largely for the sake of the following corollary:

Corollary 23.2. *Let a be a maximum critical value of a* ND $f \in C^\infty(M)$ *which has no absolute maximum on M_n. If M_n is boundedly f-compact, there exists a unique $\downarrow f$-linear deformation retracting M_n onto f_a in such a manner that f_{a+} is retracted onto f^a* (cf. Theorem 23.3) *with* $b = \sup f$, *and points of f_a are fixed.*

There is a similar corollary in which minimum and inf f, respectively, replace maximum and sup f, while f_a and f_{a^+} are interchanged. The retraction is $\uparrow f$-linear.

The Determination of Homology Groups of M_n. As will be seen in Part III, if there is a deformation retracting a topological space Z onto a subspace B, the singular homology groups on Z of the different dimensions are isomorphic to the singular homology groups on B of the corresponding dimensions. In the terminology of Corollary 23.1, the qth singular homology group on f_a is isomorphic to the qth singular homology group of $f_b - z_b$. Hence to determine the homology groups on f_b, up to an isomorphism, from those of f_a it is merely necessary to determine the effect on the homology groups of adding the critical point z_b to $f_b - z_b$. This problem will be solved in Part III.

Suppose that a ND $f \in C^\infty(M_n)$ is given and that M_n is "f-compact below." There is then a sequence of critical values $a_0 < a_1 < a_2 < \cdots$ of f which may be finite or infinite in number. Let z_r be the critical point at the f-level a_r. It will be seen in Part III that the homology groups of $f_{a_1} - z_1$ are isomorphic to those of an n-ball. Bases for the homology groups of

$$f_{a_1} - z_1, f_{a_1}, f_{a_2} - z_2, f_{a_2}, f_{a_3} - z_3, f_{a_3}, \ldots$$

are then successively determined, making use of Corollary 23.1 and theorems of Part III on the effect of adding z_r to $f_{a_r} - z_r$. If a_m is a maximum critical value but not a maximum value of f on M_n, then, up to an isomorphism, the homology groups of f_{a_m} are those of M_n by virtue of Corollary 23.2.

The condition in the theorems of this section that a critical value a of a ND $f \in C^\infty(M_n)$ be assumed at just one critical point was imposed for simplicity of statement and proof and not at all for logical necessity. In fact, the mechanism preceding this section is admirably adapted for a treatment of the case where this condition is not imposed. Corresponding to an M_n which is boundedly f-compact and to a critical value a of f we now denote by $\mathbf{z}_a$ the finite set of critical points at the f-level a. The Homotopy Theorem 23.2 is replaced by Theorem 23.4.

Hypotheses of Theorem 23.4. There is given a ND $f \in C^\infty(M_n)$ with M_n boundedly f-compact and an interval $[a, b]$ of values of f of which a and b alone are critical.

Theorem 23.4. *Under these conditions there exists a unique* $\downarrow f$*-linear deformation* $\mathbf{D}$ *retracting the set* $Z = f_{[a,b]} - \mathbf{z}_b$ *on itself onto* f^a *in such a manner that just one* f*-trajectory is retracted onto each point of* f^a *excepting the points of* $\mathbf{z}_a$. *Onto* $z_a \in \mathbf{z}_a$ *the intersection with* $f_{[a,b]} - \mathbf{z}_b$ *of the bowl ascending from* z_a *is retracted.*

A Substitute for Manifold Triangulation. The scope of theorems of this section is greatly enlarged by the following theorem:

Theorem 23.5. *Corresponding to a prescribed, connected, noncompact* C^∞*-manifold* M_n *there exists a* ND $f \in C^\infty(M_n)$ *with the following properties*:

(i) *For each value* a *of* f, f_a *is compact.*

(ii) *The function* f *has a point of absolute minimum and no other critical point of index* 0.

(iii) *The function* f *has no critical point of index* n.

Whitney's theorem in an extended form (Whitney [3], p. 113) states that M_n can be C^∞-embedded as a closed subset of some Euclidean space E_m. It is accordingly sufficient to establish Theorem 23.5 for the case in which M_n is a closed regular C^∞-manifold M_n in a Euclidean space E_m. By Theorem 6.1 there exists a point $a \in E_m$ which is neither on M_n nor a focal point of M_n. The function

$$x \to \varphi(x) = \| a - x \|, \qquad x \in M_n,$$

satisfies the conditions of Theorem 23.5 except at most for conditions (ii) and (iii).

By the methods of Morse [9] used to establish the existence of polar ND functions, φ can be modified by the elimination of critical points so that each condition of Theorem 23.5 is satisfied.

The set f_a affirmed to exist in Theorem 23.5 has no other boundary in M_n than f^a, and if a is an ordinary value of f, this boundary is a compact C^∞-manifold of dimension $n - 1$. Although the homology groups of M_n may have infinite connectivities, each subspace f_a will have finite connectivities, as we shall see.

The representation of M_n by means of the subsets f_a of M_n in which a takes on the successive critical values of f is useful in the general study of homotopy and homology on M_n. It naturally replaces the classical representation of M_n as a "triangulated" complex.

Homotopy Types. The Homotopy Theorem 23.2 has as corollary an illuminating interpretation in terms of "homotopy types." We were led to this interpretation by Theorem A of Bott [3]. Our statement of Corollary 23.3 differs from Bott's Theorem A in that the closed k-cell "attached" by Bott to the sublevel set f_{a-e} is here taken as the (e, k)-dome of the k-bowl descending from the critical point p_a. This (e, k)-dome will be defined, together with other terms which are needed.

Definition 23.3. *Homotopy Type.* In accord with Hilton ([1], p. 3), let X and Y be two nonempty Hausdorff spaces and $h : X \to Y$ and $g : Y \to X$ be two continuous maps into Y and X, respectively, such that $g \circ h : X \to X$ is deformable on X into the identity map of X onto X and $h \circ g : Y \to Y$ is deformable on Y into the identity map of Y onto Y. Such maps, h and g, are termed *homotopy equivalences*, and spaces X and Y so related are said to have the *same homotopy type*. When X and Y have the same homotopy type one writes $X \simeq Y$.

The relation $\simeq$ is reflexive, symmetric, and transitive (see Hilton [1]).

If $Y \subset X$ and if there exists a deformation $\mathbf{D}$ (Definition 23.1) whose "terminal" mapping $\mathbf{D}_1$ retracts X on X onto Y, it is trivial that $X \simeq Y$. Homotopy equivalences sufficient to establish the relation $X \simeq Y$ can be taken, respectively, as $\mathbf{D}_1$ and the inclusion map of Y into X.

Notation. Let f be given on M_n, as throughout this section. The value a of f is assumed critical, and p_a the only critical point at the f-level a. Suppose that the index k of p_a is such that $0 < k < n$. Suppose $e > 0$ and that with $a - e < a < c$ the interval $[a - e, c]$

contains no critical values other than a. It follows from Theorem 23.2 that there exists a deformation retracting f_c onto f_a, so that

$$f_c \simeq f_a . \tag{23.24}$$

Definition 23.4. *The (e, k)-Dome with Summit p_a.* Turning to Definition 22.2, let $B_{-}(p_a, k)$ be the k-bowl descending from p_a. This bowl has a subspace

$$\eta = B_{e-}(p_a, k) = \{q \in B_{-}(p_a, k) \mid a \geqslant f(q) \geqslant a - e\}, \tag{23.25}$$

which will be called the *(e, k)-dome with summit p_a*. The (e, k)-dome with summit p_a is a topological k-ball η on which f assumes an absolute maximum a at p_a. The topological $(k - 1)$-sphere $\beta\eta$ which is the geometric boundary of η is the set

$$\beta\eta = \eta \cap f^{a-e}. \tag{23.26}$$

One terms the set

$$f_{a-e} \cup \eta \tag{23.27}$$

the sublevel set f_{a-e} with the (e, k)-dome η "attached" to f_{a-e} along $\beta\eta$ by the inclusion map of $\beta\eta$ into f_{a-e} (see Bott [3]).

Corollary 23.3. *Let the constants $a - e < a < c$, the critical point p_a, and the index k of p_a be conditioned as above. One then has the homotopy relation*

$$f_{a-e} \cup \eta_e^k \simeq f_c, \qquad 0 < k < n, \tag{23.28}$$

where η_e^k is the (e, k)-dome with summit p_a.

Proof. The relation (23.24) holds as a consequence of Theorem 23.2. Because of the transitivity of the relation $\simeq$, (23.28) will follow if (23.24) is supplemented by the relation

$$f_{a-e} \cup \eta_e^k \simeq f_a . \tag{23.29}$$

We shall see that in a proper notational context (23.29) is an elementary consequence of Theorem 21.1.

The relation (23.29) may be inferred from the following lemma:

Lemma 23.3. *If constants* $a - e < a < c$ *are conditioned as above, and if* $0 < \epsilon < e$, *then the following is true.*

(i) *There exists a deformation* $\mathbf{D}'$ *of* $f_{a-\epsilon} \cup \eta_\epsilon^k$ *onto* $f_{a-e} \cup \eta_e^k$.

(ii) *If* $\epsilon > 0$ *is sufficiently small there exists a deformation* $\mathbf{D}''$ *retracting* f_a *onto* $f_{a-\epsilon} \cup \eta_\epsilon^k$.

Notation for the proof of Lemma 23.3 (ii). Use will be made of the special local coordinates $u_1, \ldots, u_n$ introduced in Theorem 22.2. The critical point q of Theorem 22.2 is identified with the point p_a of Lemma 23.3. We refer to the presentation (from (22.12))

$$(F : D_\sigma, X) \in \mathscr{D}M_n \tag{23.30}$$

of the neighborhood X on M_n of $p_a = q$, and to the quadratic form

$$Q(u) = -u_1^2 - \cdots - u_k^2 + u_{k+1}^2 + \cdots + u_n^2 \tag{23.31}$$

on the right of (22.13). For simplicity we suppose that $f(q) = 0$, so that for $u \in D_\sigma$, $(f \circ F)(u) = Q(u)$.

In the n-plane E_n of the coordinates $u_1, \ldots, u_n$ we introduce the solid n-cone Λ on which

$$-u_1^2 - \cdots - u_k^2 + 2(u_{k+1}^2 + \cdots + u_n^2) \leqslant 0. \tag{23.32}$$

On the $(n - 1)$-cone Λ' bounding Λ,

$$-u_1^2 - \cdots - u_k^2 + 2(u_{k+1}^2 + \cdots + u_n^2) = 0. \tag{23.33}$$

For arbitrary positive e the sets

$$\Lambda_e = \{u \in \Lambda \mid 0 \geqslant Q(u) \geqslant -e\} \tag{23.34}$$

$$\Lambda_e' = \{u \in \Lambda' \mid 0 \geqslant Q(u) \geqslant -e\} \tag{23.35}$$

are well-defined.

The geometric boundary $\beta\Lambda_e$ *of* Λ_e. Let Q^{-e} be the nonsingular $(n - 1)$-dimensional quadric manifold on which $Q(u) = -e$. We introduce the subset,

$$\Lambda_e'' = \Lambda \cap Q^{-e} \tag{23.36}$$

of Q^{-e}. One then has

$$\beta\Lambda_e = \Lambda_e' \cup \Lambda_e''. \tag{23.37}$$

An ample conception of these sets can be obtained by diagramming these sets in the special cases in which $n = 3$ and $k = 1$ or 2.

A First Retracting Deformation. Let π_k be the coordinate k-plane of E_n on which $u_{k+1} = \cdots = u_n = 0$. We introduce the origin-centered k-disk

$$d_e^k = \{u \in \pi_k \mid 0 \geqslant Q(u) \geqslant -e\} \tag{23.38}$$

on π_k. There exists a retracting deformation

$$\Lambda_e \to \Lambda_e'' \cup d_e^k \tag{23.39}$$

of Λ_e onto the right side of (23.39). The trajectories of this deformation can be taken as subarcs of straight lines orthogonal to π_k. Under the deformation (23.39) each point of $\Lambda_e'' \cup d_e^k$ is fixed, while each point of Λ_e, not on $\Lambda_e'' \cup d_e^k$, moves on its trajectory to a point of $\Lambda_e'' \cup d_e^k$ at a velocity equal to the distance to be traversed.

If $\epsilon > 0$ is sufficiently small,

$$\Lambda_\epsilon \subset D_\sigma \tag{23.40}$$

[D_σ from (23.30)]. As a consequence $F(d_\epsilon^k)$ is well-defined, and

$$\eta_\epsilon^k = F(d_\epsilon^k). \tag{23.41}$$

It follows then from (23.39) that there exists a retracting deformation

$$F(\Lambda_\epsilon) \to F(\Lambda_\epsilon'') \cup \eta_\epsilon^k \tag{23.42}$$

of $F(\Lambda_\epsilon)$ onto the right side of (23.42).

A Second Retracting Deformation. We affirm that there exists a retracting deformation $\mathbf{D}^*$

$$f_a \to f_{a-\epsilon} \cup F(\Lambda_\epsilon) \tag{23.43}$$

of f_a onto the right side of (23.43).

Under $\mathbf{D}^*$ the point p_a is fixed. Apart from this condition the deformation $\mathbf{D}^*$ is definable in terms of f-trajectories. The definition is rendered simple by the fact that the only arcs of f-trajectories in X with initial points on $f^a - p_a$ and which meet the boundary of $F(\Lambda_\epsilon)$

are those whose images under F^{-1} are hyperbolas orthogonal to the level manifolds of Q (see §22).

Since $F(\Lambda_\epsilon'')$ in (23.42) is a subset of $f^{a-\epsilon}$ and so is pointwise fixed under the deformations (23.42) and (23.43), the existence of the deformation $\mathbf{D}''$ of Lemma 23.3(ii) may be inferred from the existence of the deformations (23.42) and (23.43). The figure when $n = 2$ and $k = 1$ is helpful.

This completes the proof of Lemma 23.3(ii).

Proof of Lemma 23.3(i). A deformation $\mathbf{D}'$ of the first of the sets

$$A = f_{a-\epsilon} \cup \eta_\epsilon^k, \qquad B = f_{a-e} \cup \eta_e^k$$

onto the second will now be defined. One should refer to Theorem 21.1 for essential background.

The value of f at an ordinary point p of f will be called the *f-coordinate* of p. Under $\mathbf{D}'$ the point $p_a = q$ shall remain fixed. A point $p \in f_{a-\epsilon}$ shall be deformed on the f-trajectory λ_p meeting p, moving in the sense of decreasing f. Let p_1 be the point in which λ_p meets f_{a-e}. As the time τ increases from 0 to 1 the f-coordinate of the deform of p shall decrease at a constant rate such that the deform of p reaches p_1 when the time $\tau = 1$.

A point $p \in \eta_\epsilon^k$ at which $f(p) = Q(-\epsilon)$ is in $f_{a-\epsilon}$, so that the deformation arc of p is already defined. A point $p \in \eta_\epsilon^k$ such that $f(p)$ divides the interval $[Q(-\epsilon), 0]$ in a ratio ρ between 0 and 1 shall be deformed on the f-trajectory λ_p meeting p into the point p_1 on λ_p whose f-coordinate divides the interval $[Q(-e), 0]$ in the above ratio ρ. The rate of decrease of the f-coordinate of the deform of p on λ_p shall be constant and such that p_1 is reached when $\tau = 1$.

It is seen that $\mathbf{D}'$ is a continuous deformation retracting A onto B.

The proof of Lemma 23.3 is complete.

Corollary 23.3 *follows.*

"Handlebodies." This is a term associated with a process P used in §13 of Morse [1] in 1925 to pass from f_{a-e} to f_{a+e} when $e > 0$ is sufficiently small and p_a is the only critical point of f with critical value on the interval $[a - e, a + e]$. We shall show how the process P is related to the process P_0 of attachment of the (e, k)-dome η of (23.25) to f_{a-e} to form $f_{a-e} \cup \eta$ in (23.27). Briefly, P attaches a "thickened" η, say η^*, to f_{a-e} along $f^{a-e} \cap \eta^*$. We shall be more explicit.

Recall that in the process P_0, η was attached to f_{a-e} along the boundary $\beta\eta = \eta \cap f^{a-e}$ of η. To "thicken" the (e, k)-dome the process P replaces η by a cellular neighborhood N of p_a relative to $\text{Cl}(f_{a+e} - f_{a-e})$ as follows. The neighborhood N is the image on M_n under a homeomorphism Φ into M_n of a product $X^k \times Y^{n-k}$ of closed Euclidean balls of dimensions k and $n - k$, respectively. Thus

$$N = \Phi(X^k \times Y^{n-k}), \qquad \Phi(\text{origin}) = p_a .$$

If $\mathbf{0}$ is the center of Y^{n-k}, Φ is to be such that

$$\eta = \Phi(X^k \times \{\mathbf{0}\}) \subset N.$$

N is to be "attached" to f_{a-e} along the set

$$Z^{n-1} = N \cap f^{a-e} = \Phi(\beta X^k \times Y^{n-k})$$

by the inclusion map of Z^{n-1} into f_{a-e}, and be such that

$$\beta\eta = \Phi(\beta X^k \times \{\mathbf{0}\}) \subset Z^{n-1}.$$

The boundary of $f_{a-e} \cup N$ is an $(n - 1)$-manifold. Without modifying f_{a-e} and without altering N or $f_{a-e} \cup N$ topologically one can so choose N that the boundary of $f_{a-e} \cup N$ is a differentiable manifold and f_{a+e} admits an isotopic deformation whose deformation arcs are f-trajectories and which retracts f_{a+e} onto its homeomorph $f_{a-e} \cup N$.

In Morse [1], Section 13, on "*Incidence relations between the boundary of D_r and the remainder of the complex $f \leqslant c + e^2$*," e^2 replaces the above e and there is a detailed definition of the above cells and their incidence relations. In the abstract case which concerns us here the incidence relations are the same. The special coordinates $u_1, \ldots, u_n$ of Theorem 22.2 and the *isometry* between the domain D_σ of these coordinates and the range X of the presentation $(F : D_\sigma, X)$ of Theorem 22.2 make the above process P particularly simple.

One can extend the above construction.

Suppose that there are r ND critical points at the f-level a (cf. §39) each with index k. If $e > 0$ is sufficiently small, disjoint neighborhoods $N_1, \ldots, N_r$ of the respective critical points can be defined as above and attached to f_{a-e} along disjoint "thickened" topological $(k - 1)$-spheres, $Z_1^{n-1}, \ldots, Z_r^{n-1}$, of f^{a-e} to form what is termed a *handlebody*

$$f_{a-e} \cup (N_1 \cup \cdots \cup N_r) \subset f_{a+e}$$

of index k *based* on f_{a-e}.

Smale ([1], p. 374) introduced handlebodies based on an n-ball D^n. He uses such handlebodies in deriving his penetrating solution of the Poincaré problem, $n > 4$.

EXERCISE 23.1. If a ND $f \in C^\infty(M_n)$ has just one critical value a and if M_n is boundedly f-compact, show that there exists a deformation $\mathbf{D}$ retracting M_n onto f^a.

EXERCISE 23.2. Give a direct proof of Theorem 23.5 in the special case in which $n = 2$ and the carrier of M_2 is an arbitrary, open, connected subset of E_2.

PART III

SINGULAR HOMOLOGY THEORY

§ 24

INTRODUCTION

Vector Spaces. Let $\mathscr{K}$ be a commutative field with a unit element e. An Abelian group G is called a vector space,† G *over* $\mathscr{K}$, if for each $r \in \mathscr{K}$ and each $g \in G$ an element $rg \in G$ is defined such that for $g, g_1, g_2 \in G$ and $r, r_1, r_2 \in \mathscr{K}$

$$\begin{gathered} r(g_1 + g_2) = rg_1 + rg_2, \qquad (r_1 + r_2)g = r_1 g + r_2 g, \\ r_1(r_2 g) = (r_1 r_2)g, \qquad eg = g. \end{gathered} \tag{24.1}$$

Given a vector space G over $\mathscr{K}$, we admit only those subgroups H of G which are *linear subspaces* of G, that is, are such that $rg \in H$ whenever $r \in \mathscr{K}$ and $g \in H$.

A mapping φ of G into another vector space G' over $\mathscr{K}$ is called a *homomorphism* if

$$\varphi(g + g') = \varphi(g) + \varphi(g'), \qquad g, g' \in G, \tag{24.2}$$

and is called *linear* if in addition

$$\varphi(rg) = r\varphi(g), \qquad r \in \mathscr{K}. \tag{24.3}$$

The Quotient Group G/L. Let G be a vector space over a field $\mathscr{K}$ and L a subgroup of G. Corresponding to each $g \in G$ the subset $g + L$ of G is called the *coset* L_g of g in G. The cosets of G partition G. If $g' \in L_g$, then $L_{g'} = L_g$. The disjoint cosets of G are the elements of a group G/L, termed the *quotient* of G by L.

† See Birkhoff and Mac Lane [1], Chapter VII for a detailed treatment of vector spaces.

Addition in G/L and multiplication by $r \in \mathscr{K}$ are defined in accord with the respective set relations

$$(g_1 + L) + (g_2 + L) = (g_1 + g_2) + L, \qquad r(g + L) = rg + L, \qquad r \in \mathscr{K}.$$

The null element is $0 + L = L$. With this understood the following three lemmas are readily verified:

Lemma 24.0. *If G is a vector space over a field $\mathscr{K}$ and L a linear subgroup of G, then G/L is a vector space over $\mathscr{K}$.*

Unless otherwise stated a vector space will be understood as over $\mathscr{K}$.

Lemma 24.1. *If α is a linear homomorphism of a vector space G into a vector space G' that maps a subspace L of G into a subspace L' of G', a linear homomorphism $\alpha_* : G/L \to G'/L'$ is thereby induced under which a coset L_g of G goes into the coset $L'_{\alpha g}$ of G'.*

Lemma 24.2. *If α and γ are linear homomorphisms,* $\alpha : G \to G'$ and $\gamma : G' \to G''$, *of vector spaces G and G' into vector spaces G' and G'', respectively, and if restrictions of α and γ, respectively, define linear homomorphisms $L \to L'$ and $L' \to L''$, where L, L', and L'' are vector subspaces, respectively, of G, G', and G'', there is thereby induced a linear homomorphism*

$$(\gamma\alpha)_* = \gamma_*\alpha_* : G/L \to G''/L''.$$

Proof. Lemma 24.1 implies that for $g \in G$ the coset L_g of G goes into the coset of G'' of form $(\gamma\alpha)_*L_g = L''_{(\gamma\alpha)g} = \gamma_*L'_{\alpha g} = \gamma_*\alpha_*L_g$.

Generators. A set (u) of non-null elements in G is called *linearly independent* or *free* if no element $v \in (u)$ equals a finite linear combination with coefficients in $\mathscr{K}$ of elements in (u), v deleted. If each element of G is a finite linear combination over $\mathscr{K}$ of elements of (u), then the elements of (u) are called *generators* of g. If in addition the elements of (u) are *free*, (u) is said to be a *base* for G.

One proves readily that if G has two finite bases $(u_1, \ldots, u_m)$ and $(v_1, \ldots, v_n)$, then $n = m$. One terms the number of elements in a finite base for G the *dimension* of G. If G is not trivial and has no finite base, the dimension of G is said to be *infinite*. If G is trivial, its dimension is said to be zero and it is said to have an *empty* base.

If there is a biunique linear homomorphism α of one vector space G_1 onto another G_2, a base of G_1 is mapped by α onto a base of G_2. The inverse α^{-1} is then a biunique linear homomorphism of G_2 onto G_1 and G_1 and G_2 are termed *isomorphic*. Trivial vector spaces are regarded as isomorphic. Vector spaces over $\mathscr{K}$ with finite dimensions are isomorphic if and only if their dimensions are equal.

We record a definition:

Definition 24.0. *Reduced Representations.* A representation

$$g = r_1 u_1 + \cdots + r_\mu u_\mu, \qquad r_i \in \mathscr{K}, \quad u_i \in G, \tag{24.4}$$

of a non-null element $g \in G$ will be called *reduced* if no $r_i = 0$ and if the elements $u_1, \ldots, u_\mu$ are linearly independent over $\mathscr{K}$.

The Existence of Bases. A vector space G may be defined by giving a base for G. More generally, it can be shown by Zorn's lemma that every vector space has a base (see Bourbaki [2], p. 147). In the most important cases bases are finite and their existence follows from the definition of G. The following lemma is needed in the singular homology theory:

Lemma 24.3. *Let G be a vector space generated by a base* $\mathbf{u}$ *and g a non-null element of G. There then exists a base for G which contains g.*

By hypothesis g has a reduced representation of form (24.4) in which the elements $u_1, \ldots, u_\mu$ belong to the base $\mathbf{u}$. The elements of $\mathbf{u}$, with u_1 replaced by g, then form a base for G, as one readily verifies.

The following definition is needed:

Definition 24.1. *Direct Sums of Vector Spaces.* Let G_1 and G_2 be vector spaces with no element in common other than the null element:

(i) A vector space G with a base which is the union of bases of G_1 and G_2 is called the *direct sum*

$$G = G_1 \oplus G_2 \tag{24.5}$$

of G_1 and G_2.

(ii) Equivalently, a vector space G is the *direct sum* of vector subspaces G_1 and G_2 if each element $z \in G$ has a unique representation

$$z = z_1 + z_2, \qquad z_1 \in G_1, \quad z_2 \in G_2. \tag{24.6}$$

We turn to an abstract homology theory with a generality sufficient for our purposes:

An Abstract Homology Theory. Let K be a collection of formally distinct elements σ^q, called q-cells, each assigned an integer $q \geqslant 0$ called its *dimension*. The cells of dimension q serve as generators over $\mathscr{K}$ of a vector space denoted by $C_q(K, \mathscr{K})$. These generators are supposed *free* in that each finite subset of cells σ^q of the same dimension is supposed free. For us the field $\mathscr{K}$, although arbitrary, is invariable unless otherwise stated, and we shall denote $C_q(K, \mathscr{K})$ by $C_q(K)$. Elements in $C_q(K)$ are called *q-chains* of K. For $q < 0$ we understand that $C_q(K)$ is the trivial Abelian group.

The Boundary Operator ∂. For each nonnegative integer q there is given a linear homomorphism

$$\partial : C_q(K) \to C_{q-1}(K), \tag{24.7}$$

more explicitly denoted by ∂_q. It is required that

$$\partial_{q-1}(\partial_q x) = 0, \qquad x \in C_q(K). \tag{24.8}$$

One writes (24.8) in the form $\partial(\partial x) = 0$, and one refers to the collection of conditions (24.8) as the condition $\partial\partial = 0$.

If σ is a 0-cell, $\partial\sigma$ is the null element in $C_{-1}(K)$.

Definition 24.2. *Admissible Complexes K.* A collection of abstract cells σ^q and an operator ∂ satisfying the preceding conditions will be called a *∂-structured complex K over $\mathscr{K}$* and termed *admissible.*

We now define certain subgroups of $C_q(K)$ essential for a homology theory, admitting only those subgroups of $C_q(K)$ which are "subspaces" of $C_q(K)$.

Definition 24.3. *The Subspace $Z_q(K)$ of q-Cycles.* A q-chain $x \in C_q(K)$ is termed a q-cycle if $\partial x = 0$. Since the homomorphism ∂

is linear by hypothesis, $r(\partial x) = \partial(rx)$ for each $r \in \mathscr{K}$, so that $\partial(rx) = 0$ if $\partial x = 0$. It follows that the ensemble of q-cycles is a linear subspace $Z_q(K)$ of $C_q(K)$. Each 0-chain is a 0-cycle.

Definition 24.4. *The Subspace $B_q(K)$ of Bounding q-Cycles.* A q-cycle x^q is termed *bounding* or homologous to 0, written $x^q \sim 0$, if $x^q = \partial x^{q+1}$ for some $(q+1)$-chain x^{q+1}. The subgroup of $Z_q(K)$ of bounding q-cycles is linear and is denoted by $B_q(K)$. If q-cycles x^q and y^q are such that $x^q - y^q \sim 0$, one writes $x^q \sim y^q$, and says that x^q is *homologous* to y^q.

Definition 24.5. *The Homology Group $H_q(K)$.* The quotient group

$$H_q(K) = Z_q(K)/B_q(K) \tag{24.9}$$

is a vector space over $\mathscr{K}$ (Lemma 24.0) and is called the qth homology group of K. For $q < 0$, $H_q(K)$ is trivial. The cosets of $Z_q(K)$ are called *homology classes.*

Definition 24.6. *The qth Connectivity R_q of K.* If a homology group $H_q(K)$ has a finite dimension, this dimension is called the qth *connectivity* $R_q(K)$ of K. If the dimension of $H_q(K)$ is not finite, one says that $R_q(K) = \infty$.

Definition 24.7. *Homology Prebases for K.* Any subset of $Z_q(K)$ which contains just one q-cycle from each homology class of some base for $H_q(K)$ will be termed a *homology prebase for K of q-cycles.*

If $\mathbf{b}_q(K)$ is a homology prebase for K of q-cycles, no proper linear combination over $\mathscr{K}$ of q-cycles of $\mathbf{b}_q(K)$ is homologous to zero, or, as wc shall say, $\mathbf{b}_q(K)$ is *free* homology-wise. Moreover, each q-cycle is homologous to a linear combination over $\mathscr{K}$ of q-cycles of $\mathbf{b}_q(K)$, or, as we shall say, $\mathbf{b}_q(K)$ is *generating* homology-wise.

Any subset of $Z_q(K)$ which is both "free" and "generating" homology-wise contains just one q-cycle in each homology class of a uniquely determined base for $H_q(K)$ and so is a homology prebase for K of q-cycles. If $R_q(K)$ is finite, the number of elements in a homology prebase for K of q-cycles is $R_q(K)$.

Homology prebases should not be confused with bases for homology groups.

Induced Homomorphisms α_*. Let there be given two admissible complexes K' and K''.

Definition 24.8. *∂-Permutable Chain-Transformation.* A set of linear homomorphisms

$$\alpha : C_q(K') \to C_q(K''), \qquad q = 0, 1,..., \tag{24.10}$$

such that $\alpha\partial = \partial\alpha$ or, more explicitly,

$$\alpha_{q-1}\partial = \partial\alpha_q , \qquad q = 0, 1,..., \tag{24.11}$$

will be called a *∂-permutable chain-transformation*† $K' \to K''$. For brevity, the set of mappings (24.10) will be referred to as the chain-transformation $\alpha : K' \to K''$.

When (24.11) holds α maps a q-cycle x of K' into a q-cycle of K''. Thus α defines a linear homomorphism $Z_q(K') \to Z_q(K'')$ for each q. Moreover, α defines a linear homomorphism $B_q(K') \to B_q(K'')$ for each q, since whenever a q-cycle x of K' is bounding αx is bounding, as (24.11) implies. From Lemma 24.1 one can thus infer the following:

Theorem 24.1. *A ∂-permutable chain-transformation* $\alpha : K' \to K''$ *induces linear homomorphisms*

$$\alpha_* : H_q(K') \to H_q(K''), \qquad q = 0, 1,..., \tag{24.12}$$

under which a homology class of a q-cycle x of K' *goes into the homology class of the q-cycle* αx *of* K''.

We point out a consequence (24.13) of Lemma 24.2.

Let K', K'' and K''' be three "∂-structured abstract complexes" and let

$$\alpha : C_q(K') \to C_q(K''), \qquad q = 0, 1,...,$$

and

$$\gamma : C_q(K'') \to C_q(K'''), \qquad q = 0, 1,...,$$

† See Eilenberg [1], p. 411. Strictly, α should bear a subscript q in (24.10).

be ∂-permutable chain-transformations. Then the composite mapping $\gamma\alpha$ is ∂-permutable, since $\partial(\gamma\alpha) = \gamma\partial\alpha = (\gamma\alpha)\partial$. There are induced linear homomorphisms,

$$\alpha_* : H_q(K') \to H_q(K''), \quad \gamma_* : H_q(K'') \to H_q(K'''), \quad (\gamma\alpha)_* : H_q(K') \to H_q(K''').$$

Moreover, it follows from Lemma 24.2 that

$$(\gamma\alpha)_* = \gamma_*\alpha_* . \tag{24.13}$$

It should be observed that an identity chain-transformation $\alpha : K \to K$ induces an identity homomorphism α_* of $H_q(K)$.

Definition 24.9. **Geometric Simplices a^q.** Let

$$x^{(0)}, x^{(1)}, \ldots, x^{(q)}, \qquad q \geqslant 0, \tag{24.14}$$

be $q + 1$ points in E_n with $q \leqslant n$. If these points do not lie in a $(q - 1)$-plane, the set of points $x \in E_n$ with vectorial representations

$$x = \mu_0 x^{(0)} + \mu_1 x^{(1)} + \cdots + \mu_q x^{(q)}, \tag{24.15}$$

where the parameters $\mu_0, \mu_1, \ldots, \mu_q$ are subject to the conditions

$$\mu_0 + \mu_1 + \cdots + \mu_q = 1, \qquad 0 \leqslant \mu_i \leqslant 1, \tag{24.16}$$

is called a *geometric simplex* a^q with vertices (24.14). The simplex a^q is independent of the order in which its "vertices" (24.14) are given and is to be distinguished from an *ordered* simplex presently to be defined.

Any subset of $r + 1$ distinct vertices of a^q, $0 \leqslant r \leqslant q$, determines a geometric r-simplex, called an *r-face* of a^q. If ν is any integer in the range 0, 1,..., q and if

$$x^{(0)}, x^{(1)}, \ldots, \hat{x}^{(\nu)}, \ldots, x^{(q)}, \qquad q > 0, \tag{24.17}$$

denotes the subset of the vertices (24.14) of a^q with $x^{(\nu)}$ deleted, the geometric simplex with vertices (24.17) is called the $(q - 1)$-face of a^q *opposite* the vertex $x^{(\nu)}$. As a polyhedron (see §25), an n-simplex shall be regarded as including all of its faces.

Definition 24.10. Ordered Simplices. When $q \geqslant 0$ a geometric simplex a^q whose vertices p_i have been assigned a definite order (see Eilenberg [1], p. 420) $p_0 < p_1 < \cdots < p_q$ will be termed an *ordered q-simplex* and denoted by

$$s = p_0 p_1 \cdots p_q = s^q. \tag{24.18}$$

For notational reasons we do not ordinarily write s as s^q, although s is understood as having the dimension q. We term a^q the *carrier* of s and write $|s| = a^q$. When $q = 0$ we set $s = |s| = p_0$.

When $q > 0$ the ordered simplex

$$s(i) = p_0 \cdots \hat{p}_i \cdots p_q, \qquad 0 \leqslant i \leqslant q, \tag{24.19}$$

obtained by deleting the vertex p_i from s is called the ith *face* of s. The symbol $s(i)$ is used for no purpose other than to represent the ith face of s.

To give a definition of the algebraic boundary ∂s of an ordered simplex s, let Γ_q, $q \geqslant 0$, denote the Abelian group (with integral coefficients) generated by the set of all *distinct* ordered q-simplices in all Euclidean spaces E_n of all dimensions, subject to the following convention as to distinctness: A q-simplex given in E_n shall be identified with a simplex s' in a Euclidean space $E_{n'}$ for which $n' > n$ when s' is the image of s under the mapping

$$(x_1, \ldots, x_n) \to (x_1, \ldots, x_n; 0, \ldots, 0) : E_n \to E_{n'}.$$

With Γ_q so defined, when $\dim s = q > 0$ set[†]

$$\partial s = (-1)^j s(j) \in \Gamma_{q-1}, \tag{24.20}$$

and when $\dim s = 0$ set $\partial s = 0$. Let ∂ so defined be extended linearly to define a homomorphism

$$\partial : \Gamma_q \to \Gamma_{q-1}, \qquad q = 0, 1, \ldots, \tag{24.21}$$

understanding that Γ_m is the trivial group when $m < 0$.

A classical lemma can now be verified.

[†] Here as elsewhere a repeated index in a term indicates summation of the term over the range of the index.

Lemma 24.4. *If s is an ordered simplex of dimension $q \geqslant 0$, then $\partial(\partial s)$ is the null element in Γ_{q-2}.*

Proof. The lemma is trivial when dim s is 1 or 0. Suppose then that dim $s = q > 1$. Let i and k be restricted to the range $0, 1, \ldots, q$. For fixed i

$$\partial s(i) = \sum_{k<i} (-1)^k p_0 \cdots \hat{p}_k \cdots \hat{p}_i \cdots p_q + \sum_{k>i} (-1)^{k-1} p_0 \cdots \hat{p}_i \cdots \hat{p}_k \cdots p_q \tag{24.22}$$

by virtue of (24.19) and the definition (24.20) of ∂s. Applying ∂ to both sides of (24.20), one finds that

$$\partial(\partial s) = \partial \sum_{i=0}^{q} (-1)^i s(i) = \sum_{i=0}^{q} (-1)^i \partial s(i). \tag{24.23}$$

If one replaces $\partial s(i)$ in the right side of (24.23) by the chain on the right of (24.22), the lemma follows.

Barycentric Coordinates. Let s be an ordered simplex with vertices

$$x^{(0)}, x^{(1)}, \ldots, x^{(q)}, \qquad 0 \leqslant q \leqslant n, \tag{24.24}$$

in E_n in their given order in s. A point $x \in |s| = a^q$ of the form (24.15) with parameters μ conditioned as in (24.16) will be said to have *barycentric coordinates* μ *relative to* s. An induction relative to q suffices to prove the following lemma:

Lemma 24.5. *Points in $|s| = a^q$ with different barycentric coordinates relative to s are different.*

Theorem 24.2 below requires the following definition: A mapping H of a subset X of a Euclidean space E_n onto a subset Y of a Euclidean space E_m will be termed λ-*linear* if whenever H maps points x and x' in X, respectively, onto points y and y' in E_m, H maps the point†

† We are representing a point by a vector whose components are the coordinates of the point.

$\lambda x + (1 - \lambda)x'$ onto the point $\lambda y + (1 - \lambda)y'$ for every choice of λ in the interval [0, 1].

Theorem 24.2. *Let a^q and b^q, $q > 1$, be geometric simplices in E_n and E_m which are carriers of ordered simplices s_a and s_b, respectively.*

A mapping of a^q onto b^q in which points of a^q and b^q correspond if they have the same barycentric coordinates is a λ-linear homeomorphism of a^q onto b^q.

That this mapping is biunique follows from Lemma 24.5. That it is λ-linear is formally verified without difficulty.

To show that this mapping is bicontinuous, we introduce a model geometric simplex c^q in E_{q+1} which is the carrier of an ordered simplex s_c whose vertices have coordinates $x_1, \ldots, x_{q+1}$ in E_{q+1} given by the successive rows of the unit $(q + 1)$-square matrix. The representation of points $y \in c^q$ in terms of their barycentric coordinates relative to s_c takes the special form

$$y = (y_1, \ldots, y_{q+1}) = (\mu_0, \mu_1, \ldots, \mu_q).$$

The theorem is true if a^q is replaced by c^q. This follows from the fact that the mapping of c^q, onto b^q of the theorem, is biunique and continuous and c^q is compact. Since there is a similar λ-linear homeomorphism of c^q onto a^q, Theorem 24.2 follows as stated.

We term the mapping of a^q onto b^q affirmed to exist in Theorem 24.2 a *barycentric homeomorphism.*

§ 25

MODEL POLYHEDRAL COMPLEXES P

Singular simplexes and complexes will presently be defined. This section is concerned with certain model polyhedral complexes whose continuous images in a Hausdorff space χ will be useful in singular homology theory. Among polyhedra are prisms with the aid of which model chain homotopies will be defined.

One starts with a simplicial n-polyhedron P in E_m, $m \geqslant n \geqslant 0$, defined as the union of a finite set of geometric n-simplices in E_m no two of which intersect other than in a common face of dimension less than n. Recall the convention that a geometric simplex, as a polyhedron, includes each of its faces.

Given an n-polyhedron P we shall define a ∂-structured complex $\mathbf{P}$ over $\mathscr{K}$, based, as we shall say, on P, terming $\mathbf{P}$ a model polyhedral complex. To that end, we must define the cells of $\mathbf{P}$, and on $\mathbf{P}$ a boundary operator ∂ such that $\partial\partial = 0$.

Definition 25.1a. *The Cells* ρ^q *of* $\mathbf{P}$. A q-cell of $\mathbf{P}$ is any ordered q-simplex ρ^q whose carrier is a geometric simplex a^q of P. Two q-cells of $\mathbf{P}$ are regarded as identical if and only if their carriers are the same geometric simplex a^q and their vertices have identical orderings.

Definition 25.1b. *The Boundary Operator* ∂ *on* $\mathbf{P}$. For each q-cell $\rho^q = s$ of $\mathbf{P}$ $\partial\rho^q$ is defined as was ∂s in §24, with $\pm 1 = \pm e$ in $\mathscr{K}$, as an element in $C_{q-1}(\mathbf{P})$. The operator ∂ is extended as a linear homomorphism

$$\partial : C_q(\mathbf{P}) \to C_{q-1}(\mathbf{P}) \qquad \text{over} \quad \mathscr{K}.$$

It follows from Lemma 24.4 that $\partial(\partial x) = 0$ for each q-chain of $C_q(\mathbf{P})$. This completes the definition of $\mathbf{P}$ as an admissible ∂-structured n-complex "based" on the n-polyhedron P.

Carriers of Chains of $\mathbf{P}$. A non-null q-chain u^q in $C_q(\mathbf{P})$ admits a unique "reduced representation" (Definition 24.0)

$$u^q = r_1\rho_1{}^q + \cdots + r_\mu\rho_\mu{}^q, \qquad \rho_i{}^q \text{ a cell of } \mathbf{P}.$$

When u^q is so represented the carrier $| u^q |$ of u^q is defined as the set (see Example 25.1):

$$| u^q | = | \rho_1{}^q | \cup \cdots \cup | \rho_\mu{}^q |. \tag{25.1}$$

Definition 25.2. *The Geometric Boundary βP of P.* By βP we shall mean the $(n - 1)$-polyhedron which is the union of those geometric $(n - 1)$-simplices of P which are faces of an odd number of n-simplices of P.

Definition 25.3. *The Star P_v of a Vertex v of P.* Given a vertex v of an n-polyhedron, $n > 0$, let P_v be the closed subpolyhedron of P which is the union of the geometric simplices (closed by definition) which are incident with v. Let $\mathbf{P}_v$ be the subcomplex of $\mathbf{P}$ based on P_v.

Definition 25.4. *The Outer Complex of P_v.* The union of the geometric simplices of P_v, not incident with v, is an $(n - 1)$-polyhedron called the *outer boundary* $\hat{P}_v$ of P_v. The complex based on $\hat{P}_v$ is termed the *outer complex* of P_v. The outer complex of P_v is a subcomplex of $\mathbf{P}$ of dimension $n - 1$.

Definition 25.5. *Vertex-Joins in* $\mathbf{P}$. Let ρ^q, $0 \leqslant q < n$, be a q-cell of the "outer complex" of P_v. Let *Join* $v\ \rho^q$ denote the $(q + 1)$-cell of $\mathbf{P}$ whose ordered set of vertices are those of ρ^q preceded by v. If $c = r_i\rho_i{}^q$ is a chain of the outer complex of P_v, set

$$\text{Join } v\ c = r_i \text{ Join } v\ \rho_i{}^q, \qquad 0 \leqslant q < n. \tag{25.2}$$

Join $v\ c$ is a $(q + 1)$-chain of the complex based on P_v.

Lemma 25.1. *If P_v is the star of a vertex v of P and c a q-chain of the outer complex of P_v, then for $0 < q < n$*

$$\partial \text{ Join } v\, c = c - \text{Join } v\, \partial c. \tag{25.3}$$

Formula (25.3) is clearly valid if c is a q-cell of the outer complex of P_v. Relation (25.3) follows when c is a q-chain $r_i \rho_i^q$ of the outer complex of P_v, on taking account of the permutability of r_i with ∂ and with Join v regarded as a linear operator.

We draw the following conclusion:

Lemma 25.2. *Any q-cycle of the outer complex of P_v for which $0 < q < n$ is bounding in the complex based on P_v.*

Barycentric Subdivision. A geometric q-simplex is a simplicial q-polyhedron. Its barycentric subdivision $b(a^q)$, as we shall define it, is another simplicial q-polyhedron, essential as a model in building the singular homology theory.

Definition 25.6 $b(a^n)$. Set $b(a^0) = a^0$. Proceeding inductively, suppose that $n > 0$ and that for each integer m on the range $0, 1, 2, \ldots, n-1$ $b(a^m)$ has been so defined that the new vertices in $b(a^m)$ when $m > 0$ are the barycenters of the respective faces of a^m of positive dimensions. The induction from the case $m = n - 1$ to the case $m = n$ is as follows.

Let $f_0, f_1, \ldots, f_n$ be the faces of a^n of dimension $n - 1$. We are assuming that the polyhedra

$$b(f_0), b(f_1), \ldots, b(f_n), \qquad n > 0, \tag{25.4}$$

have been defined. Let $p_n = \text{bary } a^n$ be the barycenter of a^n and let u^{n-1} be an arbitrary one of the $(n-1)$-simplices of the polyhedra (25.4). Let $|\, p_n u^{n-1} \,|$ be the geometric simplex whose vertices are p_n and the vertices of u^{n-1}. The polyhedron $b(a^n)$ shall be the union of all geometric simplices obtained in this way as u^{n-1} ranges over the $(n-1)$-simplices of the polyhedra (25.4). One sees that this union is a simplicial n-polyhedron and is identical as a set with a^n.

The geometric operations $a^q \to b(a^q)$ have algebraic counterparts which involve complexes Join $v\ c$ with v a barycenter.

The Barycentric Operator B. Let a^n be a geometric n-simplex and $b(a^n)$ its barycentric subdivision. Both a^n and $b(a^n)$ determine simplicial polyhedra. Complexes $\mathbf{a}^n$ and $\mathbf{b}(a^n)$ are "based" on these polyhedra. Of the linear homomorphisms of the form

$$C_q(\mathbf{a}^n) \to C_q(\mathbf{b}(a^n)), \qquad q = 0, 1, \ldots, n, \tag{25.5}$$

there is one, denoted by B, which we call a *barycentric operator*. The barycentric operator B is the algebraic counterpart of the geometric operation $a^n \to b(a^n)$.

Definition 25.7. *The Operator B.* If u^0 is a 0-chain of $\mathbf{a}^n$, set $Bu^0 = u^0$. Proceeding inductively, when $q > 0$ set

$$B\rho = \text{Join } v(B\partial\rho), \qquad v = \text{bary } |\rho|, \tag{25.6}$$

for each q-cell ρ of $\mathbf{a}^n$, assuming that Bu^{q-1} has already been defined for chains u^{q-1} of $\mathbf{a}^n$. This definition for q-cells of $\mathbf{a}^n$ will be extended linearly over q-chains of $\mathbf{a}^n$. The completion of this inductive definition of B yields linear homomorphisms of the form (25.5).

We continue with the following lemma:

Lemma 25.3. *The operator B defines a set of linear homomorphisms of form* (25.5) *such that for each q-cell* ρ *of the complex* $\mathbf{a}^n$

$$\partial B\rho = B\partial\rho, \tag{25.7}$$

$$|B\rho| = |\rho|, \tag{25.8}$$

and (cf. Definition 25.2)

$$|B\partial\rho| = \beta\,|\rho| \tag{25.9}$$

Verification of (25.7). When $\dim \rho = 0$ (25.7) is trivially true. When $\dim \rho = q > 0$ it follows from (25.6) and (25.3) that

$$\partial B\rho = B\partial\rho - \text{Join } v(\partial B\,\partial\rho), \qquad v = \text{bary } \rho. \tag{25.10}$$

Proceeding inductively, we assume that $\partial Bz = B\partial z$ when z is a $(q-1)$-chain of $\mathbf{a}^n$, so that in (25.10) $\partial B(\partial \rho) = B\partial(\partial \rho) = 0$. Thus (25.10) implies (25.7).

Verification of (25.8) *and* (25.9). When dim $\rho = 0$ (25.8) and (25.9) are trivially true.

When dim $\rho = q > 0$, $\partial \rho = (-1)^i \rho(i)$, in accord with the definition (24.20) of ∂s. Because of the linearity of B

$$B\partial \rho = (-1)^i \, B\rho(i). \tag{25.11}$$

We proceed inductively. From (25.11), from (25.1), and from the truth of (25.8) when dim $\rho = q - 1 \geqslant 0$, it follows that

$$| \, B\partial \rho \, | = \bigcup_i | \, B\rho(i)| = \bigcup_i | \, \rho(i)| = \beta \, | \, \rho \, |,$$

confirming (25.9). Note that $| \, \rho \, |$ and $| \, B\rho \, |$ are convex sets.

From the definition of $B\rho$ and from (25.9) we find that in case dim $\rho > 0$

$$| \, B\rho \, | = | \, \text{Join } v \; \beta\partial\rho \, | = | \, \rho \, |, \qquad v = \text{bary } \rho,$$

confirming (25.8) and completing the proof of the lemma.

Prisms. Among the simplicial polyhedra which are most useful as models in singular homology theory are simplicially subdivided prisms.

To present such prisms properly, let $x_1, \ldots, x_m$, t be rectangular coordinates of a point (x, t) in $E_m \times R = E_{m+1}$. Corresponding to each nonempty subset w of E_m introduce the set

$$w^{\text{co}} = \{(x, t) \in E_m \times R \mid x \in w, \, t = 1\} \tag{25.12}$$

(co = congruent) in the m-plane of E_{m+1} on which $t = 1$. The set w^{co} is congruent by orthogonal projection to w in E_m. To an ordered simplex s in E_m there corresponds a congruent ordered simplex s^{co} in E_m^{co}. Congruent chains and complexes with carriers in E_m and E_m^{co} are similarly defined.

We shall consider prisms given as products $w \times I$, where w is a geometric n-simplex in E_m, $n \leqslant m$, I the interval $[0, 1]$, and

$$w \times I = \{(x, t) \in E_m \times R \mid x \in w, \, t \in I\}.$$

To the *lower base* w of the prism corresponds the *upper base* w^{co} projecting onto w. If a is a q-face of w with $q < n$, we term $a \times I$ a *lateral face* of the prism $w \times I$. The lateral face $a \times I$ has its *lower base* a and its *upper base* a^{co}.

A First Subdivision $P'(w)$ *of* $w \times I$. Given the prism $w \times I$, neither of the bases w and w^{co} is to be subdivided. The lateral faces $a \times I$ of $w \times I$ are to be simplicially subdivided in the order of their dimensions, beginning with 1-faces. Each lateral 1-face is to be divided into two geometric 1-simplices by its barycenter. Thereafter the lateral faces $a \times I$ are to be simplicially divided in the order of their dimensions by adding a barycenter p to each face $a \times I$ and dividing $a \times I$ into simplices which are radial joins of p with the already subdivided lateral faces of $a \times I$. This process extends to $w \times I$, so that $P'(w)$, the subdivided prism, is a union of n-simplices joining the barycenter of $w \times I$ to the $(n - 1)$-simplices of the subdivided boundary of $w \times I$. Neither a nor a^{co} has been subdivided.

Another Subdivision $P''(w)$ *of* $w \times I$. To define $P''(w)$, lateral faces of $w \times I$, including their upper bases but not their lower bases, are to be subdivided in the order of their dimensions, taking the barycenters of the lateral faces and their upper bases as new vertices. Finally, the upper base w^{co} is to be replaced by its subdivision $(b(w))^{co} = b(w^{co})$ and the barycenter of $w \times I$ joined to each $(n - 1)$-simplex on the already subdivided boundary of $w \times I$.

Corresponding to the $(n + 1)$-polyhedra $P'(w)$ and $P''(w)$ the complexes $\mathbf{P}'(w)$ and $\mathbf{P}''(w)$ are well-defined, and special linear homomorphisms δ and Δ, respectively, of the form

$$\delta : C_q(\mathbf{w}) \to C_{q+1}(\mathbf{P}'(w)), \qquad q = 0, 1, \ldots, \tag{25.13}$$

$$\Delta : C_q(\mathbf{w}) \to C_{q+1}(\mathbf{P}''(w)), \qquad q = 0, 1, \ldots, \tag{25.14}$$

are to be defined.

To define a homomorphism φ of either of these types, one defines φ for q-cells $\rho \in \mathbf{w}$ and extends this definition linearly over chains in $\mathbf{w}$.

The Operator δ. One defines δ inductively by setting

$$\delta\rho = v\rho^{co} - v\rho, \qquad v = \text{bary}(|\rho| \times I), \tag{25.15a}$$

for each 0-cell ρ of $\mathbf{w}$ and

$$\delta\rho = \text{Join } v(\gamma^{\text{co}} - \rho - \delta\partial\rho), \qquad v = \text{bary}(|\,\rho\,| \times I), \tag{25.15b}$$

for each cell $\rho \in \mathbf{w}$ with $\dim \rho > 0$.

The Operator Δ. One defines Δ inductively by setting

$$\Delta\rho = v\rho^{\text{co}} - v\rho, \qquad v = \text{bary}(|\,\rho\,| \times I), \tag{25.16a}$$

for each 0-cell ρ of $\mathbf{w}$ and

$$\Delta\rho = \text{Join } v((B\rho)^{\text{co}} - \rho - \Delta\partial\rho), \qquad v = \text{bary}(|\,\rho\,| \times I), \tag{25.16b}$$

for each cell $\rho \in \mathbf{w}$ with $\dim \rho > 0$.

The meaning of the above linear homomorphisms δ and Δ is indicated by the following fundamental theorem (cf. Eilenberg [1]):

Theorem 25.1. *Corresponding to an arbitrary geometric n-simplex w in E_n, there exist linear homomorphisms δ and Δ, respectively, of types* (25.13) *and* (25.14), *such that for each cell ρ in the complex* $\mathbf{w}$

$$\partial\delta\rho = \rho^{\text{co}} - \rho - \delta\partial\rho, \tag{25.17}$$

$$\partial\Delta\rho = (B\rho)^{\text{co}} - \rho - \Delta\partial\rho, \tag{25.18}$$

$$|\,\delta\rho\,| = |\,\rho\,| \times I = |\,\Delta\rho\,|, \tag{25.19}$$

and, if $\beta(|\,\rho\,| \times I)$ denotes the union of the lateral faces and upper and lower bases of $|\,\rho\,| \times I$, then

$$|\,\partial\delta\rho\,| = \beta(|\,\rho\,| \times I) = |\,\partial\Delta\rho\,|. \tag{25.20}$$

We begin by establishing (25.17). The proof of (25.18) is formally similar, Δ replacing δ, and $(B\rho)^{\text{co}}$ replacing ρ^{co}.

Proof of (25.17). In case $\dim \rho = 0$ the definition of δ in (25.15a) implies that

$$\partial(\delta\rho) = (\rho^{\text{co}} - v) - (\rho - v) = \rho^{\text{co}} - \rho,$$

as affirmed by (25.17). Proceeding inductively, we assume that (25.17) is true for $\dim \rho = 0, 1, 2, \ldots, \bar{n} - 1 < n$ and prove its truth for $0 < \dim \rho = \bar{n} \leqslant n$.

It follows from the formula (25.3) for ∂ Join $v\ c$ that when $\delta\rho$ is defined by (25.15b) and dim $\rho = \bar{n} > 0$

$$\partial\delta\rho = (\rho^{\text{co}} - \rho - \delta\partial\rho) - \text{Join } v\ \partial(\rho^{\text{co}} - \rho - \delta\partial\rho). \tag{25.21}$$

By virtue of our inductive hypothesis, (25.17) is valid when ρ is replaced by a chain in **w** of dimension on the range $0, 1,..., \bar{n} - 1$. It follows that

$$\partial\delta(\partial\rho) = (\partial\rho)^{\text{co}} - \partial\rho - \delta\partial(\partial\rho) = \partial\rho^{\text{co}} - \partial\rho \tag{25.22}$$

when dim $\rho = \bar{n}$, or, equivalently,

$$\partial(\rho^{\text{co}} - \rho - \delta\partial\rho) = 0,$$

so that (25.21) reduces to (25.17).

Proof of (25.19) *and* (25.20). We refer to the definition in (25.1) of the carrier of a chain. Using this definition and proceeding inductively with respect to dim ρ, one can establish (25.19) as a consequence of (25.15) and (25.16). One similarly establishes (25.20) as a consequence of (25.17)–(25.19).

In anticipation of the use of the relations (25.17) and (25.18) in defining "singular chain homotopies," we term the relations (25.17) and (25.18) *elementary chain-homotopies.*

Barycentric Homeomorphisms of Prisms. In applying the elementary chain homotopies of Theorem 25.1 to define singular chain homotopies in §27, we shall make use of the barycentric homeomorphism H of one ordered n-simplex s onto another s^*, as defined at the end of §24.

As a matter of permanent notation we shall set $\mathscr{s} = |s|$ and $\mathscr{s}^* = |s^*|$. The homeomorphism H of s onto s^* then admits an extension which maps the subdivided prism $P'(\mathscr{s})$ in $E_n \times R$ onto the subdivided prism $P'(\mathscr{s}^*)$. In this extension the points $(x, t) \in P'(\mathscr{s})$ and $(x^*, t^*) \in P'(\mathscr{s}^*)$ shall correspond if $t = t^*$ and $x^* = H(x)$.

So extended, H maps an ordered simplex ρ of the complex $\mathbf{P}'(\mathscr{s})$ onto a unique ordered simplex $\rho^* = H(\rho)$ of the complex $P'(\mathscr{s}^*)$.

For each q on the range 0, 1,..., n H thus determines a biunique mapping (onto)

$$C_q(\mathbf{P}'(\sigma)) \to C_q(\mathbf{P}'(\sigma^*)), \qquad \sigma = | s |; \quad \sigma^* = | s^* |. \tag{25.23}$$

It is important to note that $H \,\|\, |\rho|$ is the barycentric homeomorphism of ρ onto ρ^* determined by the mapping by H of the vertices of ρ onto those of ρ^*.

EXERCISE 25.1. Prove the following: If u^q and v^q are chains in **P**, then $| u^q + v^q | \subset | u^q | \cup | v^q |$.

§ 26

SINGULAR COMPLEXES $S(\chi)$. EILENBERG

Singular q-Cells. Following Eilenberg [1, p. 420], we start with vertex-ordered simplices (Definition 24.10)

$$s = p_0 p_1 \cdots p_q, \qquad q = 0, 1, \ldots . \tag{26.1}$$

We distinguish between *singular simplices* τ and *singular q-cells* $\mathring{\tau}$, the latter being equivalence classes of the former.

Definition 26.1. *Singular q-Simplices.* Let χ be a Hausdorff space. By a *singular q-simplex* $q \geqslant 0$ on χ is meant a continuous mapping $\tau : s \to \chi$ of an ordered q-simplex s into χ. Two singular q-simplices, $q \geqslant 0$,

$$\tau' : s' \to \chi \qquad \text{and} \qquad \tau'' : s'' \to \chi \tag{26.2}$$

are termed *equivalent*, written $\tau' \equiv \tau''$, if

$$\tau'(x') = \tau''(x'') \tag{26.3}$$

whenever x' and x'' are points of s' and s'', respectively, with the same barycentric coordinates relative to s' and s''.

One verifies readily that the relation $\tau' \equiv \tau''$ is reflexive, symmetric, and transitive in the class of singular q-simplices. Consequently, the class of singular simplices in χ is partitioned into disjoint equivalence classes.

Definition 26.2a. *Singular q-Cells.* Given a singular q-simplex τ, the equivalence class that contains τ will be denoted by $\mathring{\tau}$ and termed a *singular q-cell.*

Carriers. Given a singular q-simplex $\tau : s \to \chi$, we shall term the corresponding subset $\tau(|s|)$ of χ the *carrier* $|\tau|$ of τ. Noting that equivalent q-simplices of a singular q-cell $\mathring{\tau}$ have the same carriers, we shall denote this common set by $|\mathring{\tau}|$ or $|\tau|$, and term it the *carrier* of $\mathring{\tau}$.

Definition 26.2b. *Simply Carried Singular q-Cells.* Given a singular q-simplex in χ defined by a mapping $\tau : s \to \chi$ which is a *homeomorphism* onto $|\tau|$ as a subspace of χ, the corresponding singular q-cell $\mathring{\tau}$ will be said to be *simply carried.*

The Ensemble $S(\chi)$. Following Eilenberg, the complex of singular q-cells with carriers on χ will be denoted by $S(\chi)$.

The Group $C_q(S(\chi))$. For $q \geqslant 0$ the vector spaces over $\mathscr{K}$ generated by the singular q-cells of $S(\chi)$ will be denoted by $C_q(S(\chi))$. For a negative integer q, $C_q(S(\chi))$ shall be the trivial Abelian group.

Definition 26.3. *The Boundary Operator on* $S(\chi)$. Given the ordered q-simplex s of (26.1) with $q > 0$, recall that

$$s(i) = p_0 \cdots \hat{p}_i \cdots p_q, \qquad p_i \text{ deleted}, \tag{26.4}$$

is the ordered face of s opposite p_i, $0 \leqslant i \leqslant q$. Given a singular q-simplex $\tau : s \to \chi$, $q > 0$, we introduce the singular $(q-1)$-simplices†

$$\tau_i = \tau \mid s(i), \qquad i = 0, 1, 2, \ldots, q. \tag{26.5}$$

The *algebraic boundary* of the singular q-cell $\mathring{\tau}$ shall be the $(q-1)$-chain‡

$$\partial\mathring{\tau} = (-1)^i \mathring{\tau}_i, \qquad \text{in} \quad C_{q-1}(S(\chi)). \tag{26.6}$$

When $q = 0$, $\partial\mathring{\tau}$ shall be the null chain in $C_{-1}(S(\chi))$.

It is clear that if $\tau' \equiv \tau''$, then for each i on the range $0, 1, \ldots, q$

$$\tau_i'(x') \equiv \tau_i''(x''), \tag{26.7}$$

† The symbols $s(i)$ and τ_i of Definition 26.3 will be used exclusively in the sense of Definition 26.3. We have set $\mathfrak{s} = |s|$.

‡ A repeated index i here as elsewhere indicates summation of the term over the range of i.

where x' and x'' are points of $s'(i)$ and $s''(i)$ with the same barycentric coordinates relative to s' and s'' and hence relative to $s'(i)$ and $s''(i)$. Hence $\mathring{\tau}'_i = \mathring{\tau}''_i$, so that $\partial\mathring{\tau}$, as defined by (26.6), is independent of the choice of τ in its equivalence class.

The definition of the operator

$$\partial : C_q(S(\chi)) \to C_{q-1}(S(\chi)), \qquad q \geqslant 0, \tag{26.8}$$

is completed by linear extension of ∂ as defined on the singular q-cells $\mathring{\tau}$ of $S(\chi)$.

We shall verify a classical theorem:

Theorem 26.1. *The boundary operator acting on the chains of $S(\chi)$ is such that $\partial\partial = 0$.*

The theorem is trivial when $q = 0$ or 1.

Suppose $q > 1$. If the q-cell $\mathring{\tau}$ is denoted by $\{\tau\}$, (26.6) takes the form

$$\partial\mathring{\tau} = (-1)^i\{\tau \mid p_0 \cdots \hat{p}_i \cdots p_q\}, \qquad q > 0.$$

Following the proof of Lemma 24.4 as a model, we find that

$$\partial\mathring{\tau}_i = \sum_{k<i} (-1)^k\{\tau_i \mid p_0 \cdots \hat{p}_k \cdots \hat{p}_i \cdots p_q\} + \sum_{k>i} (-1)^{k-1}\{\tau_i \mid p_0 \cdots \hat{p}_i \cdots \hat{p}_k \cdots p_q\}.$$

By virtue of (26.6)

$$\partial(\partial\mathring{\tau}) = (-1)^i\, \partial\mathring{\tau}_i\,, \qquad \dim \mathring{\tau} > 1.$$

Proceeding formally as in the proof of Lemma 24.4, we infer Theorem 26.1.

Carriers of Singular Chains. A q-chain w^q of the complex $S(\chi)$ over $\mathscr{K}$ admits a "reduced" representation

$$w^q = r_1\sigma_1{}^q + \cdots + r_\mu\sigma_\mu{}^q, \qquad r_i \in \mathscr{K},$$

in terms of cells of $S(\chi)$ (cf. Definition 24.0). The *carrier* $|\, w^q \,|$ of w^q so represented shall be the set

$$|\, w^q \,| = |\, \sigma_1{}^q \,| \cup \cdots \cup |\, \sigma_\mu{}^q \,|. \tag{26.9}$$

We shall verify the following:

Theorem 26.2. *When χ reduces to a point p the connectivity*

$$R_k(S(\chi)) = \delta_0{}^k, \qquad k = 0, 1, \ldots . \tag{26.10}$$

Proof. When $\chi = p$ each singular k-simplex is equivalent to each other singular k-simplex, so that for each k there is just one singular k-cell σ^k. If n is a positive integer, it follows then from (26.6) that $\partial\sigma^{2n} = \sigma^{2n-1}$. There are accordingly no nontrivial singular $2n$-cycles, while each singular $(2n - 1)$-cycle is bounding. We infer the truth of (26.10) when $k > 0$.

On observing that dim $Z_0(S(\chi)) = 1$ and dim $B_0(S(\chi)) = 0$ when $\chi = p$, we infer the truth of (26.10) when $k = 0$.

This establishes Theorem 26.2.

Induced Chain-Transformations $\hat{g}$ and $\hat{\varphi}$. If P is a simplicial n-polyhedron in E_m, $m \geqslant n$, and χ and χ' Hausdorff spaces, we shall show how continuous mappings

$$g : P \rightarrow \chi \qquad \text{and} \qquad \varphi : \chi \rightarrow \chi', \tag{26.11}$$

respectively, induce ∂-permutable chain-transformations

$$\hat{g} : \mathbf{P} \rightarrow S(\chi) \qquad \text{and} \qquad \hat{\varphi} : S(\chi) \rightarrow S(\chi') \tag{26.12}$$

(cf. Definition 24.8). This notation is permanent.

Definition 26.4. $\hat{g}$. If s is an ordered simplex of the complex $\mathbf{P}$ based on P, $g \mid s$ is a singular simplex on χ. Its equivalence class is denoted by $[g \mid s]^\circ$. We set

$$\hat{g}s = [g \mid s]^\circ \tag{26.13}$$

and extend this definition of $\hat{g}$ for q-cells of $\mathbf{P}$ linearly over the q-chains of $\mathbf{P}$, thereby defining $\hat{g}$ in (26.12).

For each ordered simplex $s \in \mathbf{P}$ the carrier $\mid \hat{g}s \mid = g(\mid s \mid)$ by virtue of (26.13). For each chain u of the complex $\mathbf{P}$ we shall show that

$$\mid \hat{g}u \mid \subset g(\mid u \mid). \tag{26.14}$$

Verification of (26.14). We shall assume that the reader has confirmed the result of Example 26.2 below.

Suppose that the chain u of $\mathbf{P}$ given in (26.14) has a "reduced" representation

$$u = r_1\rho_1 + \cdots + r_\mu\rho_\mu , \qquad r_i \in \mathscr{K},$$

in accord with (25.0) in terms of q-cells ρ_i of $\mathbf{P}$, so that by (25.1)

$$| u | = | \rho_1 | \cup \cdots \cup | \rho_\mu |.$$

It follows from Example 26.2 that

$$| \hat{g}u | \subset | r_1\hat{g}\rho_1 | \cup \cdots \cup | r_\mu\hat{g}\rho_\mu |,$$

so that by virtue of (26.9)

$$| \hat{g}u | \subset g(| \rho_1 |) \cup \cdots \cup g(| \rho_\mu |) = g(| u |),$$

thereby confirming (26.14).

An equality in (26.14) would be incorrect, as examples would show.

We shall verify the following theorem:

Theorem 26.3a. *The set of linear homomorphisms*

$$\hat{g} : C_q(\mathbf{P}) \to C_q(S(\chi)), \qquad q = 0, 1, \ldots,$$

is a ∂-permutable chain-transformation $\hat{g} : \mathbf{P} \to S(\chi)$.

Let s be a cell of $\mathbf{P}$. If dim $s = 0$, by definition $\partial s = 0$ and $\partial\hat{g}s = 0$, so that $\partial\hat{g}s = \hat{g}\partial s$. If dim $s > 0$, by the definition of $\hat{g}s$ as $[g \mid s]^\circ$ and of $\partial\mathring{\tau}$ in (26.6),

$$\partial\hat{g}s = (-1)^i[g \mid s(i)]^\circ,$$

(summing with respect to i), so that

$$\partial\hat{g}s = \hat{g}((-1)^i\, s(i)) = \hat{g}\partial s, \tag{26.15}$$

thereby completing the proof of Theorem 26.3a.

Definition 26.5. $\widehat{\varphi}$. Let φ be given as in (26.11). Let $\tau : s \to \chi$ be a singular q-simplex. We set

$$\widehat{\varphi}\mathring{\tau} = [\varphi \circ \tau]^\circ \tag{26.16}$$

and extend $\widehat{\varphi}$ linearly over the vector spaces $C_q(S(\chi))$ for $q \geqslant 0$.

It follows from (26.16) that the carrier of $\hat{\varphi}\mathring{\tau}$ is $\varphi(\tau(|\, s \,|))$, so that if u is a q-chain of $S(\chi)$

$$|\, \hat{\varphi} u \,| \subset \varphi(|\, u \,|) \tag{26.17}$$

[using (26.9) and Example 26.2].

Definition 26.5 of $\hat{\varphi}$ requires the following justification:

Lemma 26.1. *The equivalence class of a singular simplex $\varphi \circ \tau$ is independent of the choice of τ in $\mathring{\tau}$.*

Proof. There are given equivalent singular simplices $\tau' : s' \to \chi$ and $\tau'' : s'' \to \chi$. By hypothesis $\tau'(x') = \tau''(x'')$ for $x' \in s'$ and $x'' \in s''$ provided x' and x'' have the same barycentric coordinates relative, respectively, to s' and s''. Hence $(\varphi \circ \tau')(x') = (\varphi \circ \tau'')(x'')$ and the lemma follows.

We verify a basic theorem:

Theorem 26.3b. *The linear homomorphisms*

$$\hat{\varphi} : C_q(S(\chi')) \to C_q(S(\chi'')), \qquad q = 0, 1, \ldots,$$

induced by a continuous map $\varphi : \chi' \to \chi''$ are ∂-permutable chain-transformations.

To establish this theorem, it will be sufficient to show that if $\tau : s \to \chi$ is a singular q-simplex, $q > 0$, then $\partial \hat{\varphi} \mathring{\tau} = \hat{\varphi} \partial \mathring{\tau}$. To verify this relation recall that $\hat{\varphi}\mathring{\tau} = [\varphi \circ \tau]^\circ$ by definition, and note that in accord with (26.6)

$$\begin{aligned} \partial \hat{\varphi} \mathring{\tau} &= (-1)^i [(\varphi \circ \tau) \mid s(i)]^\circ = (-1)^i [\varphi \circ \tau_i]^\circ \\ &= (-1)^i \, \hat{\varphi} \mathring{\tau}_i = \hat{\varphi}((-1)^i \, \mathring{\tau}_i) = \hat{\varphi} \partial \mathring{\tau}. \end{aligned}$$

The theorem follows.

The Composition of Induced Homomorphisms. Let χ', χ'', χ''' be three Hausdorff spaces and $\varphi : \chi' \to \chi''$ and $\psi : \chi'' \to \chi'''$ continuous maps.

The induced ∂-permutable chain-transformations

$$\widehat{\varphi} : C_q(S(\chi')) \to C_q(S(\chi'')), \qquad q = 0, 1, \ldots,$$
$$\widehat{\psi} : C_q(S(\chi'')) \to C_q(S(\chi''')), \qquad q = 0, 1, \ldots,$$
$$\widehat{\psi\varphi} : C_q(S(\chi')) \to C_q(S(\chi''')), \qquad q = 0, 1, \ldots,$$

are well-defined. We shall verify the lemma:

Lemma 26.2. $\widehat{\psi\varphi} = \widehat{\psi}\widehat{\varphi}$.

Proof. If $\mathring{\tau}$ is a singular cell of $S(\chi')$, then by definition (26.16)

$$\widehat{\psi}(\widehat{\varphi}\mathring{\tau}) = [\psi \circ (\varphi \circ \tau)]^\circ = [(\psi \circ \varphi) \circ \tau]^\circ = \widehat{\psi\varphi}\mathring{\tau}, \tag{26.18}$$

establishing the lemma.

Turning to singular homology groups, it follows from Lemmas 26.2 and 24.2 that

$$(\widehat{\psi\varphi})_* = (\widehat{\psi}\widehat{\varphi})_* = \widehat{\psi}_*\widehat{\varphi}_* \,. \tag{26.19}$$

Hausdorff spaces which are homeomorphic images of one another are said to be *topologically equivalent*. Topologically equivalent Hausdorff spaces have isomorphic homology groups. This is a consequence of the following theorem:

Theorem 26.4. *Given a homeomorphism $\Phi : \chi' \to \chi''$ of a Hausdorff space χ' onto a Hausdorff space χ'', let Θ be the inverse of Φ. The induced chain-transformations*

$$\begin{aligned} \widehat{\Phi} &: C_q(S(\chi')) \to C_q(S(\chi'')), \qquad q = 0, 1, \ldots, \\ \widehat{\Theta} &: C_q(S(\chi'')) \to C_q(S(\chi')), \qquad q = 0, 1, \ldots, \end{aligned} \tag{26.20}$$

are then inverses, as are the homomorphisms

$$\begin{aligned} (\widehat{\Phi})_* &: H_q(S(\chi')) \to H_q(S(\chi'')), \qquad q = 0, 1, \ldots, \\ (\widehat{\Theta})_* &: H_q(S(\chi'')) \to H_q(S(\chi')), \qquad q = 0, 1, \ldots . \end{aligned} \tag{26.21}$$

Proof. $\Theta\Phi$ and $\Phi\Theta$ are by hypothesis identity maps of χ' and χ'', respectively. It follows from Lemma 26.2 that $\widehat{\Theta}\widehat{\Phi}$ and $\widehat{\Phi}\widehat{\Theta}$ define identity homomorphisms of the chain groups of $S(\chi')$ and $S(\chi'')$, respectively. From (26.19) we then infer that $(\hat{\Theta})_*(\hat{\Phi})_*$ and $(\hat{\Phi})_*(\hat{\Theta})_*$ define identity homomorphisms of the homology groups of $S(\chi')$ and $S(\chi'')$, respectively (cf. Exercise 26.1).

Theorem 26.4 follows.

Note. The chain-transformations $\hat{g}$ and $\hat{\varphi}$ induced by the mappings g and φ of (26.11) are aids in deriving the chain-homotopies on $S(\chi)$ so essential in singular homology theory and its applications.

Chain-Homotopies on an Abstract Complex K. In the next paragraphs we shall introduce concepts to be applied in the singular theory, in particular when K is a singular complex $S(\chi)$.

Let there be given an abstract ∂-structured complex K and a ∂-permutable chain-transformation ω defined by homomorphisms

$$\omega : C_q(K) \to C_q(K), \qquad q = 0, 1,..., \tag{26.22}$$

together with linear homomorphisms

$$\Omega : C_q(K) \to C_{q+1}(K), \qquad q = 0, 1,.... \tag{26.23}$$

Definition 26.6. *Chain-Homotopies on K.* We say that the above chain-homomorphisms ω are *chain-homotopic* to the identity *under Ω* if for each q and for arbitrary q-chains $z \in C_q(K)$

$$\partial\Omega z = \omega z - z - \Omega\partial z. \tag{26.24}$$

For each q we term Ω the *homotopy mapping* of the chain-homotopy (26.24) and ω the *terminal homomorphism.*

If z is a cycle and ω is chain-homotopic to the identity, (26.24) shows that

$$\omega z - z = \partial\Omega z, \tag{26.25}$$

thus implying the homology $\omega z \sim z$.

Definition 26.7. *Carriers of Chain-Homotopies.* In case K is a singular complex $S(\chi)$ the carrier $|\Omega z|$ of Ωz is well-defined [see (26.9)], and we say that the chain-homotopy (26.24) is *on* $|\Omega z|$ or *carried* by $|\Omega z|$. If z is a cycle so that $\omega z \sim z$, we say that this homology is on $|\Omega z|$. The *carriers* of chain-homotopies are of basic importance in the topology of ND functions.

Arc-Wise Separate Spaces. If χ is a Hausdorff space of the form

$$\chi = \chi_1 \cup \chi_2 \qquad \text{with} \quad \chi_1 \cap \chi_2 = \varnothing, \tag{26.26}$$

where χ_1 and χ_2 are proper subspaces of χ such that no point of χ_1 is arc-wise connected to a point of χ_2, we say that χ is the union of *arc-wise separate* subspaces χ_1 and χ_2. We shall show how the homology groups of $S(\chi)$ over $\mathscr{K}$ are determined by the homology groups of $S(\chi_1)$ and $S(\chi_2)$.

Referring to Definition 24.1 of a "direct sum" of two vector spaces, one finds that for $q \geqslant 0$

$$\begin{aligned} C_q(S(\chi)) &= C_q(S(\chi_1)) \oplus C_q(S(\chi_2)) \\ Z_q(S(\chi)) &= Z_q(S(\chi_1)) \oplus Z_q(S(\chi_2)) \\ B_q(S(\chi)) &= B_q(S(\chi_1)) \oplus B_q(S(\chi_2)) \end{aligned} \tag{26.27}$$

(see Definitions 24.3 and 24.4) when χ is the union of arc-wise separate subspaces χ_1 and χ_2. One verifies these relations in the order written, making use of each relation to prove its successors.

The qth homology group $H_q(S(\chi))$ is not in general the direct sum of the corresponding groups of $S(\chi_1)$ and $S(\chi_2)$ because the latter homology groups are not strictly subgroups of $H_q(S(\chi))$. However, on making use of the relations (26.27) and Definition 24.7 of homology prebases of q-cycles, one has the following theorem:

Theorem 26.5. *If χ is a Hausdorff space which is the union of arc-wise separate Hausdorff spaces χ_1 and χ_2, then for each integer $q \geqslant 0$ the union of homology prebases for $S(\chi_1)$ and $S(\chi_2)$ of q-cycles is a homology prebase for $S(\chi)$ of q-cycles.*

Theorem 26.5 has a corollary of importance. It affirms that under the hypotheses of Theorem 26.5 there exists an isomorphism

$$H_q(S(\chi)) \approx H_q(S(\chi_1)) \oplus H_q(S(\chi_2)). \tag{26.28}$$

Statement (ii) of Theorem 26.6 is a consequence of statement (i) of Theorem 26.6 and of Theorem 26.5. A proof of Theorem 26.6 is left to the reader.

Theorem 26.6. (i) *If* χ *is an arc-wise connected Hausdorff space, a homology prebase of singular* 0*-cells for* $S(\chi)$ *is provided by an arbitrary* 0*-cell of* $S(\chi)$.

(ii) *If in Theorem* 26.5 χ_1 *and* χ_2 *are each arc-wise connected, a homology prebase of singular* 0*-cells for* $S(\chi)$ *is provided by the union of an arbitrary* 0*-cell in* $S(\chi_1)$ *and an arbitrary* 0*-cell in* $S(\chi_2)$.

EXERCISE 26.1. Show that the identity map φ of a Hausdorff space χ onto χ induces the identity isomorphism $\hat{\varphi}$ of $S(\chi)$ onto $S(\chi)$.

EXERCISE 26.2. Using (26.9) show that if x^q and y^q are q-chains of $S(\chi)$, then

$$| x^q + y^q | \subset | x^q | \cup | y^q |.$$

EXERCISE 26.3. Let χ be a Hausdorff space which is the union of r subspaces $\chi_1, \ldots, \chi_r$ each of which is arc-wise connected, but which are pair-wise arc-wise separate. Show that $R_0(S(\chi)) = r$.

§ 27

CHAIN-HOMOTOPIES ON $S(\chi)$

In §26 we have introduced an abstract complex K and on K have defined an abstract chain-homotopy

$$\partial\Omega z = \omega z - z - \Omega\partial z, \qquad z \in C_q(K), \quad q = 0, 1, \ldots, \tag{27.0}$$

of the terminal chain-transformation ω into the identity chain-transformation of K under a "homotopy mapping" Ω.

If the complex K is $S(\chi)$, there are two types of chain-homotopies of importance for us. One is induced by deformations d of χ on χ and the other by algebraic subdivisions of the chains of $S(\chi)$.

Chain-Homotopies Induced by Deformations. Let d be a continuous deformation

$$(p, t) \to d(p, t) : \chi \times I \to \chi, \qquad I = [0, 1], \tag{27.1}$$

of χ on χ, with an initial mapping

$$p \to d(p, 0) = d_0(p) : \chi \to \chi, \qquad d_0(p) = p,$$

which reduces to the identity, and a terminal mapping

$$p \to d(p, 1) = d_1(p) : \chi \to \chi \tag{27.2}$$

denoted by d_1.

Given d, and thereby d_1, a terminal chain-transformation $\omega = \widehat{d}_1$ (cf. Definition 26.5), given by the set of homomorphisms

$$\widehat{d}_1 : C_n(S(\chi)) \to C_n(S(\chi)), \qquad n = 0, 1, \ldots, \tag{27.3}$$

is chain-homotopic, as we shall see, to the identity chain-transformation under a homotopy mapping $\Omega = \mathbf{d}$, where each homomorphism

$$\mathbf{d} : C_q(S(\chi)) \to C_{q+1}(S(\chi)), \qquad q = 0, 1, \ldots, \tag{27.4}$$

is uniquely determined by d.

We shall defined $\mathbf{d}$ in terms of d, and verify the corresponding chain-homotopy (27.0). We start with a continuous mapping of a prism into χ.

A mapping $d_\tau : \mathfrak{s} \times I \to \chi$. Given an arbitrary singular n-simplex $\tau : s \to \chi$, to define $\mathbf{d}$ it is sufficient to define $\mathbf{d}\mathring{\tau}$ for each $\mathring{\tau} \in S(\chi)$.

To that end, set $\mathfrak{s} = | s |$ and introduce the continuous mapping

$$(x, t) \to d_\tau(x, t) = d(\tau(x), t) : ((x, t) \in \mathfrak{s} \times I) \tag{27.5}$$

of the prism $\mathfrak{s} \times I$ into χ. Let $P'(\mathfrak{s})$ be the first subdivision (see §25) of the prism $\mathfrak{s} \times I$. The mapping d_τ induces a chain-transformation

$$\hat{d}_\tau : \mathbf{P}'(\mathfrak{s}) \to S(\chi) \tag{27.5'}$$

(cf. Definition 26.4 of ^).

In the equivalence class of singular n-simplices of $\mathring{\tau}$, let $\tau^* : s^* \to S(\chi)$ be a second singular simplex. Let the barycentric homeomorphism H of s onto s^* be extended, as at the end of §25, to a barycentric homeomorphism of $P'(\mathfrak{s})$ onto $P'(\mathfrak{s}^*)$, where $\mathfrak{s}^* = | s^* |$. As shown in §25, a barycentric homeomorphism of each ordered q-simplex $\rho \in \mathbf{P}'(\mathfrak{s})$ onto a corresponding ordered simplex $\rho^* \in \mathbf{P}'(\mathfrak{s}^*)$ is defined by H. If ρ and ρ^* so correspond,

$$\hat{d}_\tau \mid \rho = \hat{d}_{\tau^*} \mid \rho^* \in C_{q+1}(S(\chi)), \tag{27.6}$$

in accord with the definition of ^.

Definition 27.1. *The Operator* $\mathbf{d}$. Given a singular n-simplex $\tau : s \to \chi$ defining the singular n-cell $\mathring{\tau} \in C_n(S(\chi))$, we set

$$\mathbf{d}\mathring{\tau} = \hat{d}_\tau(\delta s), \tag{27.7}$$

where the linear homomorphisms

$$\delta : C_q(\mathbf{w}) \to C_{q+1}(\mathbf{P}'(w)), \qquad q = 0, 1, \ldots, n, \quad w = | s |,$$

are defined in §25. It follows from (27.6) that $\mathbf{d}\mathring{\tau}$, as defined by (27.7), is a singular $(n+1)$-chain of $S(\chi)$, a chain independent of the choice of the singular n-cell $\tau \in \mathring{\tau}$.

With $\mathbf{d}$ so defined for n-cells $\mathring{\tau}$ of $S(\chi)$, $\mathbf{d}$ is extended linearly over $C_n(S(\chi))$.

Heuristically, one can regard the chain $\mathbf{d}\mathring{\tau}$ as a "singular prism" with a "singular base" $\mathring{\tau}$. We are led to a chain-homotopy.

Theorem 27.1. *Given the deformation* d *as in* (27.1), *the resultant linear operator* $\mathbf{d}$ *and the terminal chain-transformation* $\widehat{d}_1$ *of* (27.3) *satisfy the chain-homotopy*

$$\partial \mathbf{d} z = \widehat{d}_1 z - z - \mathbf{d}\partial z, \qquad z \in C_n(S(\chi)), \tag{27.8}$$

for $n = 0, 1, \ldots$.

Proof. To prove (27.8), it will be sufficient to establish (27.8) in the case in which z is the equivalence class $\mathring{\tau}$ of a singular n-simplex $\tau : s \to \chi$.

By virtue of Theorem 25.1

$$\partial \delta s = s^{\mathrm{co}} - s - \delta \partial s, \tag{27.9}$$

where it is understood that $|s|$ is in E_n and s^{co} is the congruent image of s in the product $E_n \times \{1\}$. In the notation of §25, with $\mathfrak{s} = |s|$, the chains in (27.9) are in the model complex $\mathbf{P}'(\mathfrak{s})$, so that if one sets $g = d_\tau$, the ∂-permutable chain-transformation $\hat{g}$ of (27.5)′ can be applied to the terms of (27.9), giving the relation

$$\partial \hat{g}(\delta s) = \hat{g} s^{\mathrm{co}} - \hat{g} s - \hat{g}(\delta \partial s), \qquad g = d_\tau . \tag{27.10}$$

The proof of Theorem 27.1 will be completed by showing that (27.10) reduces, term by term, to the form

$$\partial \mathbf{d}\mathring{\tau} = \widehat{d}_1 \mathring{\tau} - \mathring{\tau} - \mathbf{d}\partial\mathring{\tau}. \tag{27.11}$$

Proof of (27.11). The chains on the left of (27.10) and (27.11) are equal by virtue of the definition of $\mathbf{d}\mathring{\tau}$ in (27.7) and the notation $g = d_\tau$.

Evaluation of $\hat{g}s^{co}$. This chain equals

$$[g \mid s^{co}]^o = [(d_1 \circ \tau) \mid s]^o = \overset{\frown}{d_1} \mathring{\tau}$$

by virtue, respectively, of Definition 26.4 of ^, by definition of d_τ in (27.5) and by Definition 26.5 of ⌒.

Evaluation of $\hat{g}s$. This term is equal to $[(d_0 \circ \tau) \mid s]^o$ by the definition of d_τ in (27.5), and so reduces to $\mathring{\tau}$.

Evaluation of $\hat{g}\delta\partial s$. With i summed on the range $0, 1, \ldots, n$, $\hat{g}\delta\partial s$ equals

$$(-1)^i \, \hat{g}(\delta s(i)) = (-1)^i \, \mathbf{d}\mathring{\tau}_i = \mathbf{d}\partial\mathring{\tau}$$

by virtue, respectively, of the definition of ∂s in (24.20), by Definition 27.1 of $\mathbf{d}$, and by definition of $\partial\mathring{\tau}$ in (26.6).

Thus (27.11) holds and Theorem 27.1 follows.

To apply Theorem 27.1, the following analysis of carriers of its chains is necessary.

For each subset η of χ we introduce the subset

$$d\text{-traj } \eta = \underset{0 \leqslant t \leqslant 1}{\text{Union}} \; d(\eta, t) \tag{27.12}$$

of χ and verify the following lemma:

Lemma 27.1. *Given the deformation d of* (27.1) *and a chain z of $S(\chi)$, one has the following inclusions*:

$$\mid \mathbf{d}z \mid \subset d\text{-traj} \mid z \mid, \qquad \mid \overset{\frown}{d_1} z \mid \subset d_1(\mid z \mid). \tag{27.13}$$

Proof. Turning to the definition (27.7) of $\mathbf{d}\mathring{\tau}$, recall that $\mid \delta s \mid = \mid s \mid \times I$ by (25.19). It follows from (27.5), (26.14) and (27.12) that

$$\mid \mathbf{d}\mathring{\tau} \mid \subset d_\tau(\mid s \mid \times I) = d\text{-traj} \mid \mathring{\tau} \mid. \tag{27.14}$$

The first inclusion in (27.13) is implied. The second inclusion follows from (26.17).

The following corollary of Theorem 27.1 is fundamental. Its proof with the aid of Lemma 27.1 is immediate.

Corollary 27.1. *Corresponding to the deformation d of (27.1), a q-cycle z of $S(\chi)$ is homologous on d-traj $|z|$ to the terminal cycle $\hat{d}_1 z$ on $d_1(|z|)$.*

Corollary 27.1 and Theorems 26.2 and 26.6(i) imply the following:

Corollary 27.2. *If a Hausdorff space χ can be deformed on itself into some one of its points, then*

$$R_q(S(\chi)) = \delta_0^q, \qquad q = 0, 1, \ldots .$$

Chain-Homotopies Induced by Subdivisions. Given χ, we shall define a subdivision inducing a ∂-permutable chain-transformation

$$z \to \pi z : C_q(S(\chi)) \to C_q(S(\chi)), \qquad q = 0, 1, 2, \ldots .$$

By abuse of language, πz is called the "singular barycentric subdivision" of z. One also defines a set of linear homomorphisms

$$z \to \Pi z : C_q(S(\chi)) \to C_{q+1}(S(\chi)), \qquad q = 0, 1, 2, \ldots, \tag{27.15}$$

such that the homotopy

$$\partial \Pi z = \pi z - z - \Pi \partial z, \qquad z \in C_q(S(\chi)), \tag{27.16}$$

is valid.

The mappings π and Π are singular counterparts, respectively, of the operators B and Δ of Theorem 25.1. We shall set up the chain homotopy (27.16) with the aid of the elementary homotopy (25.18). The chain-homotopy (27.16) has the following fundamental consequence:

Corollary 27.3. *Each singular q-cycle z of $S(\chi)$ is homologous on its carrier $|z|$ to its singular barycentric subdivision πz.*

This corollary will be established after the proof of (27.16).

Definition 27.2. *The "Singular Barycentric Subdivision" πz of z.* Let $\tau : s \to \chi$ be a singular simplex. To the ordered q-simplex s there corresponds a q-chain Bs, defined in §25, barycentrically subdividing

s when $q > 0$. According to Lemma 25.3 $| Bs | = | s |$. On $| Bs |$, τ is thus defined. We can accordingly define $\pi\mathring{\tau}$ by setting

$$\pi\mathring{\tau} = \hat{\tau}(Bs) \tag{27.17}$$

(see Definition 26.4 of ^) and extend π linearly over $C_q(S(\chi))$ for each integer $q \geqslant 0$.

We affirm that $\pi\mathring{\tau}$, as defined in (27.17), is *independent* of the choice of $\tau \in \mathring{\tau}$. This affirmation is implied by the following: If s and s' are ordered q-simplices and if H is the corresponding barycentric homeomorphism of s onto s', then H maps each ordered simplex u of Bs onto a unique ordered simplex u' of Bs' and $H \mid u$ is the unique barycentric homeomorphism of u onto u'.

That π is *∂-permutable* is seen as follows. One starts with (27.17) and shows that

$$\partial\pi\mathring{\tau} = \hat{\tau}B\partial s = \pi\partial\mathring{\tau},$$

using Theorem 26.3a and (25.7) to justify the first equality. The relation (25.11), with the Definitions (27.17) and (26.6), imply the second equality.

Notation for the Definition of Π. Let (x, t) be rectangular coordinates in $E_n \times R$. Let $\tau : s \to \chi$ be a singular q-simplex with s in E_n and suppose that, as in §25,

$$\mathscr{s} \times I = \{(x, t) \in E_n \times R \mid x \in \mathscr{s};\ t \in I\}, \qquad \mathscr{s} = | s |, \tag{27.18}$$

Let τ be given an extension τ_e over $| s | \times I$ defined by setting

$$\tau_e(x, t) = \tau(x), \qquad (x, t) \in | s | \times I,$$

mapping $| s | \times I$ into χ. Recall the inductively defined linear homomorphisms

$$\Delta : C_r(\mathbf{w}) \to C_{r+1}(\mathbf{P}''(w)), \qquad r = 0, 1, \ldots,$$

of Theorem 25.1. One takes w as $\mathscr{s}$. The carrier $| \Delta s | = \mathscr{s} \times I$.

Definition 27.3. *Π.* We begin by setting

$$\Pi\mathring{\tau} = \hat{\tau}_e(\Delta s) \tag{27.19}$$

(^ applied to τ_e) obtaining thereby a "singular prismatic chain" in $C_{q+1}(S(\chi))$. So defined on q-cells $\mathring{\tau}$ of $C_q(S(\chi))$, Π shall be linearly extended over $C_q(S(\chi))$ to define a homomorphism of form (27.15).

As in the definition of $\pi\mathring{\tau}$, one can readily verify the fact that $\Pi\mathring{\tau}$ as defined, is independent of the choice of $\tau \in \mathring{\tau}$. [use (25.23)].

Theorem 27.2, supplemented by the carrier inclusions below, shows how πz and Πz are related for an arbitrary chain z of $S(\chi)$.

Theorem 27.2. *The mapping* $z \to \pi z$ *and the identity mapping* $z \to z$ *of* $C_q(S(\chi))$ *admit the chain-homotopy*

$$\partial \Pi z = \pi z - z - \Pi \partial z, \qquad q = 0, 1, \ldots; \quad z \in C_q(S(\chi)). \tag{27.20}$$

Proof. It is sufficient to show that

$$\partial \Pi \mathring{\tau} = \pi \mathring{\tau} - \mathring{\tau} - \Pi \partial \mathring{\tau} \tag{27.21}$$

where $\mathring{\tau}$ is given as above. To that end, we start with the "elementary homotopy"

$$\partial \Delta s = (Bs)^{\text{co}} - s - \Delta \partial s \tag{27.22}$$

of Theorem 25.1. It follows from Lemma 25.3 and Theorem 25.1 that each term of (27.22) is a chain of $\mathbf{P}''(\sigma)$. Since $|s| \times I$ is the domain of τ_e, a ∂-permutable chain-transformation $\hat{\tau}_e$ is applicable to the complex $\mathbf{P}''(\sigma)$. Hence

$$\partial \hat{\tau}_e(\Delta s) = \hat{\tau}_e(Bs)^{\text{co}} - \hat{\tau}_e s - \hat{\tau}_e(\Delta \partial s) \tag{27.23}$$

(^ applied to τ_e). The proof of (27.20) will be completed by showing that (27.23) reduces, term by term, to (27.21).

The chains on the left of (27.21) and (27.23) are equal, in accord with the definition (27.19) of $\Pi\mathring{\tau}$.

Evaluation of $\hat{\tau}_e(Bs)^{\text{co}}$. By virtue of the relation

$$\tau_e(x, 1) = \tau(x), \qquad x \in |s|,$$

this term is equal to $\hat{\tau}(Bs)$, and so is equal to $\pi\mathring{\tau}$ by definition of $\pi\mathring{\tau}$ in (27.17).

Evaluation of $\hat{\tau}_e s$. By definition of τ_e this chain is equal to $\hat{\tau}s$ and so is equal to $\mathring{\tau}$, by definition of $\mathring{\tau}$.

Evaluation of $\hat{\tau}_e(\varDelta\partial s)$. This chain equals

$$(-1)^i\, \hat{\tau}_e(\varDelta s(i)) = (-1)^i\, \varPi\mathring{\tau}_i = \varPi\partial\mathring{\tau}$$

by virtue of the definitions of ∂s, $\varPi$, and $\partial\mathring{\tau}$ in (24.20), (27.19), and (26.6) respectively.

Thus (27.21) holds, establishing Theorem 27.2.

Proof of Corollary 27.3. This corollary follows from the chain-homotopy (27.20) once the basic inclusions

$$|\,\pi z\,| \subset |\,z\,| \quad \text{and} \quad |\,\varPi z\,| \subset |\,z\,|, \qquad z \in C_q(S(\chi)), \tag{27.24}$$

have been established. These inclusions cannot be replaced by equalities, as examples show.

Proof of (27.24). Starting with the definition (27.17) of $\pi\mathring{\tau}$, recall that $|\,Bs\,| = |\,s\,|$. It follows from (27.17) that $|\,\pi\mathring{\tau}\,| \subset |\,\mathring{\tau}\,|$ and hence $|\,\pi z\,| \subset |\,z\,|$ [use (26.9) and Example 26.2].

Similarly, starting with the definition (27.19) of $\varPi\mathring{\tau}$, recall that $|\,\varDelta s\,| = |\,s\,| \times I$. It follows from (27.19) that $|\,\varPi\mathring{\tau}\,| \subset |\,\mathring{\tau}\,|$ and hence $|\,\varPi z\,| \subset |\,z\,|$.

Corollary 27.3 now follows from (27.20) making use of (27.24).

Iterated Subdivisions of Chains. As defined, π is a ∂-permutable chain-transformation applicable to each singular chain z of $S(\chi)$. In particular, π is applicable to πz. This leads us to write π as $\pi^{(1)}$ and for each positive integer $n > 1$ to define $\pi^{(n)}$ inductively by setting

$$\pi^{(n)}z = \pi(\pi^{(n-1)}z). \tag{27.25}$$

We see that $\pi^{(n)}$ is a ∂-permutable chain-transformation, that $|\,\pi^{(n)}z\,| \subset |\,z\,|$, and that when z is a cycle

$$\pi^{(n)}z \sim z, \qquad \text{on} \quad |\,z\,|, \tag{27.26}$$

by virtue of Corollary 27.3.

The principal lemmas to be associated with the operator $\pi^{(n)}$ concern its effect on the "mesh" of a singular chain.

Definition 27.4. *Mesh* z. Let χ be a metric space, z a non-null chain of $S(\chi)$ of positive dimension, and

$$z = r_1\sigma_1 + \cdots + r_\mu\sigma_\mu , \qquad 0 \neq r_i \in \mathscr{K},$$

a "reduced" representation of z. The maximum of the diameters in χ of the carriers $|\sigma_i|$ is called the *mesh* of z. If z is a null chain, one sets mesh $z = 0$.

The following lemma will be verified:

Lemma 27.2. *If z is a singular chain of $S(\chi)$ of positive dimension, then*

$$\lim_{n\uparrow\infty} \text{mesh } \pi^{(n)}z = 0. \tag{27.27}$$

Proof. Let P be a simplicial polyhedron in a Euclidean space. By *mesh* P we mean the maximum of the diameters of the simplices of P. Let a be a geometric simplex, and for $n > 0$ let $b^n(a)$ be the nth barycentric subdivision of a. By an induction with respect to dim $a = m$ one readily proves that

$$\text{mesh } b(a) \leqslant \frac{m}{m+1} \text{ mesh } a, \qquad m > 0. \tag{27.28}$$

Hence for a given m-simplex a

$$\lim_{n\uparrow\infty} \text{mesh } b^n(a) = 0. \tag{27.29}$$

Lemma 27.2 follows.

The following lemma prepares for the fundamental "Excision Theorem" of §28 (cf. Eilenberg and Steenrod [1], p. 11). In Lemma 27.3 we make use of the following principle: Let G be a vector space over $\mathscr{K}$. If a set H of free generators of G is given a partition $H = H' \cup H''$ into disjoint subsets, then G is a direct sum $G = G' \oplus G''$, where the elements of H' generate G' and the elements of H'' generate G''.

Let $C_q^e(S(\chi))$ denote the vector subspace of $C_q(S(\chi))$ generated by the q-cells of $S(\chi)$ of mesh less than e.

Lemma 27.3. *Let χ be a metric space, A a proper subspace of χ, and A^* a subspace of A such that for some positive e*

$$(\chi - A)_e \subset \chi - A^*, \tag{27.30}$$

where $(\chi - A)_e$ is the open e-neighborhood of $\chi - A$ in χ.

Let $C_q^e(S(\chi))$ be represented (as is possible) by a direct sum

$$C_q^e(S(\chi)) = G' \oplus G'', \tag{27.31}$$

where G' is generated by the q-cells on $\chi - A^*$ of mesh $<e$. The chains of G'' are then on A.

The vector space G'' is generated by the q-cells of $S(\chi)$ of mesh $< e$ which meet A^*. These generators do not meet $\chi - A$ by virtue of (27.30). Hence the chains of G'' are on A.

§ 28

RELATIVE HOMOLOGIES

Relative cycles and homologies, as we shall define them, are needed in studying a ND $f \in C^\infty(M_n)$. In particular, we shall compare the singular homology groups of a subset

$$f_a = \{p \in M_n \mid f(p) \leqslant a\} \tag{28.1}$$

of M_n with the homology groups of $f_a - p_a$, where p_a is a critical point f at the f-level a.

Let there be given a Hausdorff space χ and a subspace A of χ. If $A \neq \chi$, we term (χ, A) an *admissible set pair* and A a *modulus* for χ. We admit the possibility that A may be empty. For q-chains u and v of $S(\chi)$ we write $u = v \bmod A$ if $u - v$ is a q-chain of $S(A)$.

Definition 28.1. (i) *Cycles on* χ mod A. A chain u^q of $S(\chi)$ will be called a q-cycle, mod A on χ, if

$$\partial u^q = 0 \bmod A. \tag{28.2}$$

(ii) *Cycles Bounding on* χ mod A. A q-cycle u^q on χ mod A will be said to *bound* mod A on χ if there exists a chain u^{q+1} of $S(\chi)$ such that

$$u^q - \partial u^{q+1} = 0 \bmod A. \tag{28.3}$$

(iii) *Homologies on* χ mod A. When (28.3) holds we write

$$u^q \sim 0 \quad \text{on} \quad \chi \bmod A \tag{28.4}$$

and say that u^q is homologous to zero on χ mod A.

Definition 28.2. *Relative Homology Groups.* The set of q-cycles on χ mod A form a vector space over the field $\mathscr{K}$ and will be denoted by $Z_q(S(\chi), S(A))$. The set of q-cycles, on χ mod A, which are bounding on χ mod A, is a vector subspace of $Z_q(S(\chi), S(A))$ and will be denoted by $B_q(S(\chi), S(A))$. The quotient group

$$Z_q(S(\chi), S(A))/B_q(S(\chi), S(A)) \tag{28.5}$$

will be called the *singular homology group* $H_q(S(\chi), S(A))$ *of* $S(\chi)$ *on* χ mod A.

The use of the following terms will abbreviate the exposition:

Ordinary cycles of $S(\chi)$ will be called *absolute cycles on* χ.
Cycles on χ mod A will be called *relative cycles on* χ.
Bounding on χ mod A will be called *relative bounding on* χ.

Conditions (28.2) and (28.3), respectively, define "relative" cycles and bounding. The same conditions respectively define absolute cycles and bounding if A is empty.

Definition 28.3. *Homology Classes on* χ mod A. Two relative k-cycles u^k and v^k are said to be in the *same relative homology class* on χ mod A if

$$u^k - v^k \in B_k(S(\chi), S(A)), \tag{28.6}$$

that is, if there exists a chain c^{k+1} of $S(\chi)$ such that

$$u^k - v^k = \partial c^{k+1} \bmod A. \tag{28.7}$$

When (28.7) holds we write

$$u^k \sim v^k \qquad \text{on} \quad \chi \bmod A. \tag{28.8}$$

We shall use *rel.* to mean relative or relatively, depending on the context.

The elements of $H_k(S(\chi), S(A))$ are rel. homology classes. The null element is the class of rel. bounding k-cycles. The property of two rel. k-cycles on χ of being in the same rel. homology class is reflexive, symmetric, and transitive.

Definition 28.4a. *Connectivities of* χ mod A. The dimension, if finite, of the rel. homology group $H_q(S(\chi), S(A))$ is called the connectivity $R_q(\chi, A)$ of χ mod A.

Definition 28.4b. *Homology Prebases of Rel. q-Cycles.* Paralleling Definition 24.7 of "homology prebases of q-cycles," any subset of $Z_q(S(\chi), S(A))$ which contains just one rel. q-cycle in each rel. homology class in a base for $H_q(S(\chi), S(A))$ will be termed a *homology prebase* $\mathbf{b}_q(\chi, A)$ on χ mod A.

Such homology prebases of rel. cycles have properties analogous to those enumerated following Definition 24.7 of homology prebases of absolute q-cycles.

The following lemma is a consequence of the definition of $R_0(\chi, A)$.

Lemma 28.1. *If each point of* χ *is arc-wise connected on* χ *to some point of the modulus* A, *then the connectivity* $R_0(\chi, A) = 0$.

Induced Homomorphisms α_* of Rel. Singular Homology Groups. Let χ' and χ'' be Hausdorff spaces with moduli, respectively, A' and A''. Let α be a ∂-permutable chain-transformation of $S(\chi')$ into $S(\chi'')$ that induces such a transformation of $S(A')$ into $S(A'')$. One sees that α defines linear homomorphisms

$$Z_q(S(\chi'), S(A')) \to Z_q(S(\chi''), S(A'')), \qquad q = 0, 1, \ldots,$$

and

$$B_q(S(\chi'), S(A')) \to B_q(S(\chi''), S(A'')), \qquad q = 0, 1, \ldots,$$

and hence by Lemma 24.1 induces linear homomorphisms

$$\alpha_* : H_q(S(\chi'), S(A')) \to H_q(S(\chi''), S(A'')), \qquad q = 0, 1, \ldots. \tag{28.9}$$

Change of Notation. For brevity we shall in the future write

$$H_q(S(\chi), S(A)) \qquad \text{as} \qquad H_q(\chi, A).$$

In accord with this we shall write (28.9) in the form

$$\alpha_* : H_q(\chi', A') \to H_q(\chi'', A''), \qquad q = 0, 1, \ldots, \tag{28.9'}$$

reading and interpreting (28.9)′ exactly as (28.9).

Let Φ now be a continuous mapping of χ' into χ'' that maps A' into A''. There is thereby induced a ∂-permutable chain-transformation $\widehat{\Phi}$ (see §26) with the properties ascribed to α in the preceding paragraph. There is accordingly induced a linear homomorphism $(\widehat{\Phi})_*$ of the nature of α_* in (28.9).

The following theorem concerns topologically equivalent pairs (χ', A') and (χ'', A''):

Theorem 28.1. *Suppose that a Hausdorff space χ' is topologically equivalent to a Hausdorff space χ'' under a homeomorphism Φ of χ' onto χ'' that maps A' onto A''. There are then induced isomorphisms*

$$(\widehat{\Phi})_* : H_q(\chi', A') \to H_q(\chi'', A''), \qquad q = 0, 1, \ldots, \tag{28.10}$$

onto, under which a rel. homology class on χ' of a rel. q-cycle z goes into the rel. homology class on χ'' of $\widehat{\Phi}z$.

This theorem follows from an obvious extension of Theorem 26.4. As in the proof of Theorem 26.4, let Θ be the inverse of the homeomorphism Φ. Then, as in the proof of Theorem 26.4, one sees that $\widehat{\Theta}_*$ and $\widehat{\Phi}_*$ are inverse linear homomorphisms onto.

Theorem 28.1 follows.

Coset-Contracting Isomorphisms. We shall prove a theorem which has several important theorems as corollaries:

Theorem 28.2. *Let (χ, A) and (χ', A') be two admissible set pairs with $\chi' \subset \chi$, $A' \subset A$. Let U be an arbitrary rel.*[†] *homology class (possibly trivial) on χ and U' the subclass of rel.*[‡] *cycles on χ'. If for each non-negative integer q: (***a***) each rel. q-cycle on χ is rel. homologous*[†] *on χ to a rel.*[‡] *q-cycle on χ', and if (***b***) each rel. q-cycle on χ' which is rel. bounding*[†] *on χ is rel. bounding*[‡] *on χ' then each set U' is a rel. homology class*[‡] *on χ' and the mapping*

$$U \to U' : H_q(\chi, A) \to H_q(\chi', A') \tag{28.11}$$

is an isomorphism onto.

[†] That is, mod A.

[‡] That is, mod A'.

The theorem follows from statements 1–5:

1. The class U' is not empty.
2. Chains in U' are in the same rel. homology class on χ'.
3. U' is a rel. homology class on χ'.
4. The class mapping $U \to U'$ is biunique and onto.
5. If V is a second rel. homology class on χ, then

$$U' + V' = (U + V)' \tag{28.12a}$$

$$rU' = (rU)', \qquad r \in \mathscr{K}. \tag{28.12b}$$

Proof of 1. The class U' is not empty, since (**a**) holds.

Proof of 2. Let x and y be chains in U'. Then x and y are in U and $x - y$ is rel. bounding on χ by definition of U. By virtue of (**b**) $x - y$ is also rel. bounding on χ', so that 2 follows.

Proof of 3. If a chain $x \in U'$ and a chain z on χ' are in the same rel. homology class on χ', then $x - z$ is rel. bounding on χ' and hence rel. bounding on χ, since $\chi' \subset \chi$ and $A' \subset A$. Thus z is in U and a rel. cycle on χ', and accordingly in U' by definition of U'.

Proof of 4. The class U' cannot be a subclass of two different classes U. That the class mapping $U \to U'$ is onto is trivial.

Proof of 5. The right and left sides of (28.12a) are rel. homology classes of $H_q(\chi', A')$. It remains to show that they are the same rel. homology class. It is sufficient to show that these two classes of $H_q(\chi', A')$ have a rel. cycle on χ' in common.

Let x and y be rel. cycles in U' and V', respectively. Then by definition of addition in $H_q(\chi', A')$, $x + y$ is a rel. cycle in $U' + V'$. But $x \in U$ and $y \in V$, so that $x + y \in U + V$. Now, $x + y$ is a rel. cycle on χ', so that $x + y \in (U + V)'$ by definition of $(U + V)'$. Thus (28.12a) is true.

The proof of (28.12b) is similar.

This establishes Theorem 28.2.

We shall call the isomorphism of Theorem 28.2 a *coset-contracting isomorphism*. With us an isomorphism of vector spaces is linear.

Note on Theorem 28.2. When a coset-contracting isomorphism (28.11) exists a "homology prebase" for $H_q(\chi', A')$ is a homology prebase for $H_q(\chi, A)$, but in general not *vice versa*.

We continue with a lemma needed in applying Theorem 28.2:

Lemma 28.2. *Corresponding to a prescribed q-cycle z on χ* mod A *and a prescribed positive integer* μ

$$z \sim \pi^{(\mu)}z \qquad \text{on} \quad \chi \bmod A. \tag{28.13}$$

Proof. On applying ∂ to the terms of the chain-homotopy (27.20) and using the inclusions (27.24), one sees that πz is a q-cycle on $\chi \bmod A$. The chain-homotopy (27.20) then implies (28.13) when $\mu = 1$. Proceeding inductively with respect to μ, one infers the truth of (28.13) for an arbitrary positive integer μ.

A first application of Theorem 28.2 is a simplified "Excision" Theorem (cf. Eilenberg and Steenrod [1], p. 11):

Theorem 28.3. *Let χ be a metric space, A a proper subspace of χ, and A^* a subspace of A such that for some positive e*

$$(\chi - A)_e \subset \chi - A^*, \tag{28.14}$$

where $(\chi - A)_e$ is the open e-neighborhood of $\chi - A$ on χ.

There then exist coset-contracting isomorphisms†

$$H_q(\chi, A) \approx H_q(\chi - A^*, A - A^*), \qquad q = 0, 1, \ldots. \tag{28.15}$$

Method of Proof. It is sufficient to show that statements (**a**) and (**b**) of Theorem 28.2 are valid under the conditions of Theorem 28.3 provided (χ', A') of Theorem 28.2 is taken as the pair

$$(\chi', A') = (\chi - A^*, A - A^*). \tag{28.16}$$

We shall employ Lemma 27.3, noting that its hypotheses are satisfied under the conditions of Theorem 28.3.

Notation. In the proofs of Theorems 28.3 and 28.4 a singular chain carried by one of the spaces χ, χ', A, A' will be denoted by a letter u, v, z, etc., with a subscript denoting the space. Note that $\chi' \cap A = A'$ by virtue of (28.16), so that a chain carried by χ' and A is carried by A'.

† A^* is "excised" from χ and A in the right side of (28.15).

Verification of (**a**) *of Theorem* 28.2. Let z_χ be a q-cycle on χ mod A. To verify (**a**), we shall show that

$$z_\chi \sim y_{\chi'} \quad \text{on} \quad \chi \bmod A \tag{28.17}$$

for a suitable cycle $y_{\chi'}$ on χ' mod A'.

For each positive integer μ, by Lemma 28.2

$$z_\chi \sim \pi^{(\mu)} z_\chi \quad \text{on} \quad \chi \bmod A. \tag{28.18a}$$

Let μ be so large that mesh $\pi^{(\mu)} z_\chi < e$. Then by Lemma 27.3 $\pi^{(\mu)} z_\chi$ is the sum of an element in G' and an element in G'', that is, of an element on $\chi' = \chi - A^*$ and an element on A. Equivalently,

$$\pi^{(\mu)} z_\chi = y_{\chi'} \bmod A \tag{28.18b}$$

for some chain $y_{\chi'}$. From this relation we infer that $y_{\chi'}$ is a cycle mod A, and hence mod A', since $\chi' \cap A = A'$. The homology (28.17) follows from the relations (28.18).

Verification of (**b**) *of Theorem* 28.2. There is given a q-cycle $z_{\chi'}$ on χ' mod A' which is rel. bounding on χ. To establish (**b**), we must prove that $z_{\chi'}$ is rel. bounding on χ' mod A'.

By hypothesis $z_{\chi'} = \partial u_\chi$ mod A for some chain u_χ. Thus ∂u_χ is a q-cycle on χ' mod A. By Lemma 28.2 and the relation $\pi\partial = \partial\pi$ one infers for each $\mu > 0$ that

$$\partial u_\chi \sim \pi^{(\mu)}\, \partial u_\chi = \partial \pi^{(\mu)} u_\chi \quad \text{on} \quad \chi' \bmod A. \tag{28.19}$$

Let μ be so large that mesh $\pi^{(\mu)} u_\chi < e$ on χ. Then by Lemma 27.3

$$\pi^{(\mu)} u_\chi = v_{\chi'} \bmod A \tag{28.20}$$

[cf. (28.18b)], so that by (28.19) and (28.20)

$$\partial u_\chi \sim 0 \quad \text{on} \quad \chi' \bmod A.$$

Since $z_{\chi'} = \partial u_\chi$ mod A, we infer that

$$z_{\chi'} \sim 0 \quad \text{on} \quad \chi' \bmod A'$$

(since $\chi' \cap A = A'$), thereby verifying (**b**).

Theorem 28.3 follows from Theorem 28.2.

A second application of Theorem 28.2 concerns an isomorphism induced by a deformation retracting χ onto χ' (see Definition 23.1).

Theorem 28.4. *Let* (χ, A) *and* (χ', A') *be admissible set pairs with* $\chi' \subset \chi$ *and* $A' \subset A$, *and let* d *be a deformation retracting* χ *onto* χ' *and* A *onto* A'. *There then exist coset-contracting isomorphisms*

$$H_q(\chi, A) \approx H_q(\chi', A'), \qquad q = 0, 1, \ldots, \tag{28.21}$$

under which the rel. homology class on χ *of a rel.* q*-cycle* z_χ *goes into the rel. homology class on* χ' *of* $\overset{\frown}{d}_1 z_\chi$, *where* d_1 *is the terminal mapping of* d.

We shall show that under the hypotheses of the theorem conditions (**a**) and (**b**) of Theorem 28.2 are satisfied. The rel. q-cycle z_χ is given.

Verification of (**a**). That $\overset{\frown}{d}_1 z_\chi$ is a q-cycle on χ' mod A' follows from the inclusions $| \overset{\frown}{d}_1 z_\chi | \subset \chi'$ and $| \overset{\frown}{d}_1 \partial z_\chi | \subset A'$ implied by Definition 26.5 of $\frown$, and from the ∂-permutability of the operator $\overset{\frown}{d}_1$. (Theorem 26.3b).

To establish a homology $d_1 z_\chi \sim z_\chi$ on χ mod A, we turn to the chain-homotopy (27.8) induced by d. Recall that [see (27.13)]

$$| \mathbf{d} z_\chi | \subset d\text{-traj } | z_\chi | \subset \chi \qquad \text{and} \qquad | \mathbf{d} \partial z_\chi | \subset d\text{-traj } | \partial z_\chi | \subset A.$$

The chain homotopy (27.8) is accordingly on χ and takes the form

$$\partial \mathbf{d} z_\chi = \overset{\frown}{d}_1 z_\chi - z_\chi, \qquad \text{mod } A, \tag{28.22}$$

thereby establishing (**a**) for the rel. q-cycle z_χ.

Verification of (**b**). By hypothesis of (**b**) there is given a q-cycle $u_{\chi'}$ mod A' and a $(q + 1)$-chain y_χ such that $\partial y_\chi = u_{\chi'}$ mod A. We shall show that $u_{\chi'} \sim 0$ on χ' mod A'.

By hypothesis $\partial y_\chi = u_{\chi'} + v_A$, introducing the chain v_A. Recall that $\overset{\frown}{d}_1$ is ∂-permutable, that d_1 reduces to the identity on χ', and that $d_1 A \subset A'$, so that $\overset{\frown}{d}_1 u_{\chi'} = u_{\chi'}$ and $\overset{\frown}{d}_1 v_A$ is on A'. It follows that

$$\partial(\overset{\frown}{d}_1 y_\chi) = \overset{\frown}{d}_1 \partial y_\chi = \overset{\frown}{d}_1 (u_{\chi'} + v_A) = u_{\chi'} \text{ mod } A'.$$

Thus $u_{\chi'} \sim 0$ mod A', thereby verifying (**b**).

Theorem 28.4 follows from Theorem 28.2.

Basic Isomorphisms in the Critical Point Theory. The preceding theorems imply basic isomorphisms in the critical point theory.

Let a ND $f \in C^{\infty}(M_n)$ be given with a critical value a assumed at a critical point p_a. In §28 and §29 it is not necessary to assume that p_a is the *only* critical point at the f-level a.

Let D^σ be an open n-ball in E_n with center at the origin $\mathbf{0}$ and with radius σ. If p_a has the index k, there exists (Theorem 22.2) a presentation $(F : D^\sigma, X) \in \mathscr{D}M_n$ such that $F(\mathbf{0}) = p_a$, and for $u \in D^\sigma$

$$(f \circ F)(u) = a - u_1^2 - \cdots - u_k^2 + u_{k+1}^2 + \cdots + u_n^2 = \Phi^k(u), \quad (28.23)$$

introducing Φ^k. When (28.23) holds set

$$D_k^\sigma = \Phi_a^k \cap D^\sigma \subset E_n \quad (28.24)$$

and

$$X_k = f_a \cap X \subset M_n . \quad (28.25)$$

Let $\dot{D}_k^\sigma$ denote D_k^σ with the origin deleted. If Y is any subset of M_n which contains p_a, $\dot{Y}$ shall denote $Y - p_a$, except as noted.

Theorem 28.1 has the following corollary:

Theorem 28.1′. *The topological equivalence under F of the set pairs $(D_k^\sigma, \dot{D}_k^\sigma)$ and $(X_k, \dot{X}_k)$ implies the isomorphisms* (onto)

$$H_q(X_k, \dot{X}_k) \approx H_q(D_k^\sigma, \dot{D}_k^\sigma), \qquad q = 0, 1, \ldots . \quad (28.26)$$

Theorem 28.3′ is a corollary of Theorem 28.3 and concerns the subset f_a of M_n defined in (28.1):

Theorem 28.3′. *There exist coset-contracting isomorphisms*

$$H_q(f_a, \dot{f}_a) \approx H_q(X_k, \dot{X}_k), \qquad q = 0, 1, \ldots, \quad (28.27)$$

where $\dot{f}_a = f_a - p_a$ and $\dot{X}_k = X_k - p_a$.

Proof. One identifies (χ, A) of Theorem 28.3 with $(f_a, \dot{f}_a)$ and sets $f_a - X_k = A^*$, so that $A^* \subset A$, and $X_k = \chi - A^*$ and $\dot{X}_k = A - A^*$. The Excision Condition (28.14) of Theorem 28.3 is satisfied; if $e > 0$ is sufficiently small, an e-neighborhood on f_a of $\chi - A = p_a$ is included in $\chi - A^* = X_k$.

Theorem 28.3′ follows from Theorem 28.3.

Note. The proof is valid even in the special case in which the index $k = 0$. In this case $X_0 = p_a$, $\dot{X}_0 = \varnothing$, p_a is an isolated point of f_a, and the right member of (28.27) reduces to $H_q(p_a)$.

To apply Theorem 28.4, new notation is required.

Let E_k be the coordinate plane of E_n on which $u_{k+1} = \cdots = u_n = 0$. For $0 < k \leqslant n$ we introduce the k-disk (closed)

$$\Delta_k = E_k \cap \text{Cl}\, D^\sigma = E_k \cap \text{Cl}\, D_k{}^\sigma, \qquad k > 0, \tag{28.28a}$$

and remove the origin $\mathbf{0}$ from Δ_k to form the "centerless" disk $\dot{\Delta}_k$. Note that $\Phi^k(\mathbf{0}) = a$ and that on $\dot{\Delta}_k$ $\Phi^k(u) < a$.

One denotes the geometric boundary of Δ_k by S_{k-1} and notes that

$$\text{Int}\, \Delta_k = \Delta_k - S_{k-1}, \qquad \text{Int}\, \dot{\Delta}_k = \dot{\Delta}_k - S_{k-1}.$$

The Excision Theorem 28.3 leads to the following lemma:

Lemma 28.3. *There exist coset-contracting isomorphisms*

$$H_q(\Delta_k, \dot{\Delta}_k) \approx H_q(\text{Int}\, \Delta_k, \text{Int}\, \dot{\Delta}_k) \qquad q = 0, 1, \ldots. \tag{28.28b}$$

Proof. One identifies (χ, A) of Theorem 28.3 with $(\Delta_k, \dot{\Delta}_k)$ and sets $A^* = S_{k-1}$, the geometric boundary of Δ_k. Then

$$\text{Int}\, \Delta_k = \chi - A^*, \qquad \text{Int}\, \dot{\Delta}_k = A - A^*.$$

Moreover, condition (28.14) is satisfied, so that (28.15) implies (28.28b).

Theorem 28.4′. *If the index k of p_a is positive, there exists a deformation d_k retracting $D_k{}^\sigma$ onto* Int Δ_k *and thereby $\dot{D}_k{}^\sigma$ onto* Int $\dot{\Delta}_k$, *implying coset-contracting isomorphisms*

$$H_q(D_k{}^\sigma, \dot{D}_k{}^\sigma) \approx H_q(\text{Int}\, \Delta_k, \text{Int}\, \dot{\Delta}_k) \approx H_q(\Delta_k, \dot{\Delta}_k), \qquad q = 0, 1, \ldots. \tag{28.29}$$

The deformation d_k is taken as a mapping of $D_k{}^\sigma \times I$ onto $D_k{}^\sigma$ of the form

$$(u, t) \to d_k(u, t) = (u_1, \ldots, u_k, (1 - t)\, u_{k+1}, \ldots, (1 - t)u_n), \tag{28.29′}$$

where u is in $D_k{}^\sigma$ and $0 \leqslant t \leqslant 1$. One replaces d of Theorem 28.4 by d_k.

Theorem 28.4′ follows from Theorem 28.4 and Lemma 28.3.

We record a corollary of Theorems 28.1′, 28.3′, 28.4′ obtained by an appropriate composition of their isomorphisms:

Corollary 28.1. *When the index k of the critical point p_a is positive*

$$H_q(f_a, \dot{f}_a) \approx H_q(\Delta_k, \dot{\Delta}_k), \qquad q = 0, 1, \ldots. \tag{28.30}$$

We shall give a second basic application of Theorem 28.4.

Theorem 28.4″. *Let $f \in C^\infty(M_n)$ be* ND *and M_n boundedly f-compact* (Definition 21.1). *Let (a, b) be an open interval of ordinary values of f, with a and b critical values of which b is taken on at a unique critical point p_b. There then exist coset-contracting isomorphisms*

$$H_q(\dot{f}_b, \dot{f}_a) \approx H_q(f_a, \dot{f}_a), \qquad q = 0, 1, \ldots; \quad \dot{f}_b = f_b - p_b .$$

The isomorphisms of Theorem 28.4″ exist in accord with Theorem 28.4 because there exists (Corollary 23.1) a deformation retracting $\dot{f}_b$ onto f_a.

EXERCISE 28.1. Verify the following:

Corollary 28.2. *Let d be a deformation rectracting a Hausdorff space χ onto a Hausdorff subspace χ'. A q-cycle z^q on χ' which is bounding on χ is bounding on χ'.*

Suggestion. Apply Theorem 28.4, taking A and A' as empty sets, and observe that $\hat{d}_1 z^q = z^q$.

§ 29

COMPARISON OF THE HOMOLOGY GROUPS ON f_a AND ON $f_a - p_a$

As in §28, a ND $f \in C^\infty(M_n)$ is given with a critical value a assumed at a critical point p_a. To make the desired comparisons, the isomorphisms established in §28 are utilized. To that end, it is necessary to determine the singular homology groups on a k-disk Δ_k, the centerless k-disk $\dot{\Delta}_k$, and the k-sphere S_k.

It is understood that Δ_k is a closed Euclidean k-disk in E_n, reducing to a point when $k = 0$. The set $\dot{\Delta}_k$ is defined only when $k > 0$, and is Δ_k with its center removed. S_k is a Euclidean k-sphere. We understand that S_0 is a pair of distinct points. The geometric boundary of Δ_k when $k > 0$ is identified with a sphere S_{k-1}.

Change of Notation. As in the relative homology theory, we shall make the replacements: $H_q(S(\chi))$ by $H_q(\chi)$ and $R_q(S(\chi))$ by $R_q(\chi)$ for a Hausdorff space χ.

For $k > 0$, Δ_k admits a deformation retracting Δ_k onto its center. It follows from Corollary 27.2 that for q and k nonnegative integers

$$R_q(\Delta_k) = \delta_0{}^q. \tag{29.1}$$

We shall show that the connectivities

$$R_q(S_k) \qquad (k = 0, 1, \ldots;\ q = 0, 1, \ldots;\ S_k \text{ a } k\text{-sphere}) \tag{29.2}$$

are given as in Table I, properly extended, while the connectivities

$$R_q(\dot{\Delta}_k) \qquad (k = 1, 2, \ldots;\ q = 0, 1, \ldots;\ \Delta_k \text{ a } k\text{-disk}) \tag{29.3}$$

TABLE I. $R_q(S_k)$

$q \backslash k$	0	1	2	3	4
0	2	1	1	1	1
1	0	1	0	0	0
2	0	0	1	0	0
3	0	0	0	1	0
4	0	0	0	0	1

are given as in Table II, properly extended. In these tables the rows and diagonals of 1's are to be extended indefinitely to the right.

TABLE II. $R_q(\varDelta_k)$

$q \backslash k$	0	1	2	3	4
0	—	2	1	1	1
1	—	0	1	0	0
2	—	0	0	1	0
3	—	0	0	0	1
4	—	0	0	0	0

The entries not otherwise defined are to be 0. There are no entries in the column of Table II headed by $k = 0$.

Notation. Let the tables consisting of the columns headed by $k = 0, 1, 2, \ldots, \mu$ in Tables I and II be termed, respectively, Tables I(μ) and Tables II(μ).

Before turning to the verification of these tables we prove an essential lemma.

Lemma 29.0. *Given a rel. q-cycle y^q on $\varDelta_k$* mod $\dot{\varDelta}_k$, $k > 0$, $\partial y^q \sim 0$ *on $\dot{\varDelta}_k$ if and only if $y^q \sim 0$ on $\varDelta_k$* mod $\dot{\varDelta}_k$.

If $\partial y^q \sim 0$ on $\dot{\varDelta}_k$, then $\partial y^q = \partial u^q$ for some chain u^q of $S(\dot{\varDelta}_k)$, so that $y^q - u^q$ is an absolute q-cycle on $\varDelta_k$. Such a cycle is bounding on $\varDelta_k$, implying $y^q \sim 0$ on $\varDelta_k$ mod $\dot{\varDelta}_k$.

A homology $y^q \sim 0$ on Δ_k mod $\dot{\Delta}_k$ implies the existence of a chain w^{q+1} on Δ_k and a chain u^q on $\dot{\Delta}_k$ such that $y^q = \partial w^{q+1} + u^q$ and hence implies that $\partial y^q \sim 0$ on $\dot{\Delta}_k$.

The First Row and Column of Table I. That the entry 2 in Table I is correct follows from Theorem 26.6(ii). That the entries 1 in the first row ($q = 0$) are correct follows from Theorem 26.6(i).

The 0-entries in Table I(0), that is, in the column $k = 0$ of Table I, and correct, since S_0 is the union of a pair of points p_1 and p_2, so that by (26.28)

$$H_q(S_0) \approx H_q(p_1) \oplus H_q(p_2), \qquad q = 0, 1, \ldots . \tag{29.4}$$

The groups on the right of (29.4) are trivial when $q > 0$, in accord with Theorem 26.2, so that Table I(0) is correct.

That Table I as extended is correct will be proved by a mathematical induction with the following *inductive hypotheses*:

For some integer $\mathbf{m} > 0$ *Table* I($\mathbf{m} - 1$) *is correct.*

We have already shown that this hypothesis is valid when $\mathbf{m} = 1$. At the end of this section we shall conclude that the validity of Table I($\mathbf{m} - 1$) implies the validity of Table I($\mathbf{m}$). We continue with a lemma.

Lemma 29.1$\mathbf{m}$. *If Table* I($\mathbf{m} - 1$) *is correct, Table* II($\mathbf{m}$) *is correct.*

Proof. For $k > 0$ $\dot{\Delta}_k$ admits a radial deformation retracting $\dot{\Delta}_k$ onto the outer geometric boundary S_{k-1} of Δ_k, so that by Theorem 28.4, with the moduli A and A' empty sets,

$$H_q(\dot{\Delta}_k) \approx H_q(S_{k-1}), \qquad q = 0, 1, \ldots .$$

Thus the validity of Table II($\mathbf{m}$) follows from the validity of Table I($\mathbf{m} - 1$).

We state a key theorem:

Theorem 29.1$\mathbf{m}$. *If Table* I($\mathbf{m} - 1$) *is correct,*

$$R_q(\Delta_k, \dot{\Delta}_k) = \delta_q{}^k; \qquad k = 1, \ldots, \mathbf{m}; \; q \geqslant 0. \tag{29.5}$$

Proof of (29.5) *when* $k \neq q$. Were (29.5) false when $k \neq q$, then for some rel. q-cycle z^q, $z^q \not\sim 0$ on Δ_k mod $\dot{\Delta}_k$, implying $\partial z^q \not\sim 0$ on $\dot{\Delta}_k$ (Lemma 29.0). It would follow from Table II($\mathbf{m}$) that dim ∂z^q is 0 or $k - 1$. The second alternative is contrary to the hypothesis $k \neq q$. The first alternative, $q = 1$ and $\partial z^1 \not\sim 0$ on $\dot{\Delta}_k$, implies $k = 1$, since $\dot{\Delta}_k$ is connected for $k > 1$. Thus both alternatives are contrary to the hypothesis $k \neq q$.

Proof of (29.5) *when* $1 < k \leqslant \mathbf{m}$ *and* $q = k$. Under these conditions Table II($\mathbf{m}$) shows that there exists an absolute $(k - 1)$-cycle ω^{k-1} on $\dot{\Delta}_k$ which is a "homology prebase" for $(k - 1)$-cycles on $\dot{\Delta}_k$.

Let y^k be a k-chain on Δ_k such that $\partial y^k = \omega^{k-1}$, in accord with (29.1). Since $\partial y^k \not\sim 0$ on $\dot{\Delta}_k$, $y^k \not\sim 0$ on Δ_k mod $\dot{\Delta}_k$ by Lemma 29.0. If u_k is an arbitrary rel. k-cycle on Δ_k mod $\dot{\Delta}_k$, then for some r in the field $\mathscr{K}$, by Table Table II($\mathbf{m}$),

$$\partial u^k \sim r\omega^{k-1} = r\partial y^k \qquad \text{on} \quad \dot{\Delta}_k. \tag{29.6}$$

Hence by Lemma 29.0 $u^k \sim ry^k$ on Δ_k mod $\dot{\Delta}_k$, completing the proof of (29.5) when $1 < k \leqslant \mathbf{m}$ and $q = k$.

Proof of (29.5) *when* $q = k = 1$. The preceding paragraph gives the proof for this case provided one takes ω^0 as $u^0 - v^0$, with u^0 and v^0 0-cells on $\dot{\Delta}_1$ whose carriers are the two endpoints of Δ_1. Relations (29.6) hold for some r when $k = 1$, even though ω^0 is not a "homology prebase" for 0-cycles on $\dot{\Delta}_1$.

This completes the proof of Theorem 29.1$\mathbf{m}$.

By virtue of the isomorphism of Corollary 28.1, Theorem 29.1$\mathbf{m}$ has the following corollary [see (28.1) for definition of f_a]:

Corollary 29.1$\mathbf{m}$. *If Table* I($\mathbf{m} - 1$) *is correct and the index k of the critical point p_a is on the range* 1,..., $\mathbf{m}$, *then if one sets* $\dot{f}_a = f_a - p_a$,

$$R_q(f_a, \dot{f}_a) = \delta_k{}^q, \qquad q = 0, 1, \ldots . \tag{29.7}$$

Linking and Nonlinking Critical Points p_a. As in §28, p_a is a ND critical point at the f-level a. The index of p_a is denoted by k. We have set $f_a - p_a = \dot{f}_a$. It is not assumed that p_a is the only critical point at the f-level a or that f_a is compact.

Definition 29.1. *k-Caps of* p_a. A rel. q-cycle (on f_a mod $\dot{f}_a$) which is nonbounding on f_a mod $\dot{f}_a$ will be called a *q-cap* of p_a. Under the conditions† of Corollary 29.1**m** (29.7) holds and there are no q-caps of p_a other than k-caps, and any such k-cap of p_a is a "homology prebase" on f_a mod $\dot{f}_a$ for rel. k-cycles on f_a mod $\dot{f}_a$. If $k = 0$, a 0-cell with carrier p_a is a 0-cap.

Among critical points p_a of index k we distinguish three types, of which the third will be proved to be nonexistent. The third type is introduced in order that we can say, *a priori*, that each critical point p_a is of one of these three types.

Linking Type. The class of k-caps of p_a is not empty, and each k-cap ζ^k of p_a is *linkable*, that is, $\partial\zeta^k$ is null or bounding on $\dot{f}_a$.

Nonlinking Type. The class of k-caps of p_a is not empty and each k-cap ζ^k of p_a is *nonlinkable*, that is, $\partial\zeta^k$ is neither 0 nor bounding on $\dot{f}_a$.

Neutral Type. The point p_a is not of linking or nonlinking type.

We note that a critical p_a of index 0 is of linking type.

At the end of the inductive process of this section we can conclude that there are no critical points of neutral type. At the present stage of the induction we are limited to the following lemma:

Lemma 29.2**m**. *Under the conditions of Corollary* 29.1**m**, *including the condition that the index k of* p_a *is on the range* 1,..., **m**, *there are no critical points* p_a *of neutral type.*

For each critical point p_a here admitted there exist k-caps because of (29.7). If one such k-cap ζ^k is linkable, each such k-cap η^k is linkable; when k is on the range 1,..., **m** it follows from (29.7) that for some non-null $r \in \mathscr{K}$, $\zeta^k \sim r\eta^k$ on f_a mod $\dot{f}_a$, implying that $\partial\zeta^k \sim r\partial\eta^k$ on $\dot{f}_a$. We conclude that if ζ^k is linkable (or nonlinkable), then η^k is linkable (or nonlinkable).

Lemma 29.2**m** follows.

The following lemma gives an essential characterization of critical points p_a of linking type:

Lemma 29.3**m**. *Under the conditions of Corollary* 29.1 **m** *a necessary and sufficient condition that* p_a *be of linking type is that the rel.*

† Including the condition that k be on the range 1, 2,..., **m**.

homology class on f_a mod $\mathring{f}_a$ *of each k-cap of* p_a (*k on the range* $1,\ldots,$ **m**) *contain a nonempty class of k-caps which are absolute k-cycles* λ^k *on* f_a .

The condition of the lemma is sufficient in accord with the characterization of critical points p_a of linking type.

The condition of the lemma is also necessary. Given a k-cap ζ^k of a critical point p_a of linking type, there exists, by hypothesis, a k-chain u^k on $\mathring{f}_a$ such that $\partial\zeta^k = \partial u^k$. The k-chain $\zeta^k - u^k$ is a k-cap of p_a and an absolute k-cycle λ^k on f_a . Moreover, $\zeta^k \sim \lambda^k$ on f_a mod $\mathring{f}_a$, since $\zeta^k - \lambda^k = u^k$.

Definition 29.2. *Linking k-Cycles of* p_a . If p_a has the index k, a k-cap of p_a which is an absolute k-cycle λ^k will be called a *linking k-cycle of* p_a .

We note that a linking k-cycle λ^k of p_a is nonbounding on f_a , since a homology $\lambda^k \sim 0$ on f_a would be interpretable as a rel. homology $\lambda^k \sim 0$ on f_a mod $\mathring{f}_a$,

With a critical point p_a we associate the differences

$$\varDelta R_q = R_q(f_a) - R_q(\mathring{f}_a) \qquad q = 0, 1, \ldots, \tag{29.8}$$

whenever the numbers differenced in (29.8) are finite, and state a basic theorem:

Theorem 29.2. *Corresponding to a critical point* p_a *of index k for which the differences* (29.8) *exist, the differences* $\varDelta R_q$ *are all zero except that*

$$\varDelta R_k = 1 \qquad \text{when } p_a \text{ is of linking type}$$

$$\varDelta R_{k-1} = -1 \qquad \text{when } p_a \text{ is of nonlinking type.}$$

Hypotheses Reviewed. There is no assumption in Theorem 29.2 that the manifold M_n is "boundedly f-compact" (Definition 21.1). The assumption that the numbers differenced in (29.8) are finite is implied (as we shall see in §30) by the hypothesis that f_a is a compact subset of the manifold M_n .

Theorem 29.2**m**. The first step in the proof of Theorem 29.2 will be to prove Theorem 29.2**m**, that is, Theorem 29.2 under the conditions of Corollary 29.1**m**. Theorem 29.2**m** is a corollary of a much stronger theorem, Theorem 29.3**m**. Theorem 29.3**m** compares homology groups on f_a and on $\dot{f}_a$ rather than dimensions of such groups, and is valid regardless of whether or not these dimensions are finite. However, the comparison of dimensions continues to play a fundamental role.

Comparison of Homology Prebases on f_a and on $\dot{f}_a$. Referring to Definition 24.7, for each integer $q > 0$ let

$$\mathbf{b}_q(f_a) \qquad \text{and} \qquad \mathbf{b}_q(\dot{f}_a), \tag{29.9}$$

respectively, denote "homology prebases" (possibly empty) of q-cycles on f_a and on $\dot{f}_a$. The existence of such prebases follows from the existence of bases for the homology groups $H_q(f_a)$ and $H_q(\dot{f}_a)$ over $\mathscr{K}$. Such existence will be verified directly in case f_a is a compact subset of M_n.

The principal theorem follows.

Theorem 29.3. (i) *If the critical point p_a has the index k and if one sets $\dot{f}_a = f_a - p_a$, then a homology prebase $\mathbf{b}_q(\dot{f}_a)$ is a homology prebase $\mathbf{b}_q(f_a)$ except in the following two cases: Case* 1. $q = k$ *and p_a is of linking type. Case* 2. $q = k - 1$ *and p_a is of nonlinking type,* $k > 0$.

(ii) *In case* 1 *any set of absolute k-cycles of the form*

$$\mathbf{b}_k(\dot{f}_a) \cup \lambda^k \tag{29.10}$$

(*where λ^k is a linking k-cycle of p_a*) *is a homology prebase* $\mathbf{b}_k(f_a)$.

(iii) *In case* 2 *any set of absolute $(k - 1)$-cycles, $k > 0$, of the form*

$$\mathbf{b}_{k-1}(\dot{f}_a) - w^{k-1} \tag{29.11}$$

(*in which w^{k-1} is the boundary of a k-cap of p_a and $\mathbf{b}_{k-1}(\dot{f}_a)$ contains w^{k-1}*) *is a homology prebase* $\mathbf{b}_{k-1}(f_a)$.

Continuing our induction with respect to **m**, let Theorem 29.3 subject to the conditions of Corollary 29.1**m** on **m** and k be denoted by Theorem 29.3**m**. We shall prove Theorem 29.3**m**. It will then follow that Table I(**m**) is correct and hence Tables I and II are correct. Theorems 29.2 and 29.3 are final consequences.

Proof of Theorem 29.3**m**. Before coming to the proof proper of Theorem 29.3**m** we shall verify that the construction of the set (29.11) is possible for k on the range $1, 2, \ldots, \mathbf{m}$.

It follows from Corollary 29.1**m** that a k-cap of p_a exists. The boundary w^{k-1} of this k-cap is on $\dot{f}_a$ and nonbounding on $\dot{f}_a$, since, by hypothesis of (iii) p_a is not of linking type. By virtue of Lemma 24.3 there then exists a homology prebase $\mathbf{b}_{k-1}(\dot{f}_a)$ which contains w^{k-1}, so that the set (29.11) exists.

To prove that a given set of absolute q-cycles on f_a is a homology prebase $\mathbf{b}_q(f_a)$, it is sufficient to show that the q-cycles of the given set are both "homology-wise generating" and "homology-wise free" among absolute q-cycles on f_a.

Proof† *of* (i)**m**. In (i) p_a comes under neither case 1 nor case 2. To prove that a prebase $\mathbf{b}_q(\dot{f}_a)$ is then homology-wise generating among q-cycles on f_a is to prove that when y^q is an absolute q-cycle on f_a then $y^q \sim 0$ on f_a mod $\dot{f}_a$, or, equivalently, that y^q is homologous on f_a to an absolute q-cycle on $\dot{f}_a$.

Were $y^q \not\sim 0$ on f_a mod $\dot{f}_a$, then y^q would be a q-cap, implying that $q = k$ by (29.7) and that y^k is a linking k-cycle, contrary to the exclusion of case 1 from (i).

To show that $\mathbf{b}_q(\dot{f}_a)$ is homology-wise *free*† on f_a under the conditions of (i) it is sufficient to show that an absolute q-cycle y^q on $\dot{f}_a$ which is nonbounding on $\dot{f}_a$ is nonbounding on f_a.

Were $y^q = \partial w^{q+1}$ on f_a, then $w^{q+1} \not\sim 0$ on f_a mod $\dot{f}_a$, since $y^q \not\sim 0$ on $\dot{f}_a$. This would imply that w^{q+1} is a k-cap by (29.7) and p_a nonlinking, since $y^q \not\sim 0$ on $\dot{f}_a$. This is contrary to the exclusion of case 2 from (i).

Proof† *of* (ii)**m**. The set (29.10) of k-cycles is homology-wise *generating* on f_a, since when an absolute k-cycle y^k on f_a is prescribed, an $r \in \mathscr{K}$ exists such that $y^k \sim r\lambda_k$ on f_a mod $\dot{f}_a$ [by (29.7)].

We shall show that the set of k-cycles (29.10) is homology-wise *free* on f_a, $1 \leqslant k \leqslant \mathbf{m}$.

The set $\mathbf{b}_k(\dot{f}_a)$ of k-cycles is homology-wise free on f_a. Otherwise there would exist a k-cycle u^k on $\dot{f}_a$ such that $u^k \not\sim 0$ on $\dot{f}_a$ and $u^k = \partial y^{k+1}$ for some chain y^{k+1} on f_a. Then $y^{k+1} \not\sim 0$ on f_a mod $\dot{f}_a$, contrary to (29.7).

† When the index k of p_a is on the range $1, 2, \ldots, \mathbf{m}$ and (29.7) is true.

Finally, λ^k satisfies no relation $\lambda^k \sim u^k$ on f_a with u^k on $\dot{f}_a$, since such a relation would imply that $\lambda^k \sim 0$ on $f_a \bmod \dot{f}_a$, contrary to the fact that λ^k is a k-cap.

Thus (ii)**m** of Theorem 29.3**m** is true.

Proof† *of* (iii)**m**. The set of absolute $(k-1)$-cycles of the set (29.11) is homology-wise *generating* on $\dot{f}_a$, as we now verify. Since there are no $(k-1)$-caps,† each $(k-1)$-cycle y^{k-1} on $\dot{f}_a$ is homologous to zero on $f_a \bmod \dot{f}_a$. It follows that y^{k-1} is included among the $(k-1)$-cycles on $\dot{f}_a$ homology-wise generated by the set (29.11).

The $(k-1)$-cycles of the set (29.11) are homology-wise *free* on $\dot{f}_a$, as we now verify. We are assuming that k is on the range $1, 2, \ldots, \mathbf{m}$.

Were the cycles of the set (29.11) not free, there would exist a "reduced" form u^{k-1} in the $(k-1)$-cycles of the set (29.11) such that for some chain y^k on $\dot{f}_a$, $u^{k-1} = \partial y^k$ on $\dot{f}_a$. Such a chain y^k would be a k-cap, since $\partial y^k \not\sim 0$ on $\dot{f}_a$. By hypotheses of (iii), w^{k-1} is the boundary of a k-cap v^k. Since (29.7) holds, $y^k \sim r v^k$ on $f_a \bmod \dot{f}_a$ for some $r \in \mathscr{K}$. Hence $\partial y^k - r\,\partial v^k \sim 0$ on $\dot{f}_a$, or, explicitly,

$$u^{k-1} \sim r w^{k-1} \qquad \text{on } \dot{f}_a,$$

contrary to the nature of $\mathbf{b}_{k-1}(\dot{f}_a)$.

This completes the proof of Theorem 29.3**m**.

We can now complete the verification of Table I by proving the following theorem:

Theorem 29.4**m**. *If Table* I(**m** − 1) *is correct Table* I(**m**) *is correct.*

Set $n = \mathbf{m}$. To prove this theorem, it is sufficient to show that if Table I(**m** − 1) is correct and $n = \mathbf{m}$, then

$$R_0(S_n) = 1, \qquad n > 0, \tag{29.12}$$

and

$$R_q(S_n) = \delta_n{}^q, \qquad q = 1, 2, \ldots. \tag{29.13}$$

Relation (29.12) has already been established. To establish (29.13), suppose that S_n is an n-sphere in a Euclidean $(n+1)$-space of

---

† When the index k is on the range $1, \ldots, \mathbf{m}$ and (29.7) is true.

coordinates $x_0, x_1, \ldots, x_n$ and let $p \to f(p)$ be the ND C^∞-function defined on S_n by assigning the value x_n to the point p. One sees that f has two critical values, a and b, taken on by f at critical points p_a and p_b with indices, respectively, 0 and n. The set $f_b = \Sigma_n$. The set $\dot{f}_b = f_b - p_b$ has the connectivities of the point p_a, since $\dot{f}_b$ admits a deformation retracting $\dot{f}_b$ onto p_a (Corollary 27.2).

Since $n = \mathbf{m}$, Corollary 29.1**m** (with p_b replacing p_a) implies that there exists an n-cap v^n of p_b. The boundary ∂v^n is on $\dot{f}_b$ and is bounding on $\dot{f}_b$. The critical point p_b is accordingly of linking type. It follows from (i) and (ii) of Theorem 29.3**m** that with $f_b = \Sigma_n$

$$R_q(f_b) = R_q(\dot{f}_b) + \delta_n{}^q = \delta_n{}^q, \qquad q = 1, 2, \ldots, \tag{29.14}$$

thereby verifying (29.13).

Thus Theorem 29.4**m** is true.

Hence the extended Tables I *and* II *are correct. It follows that Theorem* 29.3 *and its corollary, Theorem* 29.2, *are true.*

A Comparison of Relative Homology Groups. In the preceding part of this section we have determined the effect on the homology groups on f_a of deleting from f_a the critical point p_a at the f-level a. If η is a value of f, $\eta < a$, there are applications in which one needs similarly to determine the effect on the homology groups on $f_a \bmod f_\eta$ of deleting p_a from f_a. We shall state modifications of Theorems 29.2 and 29.3 and indicate how the proofs of Theorems 29.2 and 29.3 lead to proofs of these modified theorems. These extensions are not used until they are applied in §33 in the study of critical chords and symmetric products.

Among critical points p_a of index k we here distinguish three types:

Linking Types mod f_η. Characterized by the condition that for each k-cap ζ^k of p_a, $\partial\zeta^k$ is null or bounding on $\dot{f}_a \bmod f_\eta$.

Nonlinking Types mod f_η. Characterized by the condition that for each k-cap ζ^k of p_a, $\partial\zeta^k$ is neither null nor bounding on $\dot{f}_a \bmod f_\eta$.

Neutral Types mod f_η. With the point p_a of neither linking nor nonlinking types mod f_η; proved nonexistent (cf. Lemma 29.2**m**).

Departing from the notation of (29.9), let

$$\mathbf{b}_q(f_a, f_\eta) \qquad \text{and} \qquad \mathbf{b}_q(\dot{f}_a, f_\eta), \qquad q > 0, \tag{29.15}$$

denote homology prebases of q-cycles, respectively, on $f_a \bmod f_\eta$ and on $\dot{f}_a \bmod f_\eta$ (cf. Definition 28.4b). Similarly, departing from the notation of (29.8), with each critical point p_a and integer $q > 0$ we associate the difference

$$\Delta\mathscr{R}_q = R_q(f_a\,, f_\eta) - R_q(\dot{f}_a\,, f_\eta) \tag{29.16}$$

(cf. Definition 28.4a) whenever the numbers differenced are finite, and state the following modification of Theorem 29.2:

Theorem 29.2*. *Corresponding to a critical point p_a of index k for which the differences* (29.16) *exist, the differences $\Delta\mathscr{R}_q$ are all zero except that*

$$\begin{aligned} \Delta\mathscr{R}_k &= 1 \qquad \textit{when } p_a \textit{ is of linking type}, \bmod f_\eta\,, \\ \Delta\mathscr{R}_{k-1} &= -1 \qquad \textit{when } p_a \textit{ is of nonlinking type}, \bmod f_\eta\,. \end{aligned} \tag{29.17}$$

Theorem 29.2* follows from a modification of Theorem 29.3 which will be stated below. The proofs of the two modified theorems are given by the proofs of Theorems 29.2 and 29.3 read with the following replacements:

"cycles" or "absolute cycles"	by	"cycles mod f_η"
"on f_a" and "on $\dot{f}_a$"	by	"on $f_a \bmod f_\eta$" and "on $\dot{f}_a \bmod f_\eta$," respectively
"linking"	by	"linking mod f_η"
"nonlinking"	by	"nonlinking mod f_η"
"$\mathbf{b}_q(f_a)$" and "$\mathbf{b}_q(\dot{f}_a)$"	by	"$\mathbf{b}_q(f_a\,, f_\eta)$" and "$\mathbf{b}_q(\dot{f}_a\,, f_\eta)$," respectively.

One should leave "on $f_a \bmod \dot{f}_a$" *unchanged.* Reference to the induction with respect to the integer $\mathbf{m}$ can be deleted because the inductive verification of Tables I and II has been completed.

The modification of Theorem 29.3 takes the following form:

Theorem 29.3*. (i) *If the critical point p_a has the index k, a homology prebase $\mathbf{b}_q(\dot{f}_a\,, f_\eta)$ is a homology prebase $\mathbf{b}_q(f_a\,, f_\eta)$ provided there occurs neither: Case* 1. *$q = k$ with p_a linking* mod f_η, *nor case* 2, *$q = k - 1$ with p_a nonlinking* mod f_η ($k > 0$).

(ii) *In case* 1 *any set of k-cycles* mod f_n *of the form*

$$\mathbf{b}_k(\dot{f}_a, f_n) \cup \lambda^k$$

is a homology prebase $\mathbf{b}_k(f_a, f_n)$ (*where* λ^k *is a linking k-cycle* mod f_n *of* p_a).

(iii) *In case* 2 *any set of* $(k-1)$-*cycles* mod f_n *of the form*

$$\mathbf{b}_{k-1}(\dot{f}_a, f_n) - \omega^{k-1}$$

is a homology prebase $\mathbf{b}_{k-1}(f_a, f_n)$ (*where* ω^{k-1} *is the boundary of a k-cap of* p_a *and* $\mathbf{b}_{k-1}(\dot{f}_a, f_n)$ *contains* ω^{k-1}).

§ 30

TYPE NUMBERS AND CONNECTIVITIES

The Compact Case. Let M_n be a compact C^∞-manifold. Given a ND $f \in C^\infty(M_n)$, for each nonnegative integer k we introduce the following symbols: m_k is the number of critical points of f of index k; a_k is the number of critical points of f of index k of linking type; b_k is the number of critical points of f of index k of nonlinking type; R_k is the kth connectivity of M_n.

We shall prove a basic theorem:

Theorem 30.1. (i) *The connectivities R_k of M_n are finite and for $k > n$, $R_k = 0$.*

(ii) *Between the numbers m_k and R_k there exist the relations*

$$\begin{aligned} m_0 &\geqslant R_0 \\ m_1 - m_0 &\geqslant R_1 - R_0 \\ m_2 - m_1 + m_0 &\geqslant R_2 - R_1 + R_0 \\ &\vdots \\ m_n - m_{n-1} + m_{n-2} \cdots (-1)^n m_0 &= R_n - R_{n-1} + R_{n-2} \cdots (-1)^n R_0 \end{aligned} \tag{30.1}$$

(iii) *The relations* (30.1) *imply the inequalities*

$$m_k \geqslant R_k, \qquad k = 0, 1, \ldots, n. \tag{30.2}$$

(iv) *If $|M_n|$ is connected, $R_n = 1$ or 0.*†

† Depending on the choice of the field $\mathscr{K}$ for some manifolds.

(v) *If* $|M_n|$ *is connected and if* z^n *is an n-cycle on* $|M_n|$ *which is nonbounding on* $|M_n|$, *then* $|z^n| = |M_n|$.

Without loss of generality in proving this theorem we can suppose that each critical value of f is assumed at just one critical point. Let $c_0 < c_1 < \cdots < c_r$ be the critical values of f. Note that $b_0 = 0$, since each critical point of index 0 is of linking type. Let $\dot{f}_{c_j}$ denote the subset f_{c_j} of M_n with the critical point of f at the f-level c_j deleted.

Proof of (i) As a subset X of $|M_n|$ becomes the successive sets

$$f_{c_0}, \dot{f}_{c_1}; f_{c_1}, \dot{f}_{c_2}; f_{c_2}, \dot{f}_{c_3}; \cdots ; f_{c_{r-1}}, \dot{f}_{c_r}; f_{c_r} \tag{30.3}$$

we shall see by an inductive procedure that $R_q(X)$ takes on integral values terminating with the value $R_q = R_q(M_n)$. There is no change in $R_q(X)$ as X changes from $f_{c_{j-1}}$ to $\dot{f}_{c_j}$, by virtue of the retraction of $\dot{f}_{c_j}$ onto $f_{c_{j-1}}$ (Corollary 23.1) and the resultant isomorphisms

$$H_q(\dot{f}_{c_j}) \approx H_q(f_{c_{j-1}}) \tag{30.4}$$

(see Theorem 28.4).

Understanding that $b_0 = 0$ and setting $a_j = b_j = 0$ for $j > n$, we shall verify the equalities

$$R_q = a_q - b_{q+1}, \qquad q = 0, 1, \ldots . \tag{30.5}$$

Relations (30.5) follow on evaluating the differences

$$R_q(f_{c_j}) - R_q(\dot{f}_{c_j}), \qquad j = 1, 2, \ldots, r, \tag{30.6}$$

by means of Theorem 29.2. This evaluation is made in the order of the integers j and shows that the numbers differenced in (30.6) are finite and that the relations (30.5) hold.

Statement (i) follows from (30.5).

Proof of (ii). For each k on the range $0, 1, \ldots, n$ set $\mathscr{E}_k = m_k - R_k$. The relations

$$m_k = a_k + b_k, \qquad k = 0, 1, \ldots, n, \tag{30.7}$$

and (30.5), with $b_0 = 0$, imply that $\mathscr{E}_k = b_k + b_{k+1}$ and

$$\mathscr{E}_k - \mathscr{E}_{k-1} + \mathscr{E}_{k-2} \cdots (-1)^k \mathscr{E}_0 = b_{k+1}, \qquad k = 0, 1, \ldots, n. \tag{30.8}$$

The inequalities in (30.1) follow from the relations (30.8). The final equality of (30.1) is a consequence of the final equality in (30.8) and the vanishing of b_{n+1}.

Proof of (iii). The relations (30.2) are a trivial consequence of (30.1).

Proof of (iv). To establish (iv), we shall make use of a result established by Morse [9] using methods of the character of those used in this book. We refer to the theorem that there exists a ND f in $C^{\infty}(M_n)$ which is "*polar-nondegenerate*," i.e., which possesses just one critical point of index 0 and just one critical point of index n. According to (30.5) $R_n = a_n$, and, since M_n admits a polar-ND f, $a_n = 1$ or 0 depending on whether the point q of absolute maximum of f on $|M_n|$ is of linking or nonlinking type.

Proof of (v). Let z^n be an n-cycle on $|M_n|$ which is nonbounding but whose carrier $|z^n|$ is not equal to $|M_n|$. We shall arrive at a contradiction.

By hypothesis there exists a point $q \in |M_n|$ which is not in $|z^n|$. The theorem on the existence of a polar-ND f on $|M_n|$ can be strengthened by the affirmation that the absolute maximum of f can be prescribed in position on $|M_n|$. We suppose then that the absolute maximum of f occurs at the above point q.

We have seen in (iv) that $R_n = 1$ or 0. Since z^n is nonbounding on $|M_n|$ by hypotheses of (v), we conclude that $R_n = 1$ in (v), and hence that $a_n = 1$ in (30.5).

Because $a_n = 1$ there is associated with the critical point q a linking n-cycle λ^n. Hence for some constant $\rho \in \mathscr{K}$

$$\lambda^n \sim \rho z^n + \partial w^{n+1}, \qquad w^{n+1} \text{ on } |M_n|. \tag{30.9}$$

By hypothesis ρz^n is on $|M_n| - q$. Hence (30.9) implies that $\lambda^n \sim 0$ on $|M_n|$, mod($|M_n| - q$), so that λ^n cannot be an n-cap of q (Definition 29.1). Hence λ^n cannot be a linking n-cycle of q (Definition 29.2).

From this contradiction we infer that the point q does not exist and that (v) is true.

This completes the proof of Theorem 30.1.

We shall derive a special consequence (30.10) of the relations (30.1).

Given f as in Theorem 30.1, set $f' = -f$. The type numbers of f' are then the respective numbers m_n, $m_{n-1}, \ldots, m_0$. Hence the relations (30.1) hold if m_k is replaced by m_{n-k} for each k on the range $0, \ldots, n$. A special consequence is that when n is odd

$$R_0 - R_1 + R_2 - \cdots (-1)^n R_n = 0. \tag{30.10a}$$

The Noncompact Case. Let M_n be a C^∞-manifold and g a ND function in $C^\infty(M_n)$. Suppose that for some *ordinary* value β of g the subset

$$g_\beta = \{p \in |M_n| \mid g(p) \leqslant \beta\} \tag{30.10b}$$

of M_n is compact. We shall verify the following concerning g_β:

Theorem 30.2. (i) *The connectivities R_k of the singular complex $S(g_\beta)$ are finite and vanish for $k > n$.*

(ii) *If m_k denotes the number of critical points of $g \mid g_\beta$ of index k, the relations* (30.1) *hold.*

(iii) *The inequalities* (30.1), *affirmed to hold in* (ii), *imply the relations $m_k \geqslant R_k$.*

The proof of this theorem is similar to the proof of (i)–(iii) of Theorem 30.1.

Proof of Theorem 9.1. We shall show that Theorem 9.1 is a corollary of Theorem 30.2.

In Theorem 9.1 there is given a "regular" C^∞-domain Z in E_{n+1} bounded by a "regular" n-manifold Σ of class C^∞ (Definition 9.1). On some open neighborhood D_f of Z there is given a real-valued function f of class C^2 which is ND on Z and ordinary on Σ. The function f of Theorem 9.1 is "admissible" relative to Z and grad f is emergent at each point of Z.

The reader will recall that once Theorem 9.1 is established its variants in §9, §10, and Appendix II admit the supplementary proofs given in Part I.

To apply Theorem 30.2 to prove Theorem 9.1, an introductory lemma is required.

Approximation Lemma 30.1. *Corresponding to a function f admissible in Theorem* 9.1 *there exists a real-valued function f' of class C^∞, defined on the domain D_f of f, with first and second partial derivatives approximating the first and second partial derivatives of f so closely on Z that* grad f' *is emergent on Σ and the critical points of $f' \mid Z$ are* ND *and correspond biuniquely to those of $f \mid Z$ with preservation of indices.*

A classical analytic or trigonometric approximation of f will suffice. The domain D_f of f' is open and the approximation is on the compact subset Z of D_f.

The Modification g of f'. We cannot apply Theorem 30.2 directly to f'. A modification g of f' is called for.

By virtue of Theorem 12.2 there exists a ND function g of class C^∞ defined on an open neighborhood D_g of Z and such that the following are true:

1. The function $g \mid Z$ is "critically equivalent" to $f' \mid Z$ in the sense of Definition 12.1.

2. The boundary Σ of Z is a level manifold of g at each point of which g is ordinary.

3. The value β of g on Σ exceeds the value of g at each point $x \in \mathring{Z}$.

4. The neighborhood D_g of Z is so small that β is less than the value of g at each point of $D_g - Z$.

We apply Theorem 30.2 to g. The function g is defined on the C^∞-manifold M_n with carrier D_g and with the Euclidean differentiable structure (Definition 13.5). As defined above $g \mid Z$ is "critically equivalent" to $f \mid Z$. Since the set $g_\beta = Z$ is compact, Theorem 30.2 applies and yields Theorem 9.1.

We verify a corollary of Theorem 30.2.

Corollary 30.1. *The connectivities R_r, $r \geqslant n$, of a connected non-compact C^∞-manifold M_n are zero.*

Proof. Let f be a ND $f \in C^\infty(M_n)$ with the properties (i)–(iii) of Theorem 23.5. Corresponding to each ordinary value a of f one has the relations

$$R_r(f_a) = 0, \qquad r > n, \tag{30.11}$$

by virtue of Theorem 30.2. Moreover, (30.11) holds when $r = n$, as we now verify.

The number m_n of critical points of $f \mid f_a$ of index n is equal to 0, as implied by (iii) of Theorem 23.5. According to Theorem 30.2 $m_n \geqslant R_n(f_a)$ [cf. (30.2)], so that (30.11) holds when $r = n$.

Since (30.11) holds for each ordinary value a of f, Corollary 30.1 follows.

We shall characterize an important special situation.

Definition 30.1. *A Lacunary Index of f.* Let M_n be a compact C^∞-manifold and f ND in $C^\infty(M_n)$. If the type numbers of f are such that for some i on the range 0, 1,..., n the type numbers m_{i+1} and m_{i-1} adjacent to m_i vanish, i will be called a *lacunary index* of f (see Morse [7], p. 151).

Corollary 30.2. *Let M_n be a compact C^∞-manifold and f* ND *in $C^\infty(M_n)$. If i is a lacunary index of f, then $m_i = R_i$ and $R_{i-1} = R_{i+1} = 0$.*

Under the conditions of Theorem 30.1 it follows from the relations (30.1) that

$$(m_{j+1} - R_{j+1}) + (m_{j-1} - R_{j-1}) \geqslant m_j - R_j\,, \qquad 0 \leqslant j \leqslant n, \quad (30.12)$$

with the understanding that m_k and R_k vanish when $k < 0$ or $k > n$. By the lacunary hypothesis $m_{i+1} = m_{i-1} = 0$. It follows from (30.2) that $R_{i+1} = R_{i-1} = 0$ and $m_i \geqslant R_i$. We infer from (30.12) that $m_i = R_i$ for the given i.

This establishes Corollary 30.2.

As we shall see, the case arises in which each of the nonvanishing type numbers of f has an even index. In such cases for each index i $R_i = m_i$.

The result of Corollary 30.2 *will be referred to as the Lacunary Principle.*

The Case of Relative Connectivities. The connectivities R_k entering into the relations of Theorem 30.1 and 30.2 have been "absolute" connectivities. With the aid of Theorem 29.3* a similar set of relations may be derived involving the relative connectivities introduced in Theorem 29.3*. These relations are needed in studying "symmetric products" and "critical chords" of differentiable manifolds in §33.

Hypotheses of Theorem 30.3. There is given a C^∞-manifold M_n, a function $f \in C^\infty(M_n)$, and an ordinary value η of f such that the subset

$$\{p \in |M_n| \mid f(p) \geqslant \eta\} = f_{\eta^+} \tag{30.13}$$

of $|M_n|$ is compact. We assume that f is ND on f_{η^+}. There are thus three conditions on the value η of f:

$$\eta \text{ is ordinary}; \qquad f_{\eta^+} \text{ is compact}; \qquad f \text{ is ND on } f_{\eta^+}. \tag{30.14}$$

Theorem 30.2 is replaced by the following theorem:

Theorem 30.3. *Given a C^∞-manifold M_n, an $f \in C^\infty(M_n)$, and a value η of f such that conditions* (30.14) *are satisfied, the following are true*:

(i) *The connectivities R_i' of* $|M_n| \bmod f_\eta$ *are finite and vanish for $i > n$.*

(ii) *Between the type numbers m_k of the critical points p of f at which $f(p) > \eta$ and the connectivities R_k' the relations* (30.1) *hold.*

Affirmations (iii)–(v) of Theorem 30.1 are also true under the hypotheses of Theorem 30.3 provided R_k is replaced by R_k' and (v) is appropriately modified. We shall not need these modifications of (iii)–(v) of Theorem 30.1.

Proof of Theorem 30.3(i). This proof is similar to the proof of Theorem 30.1(i) except for minor modifications due to the presence of the modulus f_η. We proceed with the proof following the logical order of the proof of Theorem 30.1(i).

In addition to the numbers m_k and R_k' already defined we redefine two symbols: a_k, the number of critical points of f on f_{η^+} of index k and of "linking type mod f_η" (see §29); and b_k, the number of critical points of f on f_{η^+} of index k and of "nonlinking type mod f_η" (see §29).

Without loss of generality we can suppose that each critical value $a > \eta$ is assumed at just one critical point. Let $c_0 < c_1 < \cdots < c_r$ be the critical values of f which exceed η. For each critical value $a = c_i$ let p_a be the critical point at the f-level a. As previously, set $\dot{f}_a = f_a - p_a$. Let $\mathbf{b}_q(f_a, f_\eta)$ and $\mathbf{b}_q(\dot{f}_a, f_\eta)$ be homology prebases of q-cycles mod f_η, respectively, on f_a and $\dot{f}_a$, as in Theorem 29.3*.

Departing from the proof of Theorem 30.1, note that when $a = c_0$, $\dot{f}_a$ can be retracted by a deformation onto f_η. It follows from Theorem 28.4 that for each q, $\mathbf{b}_q(\dot{f}_{c_0}, f_\eta) = \varnothing$.

As in the proof of Theorem 30.1, the integer $b_0 = 0$, since each critical point of index 0 carries a 0-cycle and so is of linking type and, in particular, of linking type mod f_η.

As a subset X of $|M_n|$ takes on the successive sets listed in (30.3), taken with present connotations, we shall see by an inductive procedure that $R_q(X, f_\eta)$ takes on finite integral values, terminating with the value R_q'. There is no change in $R_q(X, f_\eta)$ as X changes from $f_{c_{j-1}}$ to $\dot{f}_{c_j}$, by virtue of the retraction of $\dot{f}_{c_j}$ onto $f_{c_{j-1}}$ (cf. Theorem 28.4).

We understand that $b_0 = 0$ and that $a_j = b_j = 0$ for $j > n$, since there are no critical points of index exceeding n. We shall verify the equalities

$$R_q' = a_q - b_{q+1}, \qquad q = 0, 1, \ldots. \tag{30.15}$$

Relations (30.15) follow on evaluating the differences

$$R_q(f_{c_j}, f_\eta) - R_q(\dot{f}_{c_j}, f_\eta), \qquad j = 1, 2, \ldots, r,$$

by means of Theorem 29.3* (or Theorem 29.2*). Theorem 30.3(i) is implied by (30.15) and the vanishing of a_j and b_j for $j > n$.

Proof of Theorem 30.3(ii). This proof is identical in form with the proof of Theorem 30.1(ii).

Thus Theorem 30.3 is true.

Simply-Carried Separate k-Cells. We shall present a lemma on nonbounding k-cycles on a topological k-sphere. This lemma has important applications. It is based on the following definitions.

Definition 30.2. *Simply-Carried Singular k-Cells.* Let χ be a Hausdorff space. A singular k-simplex on χ which is defined by a homeomorphism $\tau : s \to \chi$ of an ordered k-simplex s into χ will be said to be *simply-carried* by χ, as will the corresponding singular k-cell $\mathring{\tau}$.

Simply-Carried n-Cells of Γ_n. Let Γ_n be a topological n-manifold, and let σ^n be a singular n-cell simply-carried by Γ_n. The cell σ^n is the equivalence class of a singular simplex $\tau : s \to \Gamma_n$, where s is an ordered n-simplex in E_n (see Definition 26.1). Hence $| \sigma^n | = \tau(| s |)$.

A Convention. We understand that E_0 is a point and Γ_0 a finite set of points, both with a discrete topology.

We shall make use of a classical theorem affirming that a homeomorphism of an open subset A of E_n into E_n' is onto an open subset A' of E_n' (see Hurewicz and Wallman [1], p. 97). By virtue of this theorem and the definition of a topological manifold the subset $\tau(\text{Int} \, | s |)$ of Γ_n is open in Γ_n, and

$$\tau(\text{Int} \, | s |) = \text{Int} \, | \sigma^n |, \tag{30.16}$$

where Int $| \sigma^n |$ is the maximal open subset of $| \sigma^n |$ relative to Γ_n. If $n = 0$, the set (30.16) is a point.

Definition 30.3. *A Simply-Carried Separate n-Cell of an n-Chain.* Let z^n be a singular n-chain on Γ_n with a reduced form

$$z^n = e_1\sigma_1{}^n + \cdots + e_m\sigma_m{}^n, \qquad m > 1, \quad e_i = \pm 1. \tag{30.17}$$

If a cell of this form, say $\sigma_1{}^n$, is simply-carried and if Int $| \sigma_1{}^n |$ is not included in the subset

$$X = | \sigma_2{}^n | \cup \cdots \cup | \sigma_m{}^n | \tag{30.18}$$

of Γ_n, then $\sigma_1{}^n$ will be called a *simply-carried separate n-cell of z^n.*

Lemma 30.2. *Let z^n be a nontrivial n-cycle on a topological n-sphere Γ_n. If there exists a simply-carried separate n-cell $\sigma_1{}^n$ of z^n, then $z^n \not\sim 0$ on Γ_n and $| z^n | = \Gamma_n$.*

We understand that Γ_0 is a pair of distinct points.

Without loss of generality in proving the lemma we can suppose that Γ_n is an origin-centered n-sphere S_n in E_{n+1}. The proof is by induction with respect to n. The lemma is trivial when $n = 0$. We shall accordingly prove the theorem true for a prescribed $n > 0$, assuming the lemma true when n is replaced by $n - 1$.

A first consequence of the hypotheses is that when $n > 0$ there exists a point p in Int $| \sigma_1{}^n |$ which is not included in the set X of (30.18) and which has an open neighborhood, relative to Γ_n, which does not meet X.

Suppose that the coordinate x_1-axis in E_{n+1} is orthogonal to S_n at p, and that on S_n, x_1 assumes a proper maximum a at p. Let f be the ND function on S_n defined by the values of x_1 on S_n. Denote p by p_a. The point p_a is in the open subset Int $| \sigma_1{}^n |$ of Γ_n.

By virtue of our inductive hypothesis $\partial \sigma_1{}^n \not\sim 0$ on the topological $(n-1)$-sphere $| \partial \sigma_1{}^n |$. As the homeomorph of a geometric n-simplex $| \sigma_1{}^n |$ is the homeomorph of the n-disk Δ_n (of Lemma 29.0 if k of Lemma 29.0 is equal to n) under a mapping in which p_a can be taken as the image of the center of Δ_n and $| \partial \sigma_1{}^n |$ as the image of the geometric boundary $\beta \Delta_n$ of Δ_n. Now, $| \sigma_1{}^n |$ is the closure of Int $| \sigma_1{}^n |$, an open neighborhood of p_a, relative to Γ_n. It follows from Lemma 29.0 that $\sigma_1{}^n$ is an n-cap associated with the critical point p_a of f.

Now, $z^n = e_1 \sigma_1{}^n \bmod f_a$, so that z^n is an n-cap associated with p_a. Since z^n is both an n-cap associated with p_a and an absolute n-cycle on S_n, it is a "linking" n-cycle on S_n associated with p_a. By Theorem 29.3 $z^n \not\sim 0$ on S_n.

That $| z^n | = S_n$ when $n > 0$ follows from Theorem 30.1(v).

We continue with a topological n-manifold Γ_n.

Definition 30.4. *n-Chains Simply-Carried on Γ_n.* An n-chain z^n on Γ_n with a reduced form (30.17) will be said to be simply-carried by Γ_n if each cell $\sigma_i{}^n$ of the reduced form for z^n is simply-carried by Γ_n and if for each pair of distinct integers i and j on the range $1,\ldots, m$

$$\text{Int } | \sigma_i{}^n | \cap \text{Int } | \sigma_j{}^n | = \varnothing. \tag{30.19}$$

The following is a corollary of Lemma 30.2:

Lemma 30.3. *If Γ_n is a topological n-sphere and z^n a simply-carried n-cycle on Γ_n, then $z^n \not\sim 0$ on Γ_n and $| z^n | = \Gamma_n$.*

In §37 it will be seen that there exists a simply-carried n-cycle on each topological n-sphere.

EXERCISE 30.1. Let M_n and Q_m be two compact C^∞-manifolds of even dimension such that there exist real-valued ND functions $\varphi \in C^\infty(M_n)$ and $\psi \in C^\infty(Q_m)$ whose nonvanishing type numbers (for a given field $\mathscr{K}$) have even indices. Spheres of even dimensions serve as examples.

Show that on the product C^∞-manifold $M_n \times Q_m$ there exists a ND function $f \in C^\infty(M_n \times Q_m)$ each of whose non-vanishing type numbers has an even index and whose type numbers are then the respective connectivities of $M_n \times Q_m$.

Suggestion. Let p and q be arbitrary points of $| M_n |$ and $| Q_m |$ and $(p, q) \in | M_n \times Q_m |$. Choose φ and ψ, as is possible, so as to be positive-valued and show that the function $(p, q) \to \varphi(p) + \psi(q)$ is a ND function on $| M_n \times Q_m |$ which satisfies the exercise.

PART IV

OTHER APPLICATIONS OF CRITICAL POINT THEORY

§ 31

NORMALS FROM A POINT TO A MANIFOLD

Let M_n be a regular compact C^∞-manifold in E_m with $0 < n < m$. In Theorem 15.3 we have seen that the focal points of M_n in E_m are nowhere dense in E_m. Let q be a point fixed in $E_m - M_n$ and p an arbitrary point in M_n. We term the mapping

$$p \to \| p - q \| = f_q(p) : M_n \to R \tag{31.1}$$

the *distance function* f_q with *pole* q and domain M_n. By definition of a focal point the function f_q is ND if and only if q is not a focal point of M_n.

Let (q, ζ) be a straight arc orthogonal to M_n at a point $\zeta \in M_n$. We term (q, ζ) an arc normal to M_n at ζ and assign this arc an index equal to the index of ζ as a critical point of f_q. Theorem 30.1 then yields the following:

Theorem 31.1. *Let q be a point of $E_m - M_n$ not a focal point of M_n, and for i on the range* 0, 1,..., *n let m_i be the number of normal arcs (q, ζ) of index i, counting arcs (q, ζ) as different if their base points ζ are different.*

If R_i is the ith *connectivity of M_n, the relations* (30.1) *are satisfied, as are the relations*

$$m_i \geqslant R_i, \qquad i = 0,\dots, n. \tag{31.2}$$

An Example. Let M_2 be a torus in E_3. The focal points of M_2 are the points on the axis of the torus and on the central circle Γ of the corresponding solid torus. For the torus $R_0 = R_2 = 1$ and $R_1 = 2$, as we shall see. If q is not a focal point of the torus, nor on the torus, there are four straight arcs (q, ζ), normal to M_2 at points ζ of M_2. If q is in the plane of Γ and exterior to the circle of M_2 of maximum length, each of these normal arcs is a subarc of the longest of these arcs.

Given a normal arc (q, ζ), we shall evaluate the index of ζ as a critical point of the distance function f_q, beginning with the case in which $m = n + 1$.

In making this evaluation we shall admit translations or orthogonal transformations of the rectangular coordinates of E_m. It is understood that a point $p \in M_n$ and the pole q in (31.1) undergo the same changes of coordinates, so that if $x = (x_1, \ldots, x_n)$ and $a = (a_1, \ldots, a_n)$ represent, respectively, p and q, then $\| x - a \|$ is invariant under any admissible change of coordinates.

The Case $m = n + 1$. Centers of Curvature of M_n. When $m = n + 1$ we shall evaluate the index of a critical point $\zeta \in M_n$ of the distance function f_q in terms of the centers of principal normal curvature of M_n on the normal to M_n at ζ. Such centers must be defined.

Given a straight arc (q, ζ) normal to M_n at ζ, let a system of rectangular coordinates x be chosen in E_m such that $\zeta \in M_n$ is represented by the origin, and the pole q is represented by the point $(0, \ldots, 0, c) \in E_m$ with $(n + 1)$st coordinate $c = \| q - \zeta \|$. Setting $(x_1, \ldots, x_n) = (v_1, \ldots, v_n)$, there will exist for $\| v \|$ sufficiently small a Monge presentation

$$x_{n+1} = (a_{ij}/2)\, v_i v_j + + \tag{31.3}$$

of a neighborhood of the point ζ on M_n, in which $a_{ij} v_i v_j$ is a symmetric quadratic form and the remainder (indicated by $++$ throughout this proof) is a function $v \to L(v)$ of class C^∞, vanishing with its first and second partial derivatives at the origin.

Let $\rho_1, \rho_2, \ldots, \rho_n$ be the characteristic roots of the matrix $\| a_{ij} \|$. There then exists an orthogonal transformation of the coordinates $v_1, \ldots, v_n$ into coordinates $u_1, \ldots, u_n$, by virtue of which a neighborhood of ζ on M_n has a Monge presentation

$$x_{n+1} = \tfrac{1}{2}(\rho_1 u_1^2 + \cdots + \rho_n u_n^2) + + \tag{31.4}$$

for $\| u \| < e$ and e sufficiently small.

If the coordinates u_i are properly numbered and the roots ρ_i correspondingly numbered, then for some r on the range 0, 1,..., n

$$\begin{aligned} \rho_h &\neq 0, \qquad h = 1, 2, \ldots, r, \\ \rho_k &= 0, \qquad k = r+1, r+2, \ldots, n. \end{aligned} \tag{31.5}$$

Corresponding to each characteristic root $\rho_h \neq 0$ we set $\mathscr{R}_h = 1/\rho_h$ and term $\mathscr{R}_h$ a radius of *principal normal curvature* belonging to the point ζ on M_n. The point P_h on the normal λ_ζ to M_n at ζ whose coordinate $x_{n+1} = \mathscr{R}_h$ will be called a *center* of principal normal curvature of M_n belonging to ζ. Such a center will be counted with a *multiplicity* equal to the multiplicity of ρ_h as a characteristic root of $\| a_{ij} \|$. One sees that the centers $P_1, \ldots, P_r$ of principal normal curvature on the normal λ_ζ to M_n at ζ are uniquely determined, except for order, by M_n and $\zeta \in M_n$.

We shall prove the following theorem:

Theorem 31.2. *Suppose that* $m = n + 1$ *and that* (q, ζ) *is a straight arc joining* $q \in E_m - M_n$ *to* $\zeta \in M_n$, *orthogonal to* M_n *at* ζ, *with* ζ *a* ND *critical point of the distance function* f_q.

The index of ζ *as a critical point of* f_q *is then the number of centers of principal normal curvature of* M_n *belonging to* ζ *on the open arc* (q, ζ), *counting these centers with their multiplicities.*

To determine the index of ζ as a critical point of the distance function f_q, we make use of the representation (31.4) of M_n near ζ, denoting the right side of (31.4) by $\varphi(u)$. In terms of the coordinates $(x_1, \ldots, x_{n+1}) = (u_1, \ldots, u_n, x_{n+1})$ employed in (31.4) the pole q has a representation $(a_1, \ldots, a_{n+1}) = (0, \ldots, 0, c)$, with $c > 0$. Hence for points $x \in M_n$ with coordinates

$$(x_1, \ldots, x_{n+1}) = (u_1, \ldots, u_n, \varphi(u)), \tag{31.6}$$

$$\begin{aligned} \| x - a \|^2 - c^2 &= \| u \|^2 + | \varphi(u) - c |^2 - c^2 \\ &= (1 - \rho_1 c)u_1^2 + \cdots + (1 - \rho_r c)u_r^2 + u_{r+1}^2 + \cdots + u_n^2 + +. \end{aligned} \tag{31.7}$$

The critical point ζ of f_q is also a critical point of f_q^2 on M_n. Since $f_q(\zeta) \neq 0$, one sees that f_q and f_q^2 have the same index at their common critical point ζ.

The index of the quadratic form (31.7) is thus the index of ζ as a critical point of f_q. This index is clearly the number of the characteristic roots $\rho_h \neq 0$ for which $1 - \rho_h c < 0$, or, equivalently, the number of the positive radii $\mathscr{R}_h < c$.

Finally, no $\mathscr{R}_h = c$, since, by hypothesis q is not a focal point of M_n, so that ζ is a ND critical point of f_q, or, equivalently, (31.7) is a ND quadratic form. Hence $1 - \rho_h c \neq 0$ for $h = 1, 2, \ldots, r$, implying that no $\mathscr{R}_h = c$.

Theorem 31.2 follows.

Remark. Suppose, contrary to the hypothesis of Theorem 31.2, that the pole q is a center of principal normal curvature based on the point $\zeta \in M_n$. Then, for some h on the range $1, \ldots, r$, $1 - \rho_h c = 0$ and the multiplicity of the center q, as defined above, is the multiplicity of ρ_h as a characteristic root. One sees that this multiplicity of ρ_h is the nullity of the form (31.7), or, equivalently, the nullity μ of the critical point ζ.

The Case $m > n$. An Extension of Theorem 31.2. A point $q \in E_m - M_n$ has been called a focal point of M_n based on a point $\zeta \in M_n$ if ζ is a degenerate critical point of the distance function f_q.

Definition 31.1. *The nullity μ of a focal point q of M_n based on a point ζ of M_n is by definition the nullity of ζ as a critical point of f_q.*

The following theorem extends Theorem 31.2. It concerns the general case $m > n$ as distinguished from the special case $m = n + 1$.

Theorem 31.3. *Suppose that $m > n$ and that (q, ζ) is a straight arc from $q \in E_m - M_n$ to $\zeta \in M_n$, with ζ a ND critical point of f_q.*

The index of ζ as a critical point of f_q is then the number of focal points of M_n belonging to ζ on the open arc (q, ζ), counting these focal points with their nullities.

Proof when $m = n + 1$. In this case Theorem 31.3 is no more than a reinterpretation of Theorem 31.2 taking account of Definition 31.1 of the nullity of a focal point q of M_n. In fact, the *Remark* following the proof of Theorem 31.2 has the following consequence:

When $m = n + 1$ a center $q \in E_{n+1} - M_n$ of principal normal curvature of M_n is "based" on a degenerate critical point ζ in M_n and so is a focal point of M_n based on ζ. The multiplicity of q as a center of principal normal curvature based on ζ is by Definition 31.1 the nullity of q as a focal point of M_n based on ζ.

Thus when $m = n + 1$ Theorems 31.2 and 31.3 are equivalent.

Method of Proof of Theorem 31.3 *in the General Case.* Theorem 31.3 belongs to a class of theorems in global variational theory (Morse [13]) capable of a proof by special methods involving "broken primary and secondary extremals." It will be proved in a sequel to the present book concerned with this variational theory. The proof involves an examination of the variation of the "index" of a "critical extremal" [here the arc (q, ζ)] as the endpoints of the extremal vary.

A Second Proof of Theorem 31.3 Consider the case in which $m > n + 1$.

Notation. Suppose that the critical point ζ is at the origin of coordinates in E_m and that the n-plane E_n of coordinates $x_1, \ldots, x_n$ is tangent to M_n at ζ. Suppose further that the pole q is on the x_{n+1}-axis with $x_{n+1} = c > 0$.

For $\| v \|$ sufficiently small a neighborhood N of ζ relative to M_n admits a Monge presentation

$$\begin{aligned} (x_1, \ldots, x_n) &= (v_1, \ldots, v_n) \\ x_{n+\alpha} &= (a_{ij}^{\alpha}/2)\, v_i v_j + +, \qquad \alpha = 1, \ldots, m - n. \end{aligned} \tag{31.8}$$

Let the coordinates $(x_1, \ldots, x_n) = (v_1, \ldots, v_n)$ be orthogonally transformed into coordinates $(u_1, \ldots, u_n)$ such that (31.4) holds as before. We adopt the notation of (31.5).

In terms of the new rectangular coordinates $u_1, \ldots, u_n$ the neighborhood N of the origin, relative to M_n, admits a Monge presentation replacing the Monge presentation (31.8). With only a slight additional complexity of reasoning one finds that at a point $p(u) \in N$, represented by u, $f_q^2(p(u)) - c^2$ is equal to the form (31.7) plus a remainder $R(u)$ of the same character as in the case $m = n + 1$. The index of ζ, as a critical point of f_q or f_q^2, is the number of roots $\rho_h \neq 0$ for which $1 - c\rho_h < 0$, counting these roots with their multiplicities as characteristic roots of the matrix $\| a_{ij}^1 \|$.

An Interpretation. Let c' be a positive number and let q' be the point on the x_{n+1} axis at which $x_{n+1} = c'$. If $1 - \rho_h c' = 0$ for some h on the range $1,\ldots, r$, the form (31.7), with c' replaced by c, is degenerate. That is, by Definition 15.1 the point q' is a focal point of M_n belonging to ζ. If ρ_h has the multiplicity μ as a characteristic root of $\| a_{ij}^1 \|$, we see that the form (31.7), with c replaced by c', has the nullity μ, since just μ of the n terms in (31.7) would then vanish. By virtue of Definition 31.1 the focal point q' is "counted" with a nullity μ.

The index of the form 31.7 is thus the number of focal points of M_n "belonging to ζ" on the open arc (q, ζ), counting these focal points with their nullities.

The interpretation of focal points in the variational theory is nearer their interpretation in geometric optics.

§ 32

EQUILIBRIUM POINTS OF AN ELECTROSTATIC POTENTIAL

We return to the electrostatic potential

$$x \to V(x) = \left(\frac{\eta_1}{r_1} + \cdots + \frac{\eta_\mu}{r_\mu}\right) + \left(\frac{\zeta_1}{\rho_1} + \cdots + \frac{\zeta_\nu}{\rho_\nu}\right); \qquad x \in E_3\,, \tag{32.1}$$

introduced in (6.15). The positive numbers $\eta_1 ,..., \eta_\mu$ represent positive charges of electricity at the respective points $p^{(1)},..., p^{(\mu)}$ in E_3, while the negative numbers $\zeta_1 ,..., \zeta_\nu$ represent negative charges of electricity at the points $q^{(1)},..., q^{(\nu)}$ in E_3. Given a point $x \in E_3$ distinct from each of the points $p^{(i)}$ or $q^{(j)}$, we have set

$$r_i = \| p^{(i)} - x \|, \qquad \rho_j = \| q^{(j)} - x \|, \qquad i = 1,..., \mu; j = 1,..., \nu. \tag{32.2}$$

We seek the *points of equilibrium* of the electrostatic force defined by V, that is, the critical points of V (see Kiang [1]).

An Example. The potential V can have degenerate critical points. This is the case when $\mu = 4$, $\nu = 0$, and the points $p^{(1)}$, $p^{(2)}$, $p^{(3)}$, $p^{(4)}$ are the respective points

$$(1, 0, 0); \qquad (-1, 0, 0); \qquad (0, 1, 0); \qquad (0, -1, 0). \tag{32.3}$$

Take charges $\eta_1 = \eta_2 = \eta_3 = \eta_4 = 1$. The only critical point of V is then the origin, and this point is degenerate, as one readily verifies.

The Hypothesis of Nondegeneracy. We shall assume that the potential V is ND. This assumption is valid for almost all choices of the charged points, in accord with Theorem 6.2.

With the aid of Theorem 6.5 the reader can verify the following. If the charges and each of the charged points are fixed, excepting one such point, say q, then the points $q \in E_3$ for which V is degenerate are nowhere dense.

V a Harmonic Function. That the potential V satisfies Laplace's equation

$$\frac{\partial^2 V}{\partial x_1 \, \partial x_1} + \frac{\partial^2 V}{\partial x_2 \, \partial x_2} + \frac{\partial^2 V}{\partial x_3 \, \partial x_3} = 0 \tag{32.4}$$

at points x at which V is defined is readily verified. It is a classical property of a nonconstant harmonic function that it assumes no relative minimum or maximum at an interior point of its domain of definition (see Kellogg [1], p. 223). A particular consequence is that there are no ND critical points of index 0 or 3.

The principal theorem of this section follows.

Theorem 32.1. *Let there be given a* ND *electrostatic potential V of form* (32.1) *such that*

$$(\zeta_1 + \zeta_2 + \cdots + \zeta_\nu) + (\eta_1 + \eta_2 + \cdots + \eta_\mu) < 0. \tag{32.5}$$

Let m_2 be the number of critical points of V of index 2 *and m_1 the number of index* 1. *Then*

$$m_2 \geqslant \mu; \qquad m_1 \geqslant \nu - 1; \qquad m_1 - m_2 = \nu - \mu - 1. \tag{32.6}$$

Proof of Theorem 32.1. The proof makes use of Theorem 9.1 and of two lemmas, of which the first follows:

Lemma 32.1. *If σ is a sufficiently large positive constant, then under the conditions of Theorem* 32.1 $\| x \| < \sigma$ *at each charged point $p^{(i)}$ and $q^{(j)}$ and at each critical point of V, while* grad V *is emergent at each point of the* 2*-sphere Σ_σ on which $\| x \| = \sigma$.*

A first condition on σ is that it be so large that $\| x \| < \sigma$ *at each charged* point $p^{(i)}$ or $q^{(j)}$. At each point x at which $\| x \| \geqslant \sigma$, grad $V = \mathbf{g}(x) = (g_1(x), g_2(x), g_3(x))$ is well-defined. Grad V is emergent at each point of Σ_σ if and only if the "dot-product"

$$\mathbf{g}(x) \cdot \mathbf{x} = g_k(x) \cdot x_k > 0, \qquad \| x \| = \sigma, \tag{32.7}$$

summing with respect to k on the range 1, 2, 3. If in addition $\mathbf{g}(x) \cdot \mathbf{x} > 0$ whenever $\| x \| \geqslant \sigma$, the critical points of V are interior to the 2-sphere Σ_σ.

It will be convenient to write $V(x)$ in the form

$$V(x) = \frac{\eta_i}{r_i(x)} + \frac{\zeta_j}{\rho_j(x)} \tag{32.8}$$

summing with respect to i and j for i and j on the ranges $1,..., \mu$ and $1,..., \nu$, respectively. If the terms indexed by k on the range $1,..., n$ are summed as in (32.7) and terms indexed by i and j are summed as in (32.8), then (32.7) takes the form

$$\mathbf{g}(x) \cdot \mathbf{x} = \frac{-\eta_i}{r_i^3(x)} (x_k - p_k^{(i)})x_k - \frac{\zeta_j}{\rho_j^3(x)} (x_k - q_k^{(j)})x_k \,. \tag{32.9}$$

It follows from (32.9) that

$$\lim_{\|x\| \uparrow \infty} \| x \| (\mathbf{g}(x) \cdot \mathbf{x}) = -(\eta_1 + \cdots + \eta_\mu) - (\zeta_1 + \cdots + \zeta_\nu) > 0, \tag{32.10}$$

in accord with (32.5).

It follows from (32.10) that if σ is sufficiently large, $\mathbf{g}(x) \cdot \mathbf{x} > 0$ for $\| x \| \geqslant \sigma$, and the lemma follows.

The Auxiliary Function f. The proof of Theorem 32.1 depends upon the replacement of V by a function f conditioned as in Lemma 32.2. Theorem 9.1 is applicable to f restricted to the closed n-ball D_σ bounded by Σ_σ.

Lemma 32.2. *Corresponding to a potential V of form* (32.1) *there exists a real-valued function f of class C^∞ in E_3 such that:*

(A_1) *$f(x) = V(x)$ except at most in sufficiently small open disjoint spherical neighborhoods*

$$N_1 ,..., N_\mu : \qquad N_1' ,..., N_\nu' \tag{32.11}$$

of the respective points $p^{(1)},..., p^{(\mu)}$; $q^{(1)},..., q^{(\nu)}$.

(A_2) *f is ordinary at each point in the respective neighborhoods* (32.11) *except for* ND *critical points of index* 3 *at the points* $p^{(1)},\ldots, p^{(\mu)}$ *and of index* 0 *at the points* $q^{(1)},\ldots, q^{(\nu)}$.

We suppose that the spherical neighborhoods (32.11) are so small that V is ordinary at each point of these neighborhoods at which it is defined.

To define f of Lemma 32.2, we shall define f separately on each of the open neighborhoods (32.11). At each point x not in one of the neighborhoods (32.11) we set $f(x) = V(x)$.

Definition of $f \mid N_1$. Let e be a positive constant so small that the following is true: Each point x for which $0 < \| x - p^{(1)}\| \leqslant e$ shall be in N_1 and be such that $V(x) > 0$ and V have a negative directional derivative on the ray emanating from $p^{(1)}$. Let η then be a constant such that $0 < \eta < e$, and set

$$\omega = \max_{x\in A_e}\{V(x) + e^2\}, \qquad \text{where} \quad A_e = \{x \in E_3 \mid \| x - p^{(1)}\| = e\}. \tag{32.12}$$

The choices of e, η, and ω are such that

$$\omega - V(x) - \| x - p^{(1)}\|^2 \geqslant \omega - V(x) - e^2 \geqslant 0, \quad \eta \leqslant \| x - p^{(1)}\| \leqslant e. \tag{32.13}$$

The Auxiliary Function φ. To define $f \mid N_1$, we shall make use of an even mapping φ of class C^∞ of the real axis onto the interval [0, 1] such that

$$\begin{aligned} \varphi(t^2) &= 1, && 0 \leqslant t \leqslant \eta, \\ \varphi(t^2) &= 0, && t \geqslant e, \\ \varphi'(t^2) &< 0, && \eta < t < e. \end{aligned} \tag{32.14}$$

With φ so defined set $r(x) = \| x - p^{(1)}\|$ and define $f \mid N_1$ by setting $f(p^{(1)}) = \omega$ and

$$f(x) = \varphi(r^2(x))(\omega - r^2(x)) + (1 - \varphi(r^2(x))\, V(x), \qquad x \in N_1 - p^{(1)}. \tag{32.15}$$

So defined, $f \mid N_1$ is such that

$$f(x) = V(x), \qquad x \in N_1, \| x - p^{(1)}\| \geqslant e, \tag{32.16}$$

$$f(x) = \omega - \| x - p^{(1)}\|^2, \qquad \| x - p^{(1)}\| \leqslant \eta, \tag{32.17}$$

and is of class C^∞. It follows from (32.17) that f has a critical point of index 3 at $p^{(1)}$ and has no other critical point on the n-ball on which $\| x - p^{(1)}\| \leqslant \eta$.

We continue by proving the following:

(α) *In N_1 f is ordinary except at the critical point $p^{(1)}$.*

To establish (α), it is sufficient to show that the directional derivative $df/ds < 0$ on each ray λ issuing from $p^{(1)}$ at points x on λ at which $s = r(x)$ and $\eta \leqslant s \leqslant e$. To that end, we note that on such a ray

$$df/ds = A(s) + B(s), \qquad \eta \leqslant s \leqslant e, \tag{32.18}$$

where, in accord with (32.15),

$$A(s) = -2s\varphi(s^2) + (1 - \varphi(s^2))\, dV/ds, \qquad \eta \leqslant s \leqslant e \tag{32.19}$$

and

$$B(s) = 2s\varphi'(s^2)(\omega - s^2 - V(x)). \tag{32.20}$$

The choice of e implies that $dV/ds < 0$ in (32.19). On using (32.14) we see that $A(s) < 0$ if $\eta \leqslant s \leqslant e$. By virtue of (32.13) and (32.14) $B(s) \leqslant 0$, so that $df/ds < 0$ for $\eta \leqslant s \leqslant e$. This establishes ($\alpha$).

We define $f \mid N_i$ for $i = 1,..., \mu$ in the same manner as $f \mid N_1$.

The definition of $f \mid N_j'$ for $j = 1,..., \nu$ must be made so that (A_2) of Lemma 32.2 holds. This means that the definition $f \mid N_j'$ will differ from that of $f \mid N_1$. One way to give the definition of $f \mid N_j'$ is first to replace V by V', where $V' = -V$, and then to replace V' on a sufficiently small N_j' by an f' related to $V' \mid N_j'$ as f was related to $V \mid N_1$. If one sets $f = -f'$ on N_j' for each j, one obtains a function f satisfying Lemma 32.2.

That f so defined over E_3 is of class C^∞ follows from the fact that f is of class C^∞ on the union U of the neighborhoods (32.11) and that $f(x) = V(x)$ not only on $E_3 - U$ but also on an open neighborhood of $E_3 - U$, in accord with the relations of the type of (32.16).

This completes the proof of Lemma 32.2.

Completion of Proof of Theorem 32.1. Given V, let σ be conditioned as in Lemma 32.1, and let f be the function of class C^∞ on E_3 replacing V in accord with Lemma 32.2. Let m_0, m_1, m_2, and m_3 be the type numbers of the critical points of f. Of these type numbers $m_0 = \nu$ and $m_3 = \mu$, while m_1 and m_2 are the type numbers of V.

We can suppose σ so large that Σ_σ of Lemma 32.1 is included in the subset of E_3 on which $f(x) = V(x)$. According to Lemma 32.1 grad V is emergent on Σ_σ. Hence grad f is emergent on Σ_σ. Theorem 9.1 is applicable to f, with Σ_σ serving as the boundary of the region Z of Theorem 9.1. The connectivities of Z are zero except that $R_0 = 1$. Applying Theorem 9.1 to $f \mid Z$, the second relation of (9.4) takes the form $m_1 \geqslant \nu - 1$, while the last of the equations (9.4), with $n + 1 = 3$, gives the relation $m_1 - m_2 = \nu - \mu - 1$. These two relations imply the third relation, $m_2 \geqslant \mu$, of Theorem 32.1.

This completes the proof of Theorem 32.1.

Minimal Sets of Equilibrium Points of V. Given a ND electrostatic potential V conditioned as in Theorem 32.1, the resultant configuration of equilibrium points will be called *minimal* if

$$m_1 = \nu - 1, \qquad m_2 = \mu, \tag{32.21}$$

and *nonminimal* if

$$m_1 + m_2 > \mu + \nu - 1. \tag{32.22}$$

It follows from the last relation in (32.6) that

$$m_1 - (\nu - 1) = m_2 - \mu, \tag{32.23}$$

that is, the *excess* over the minimum possible number of critical points of type 1 is equal to the excess of type 2.

It is not *a priori* clear that there exist both minimal and nonminimal equilibrium configurations.

A *minimal* configuration of equilibrium points is afforded by an arbitrary finite set of negative charges on a straight line, as one readily verifies.

A special nonminimal configuration of equilibrium points can be set up as in the following example:

Example. Let unit charges of positive electricity be placed at each of the six points

$$(\pm 1, 0, 0), \qquad (0, \pm 1, 0), \qquad (0, 0, \pm 1), \tag{32.24}$$

and let a charge of $\zeta = -5$ units of electricity be placed at a point q near the origin. Let the resultant potential have values denoted by

$V(x, q)$. For almost all values of q near the origin this potential will be ND. The number of critical points in a minimal configuration (if any existed) would be six. We shall prove the following:

(i) *For almost all positions of q sufficiently near the origin the potential $x \to V(x, q)$ is* ND *and has more than six critical points.*

In each coordinate 2-plane of E_3 there are four open quadrants. In the three coordinate 2-planes there are thus 12 disjoint open quadrants. Corresponding to a preferred one of these quadrants, say Q, let h be an open ray bisecting Q and tending to the origin. We shall verify the following:

(ii) *The potential $x \to V(x, \mathbf{0})$ has a unique* ND *critical point on h.*

To establish (ii), let new rectangular coordinates (y_1, y_2, y_3) be taken with h the positive y_1-axis and with the y_3-axis orthogonal to the 2-plane of the quadrant Q. There will then be unit charges at the unit points on the y_3-axis. The four remaining charges will be in the (y_1, y_2) plane at the points $(\pm a, \mp a)$ and $(\pm a, \pm a)$, where $a = \sqrt{2}/2$.

Let $y \to U(y)$ be the potential into which the potential $x \to V(x, \mathbf{0})$ goes in terms of the new coordinates. Taking account of the symmetry of the problem and the hypothesis that $\zeta = -5$, simple calculations will show that there is just one critical point of U on the y_1-axis. Further calculation will show that the quadratic form underlying this critical point is of ND diagonal type.

The potential U thus has at least 12 ND critical points. Although it can be shown that U has other critical points of degenerate type, this fact can be disregarded. In any case, for almost all values of q near the origin the potential $x \to V(x, q)$ will be ND and have ND critical points near the 12 critical points of U on the respective 12 rays h.

This establishes (ii) and hence (i).

Equilibrium Problems. There are many questions concerning the nature of equilibrium configurations which would be of interest if answered. Among these questions are the following:

1. What is the maximum (if any) of the number of points in an equilibrium configuration of a ND potential V for which $\mu + \nu$ is given and (32.5) is satisfied?

2. Is it possible for a degenerate electrostatic potential defined by a finite number of charges to have an analytic arc of critical points?

3. Among ND electrostatic potentials conditioned as is V in Theorem 32.1, are there potentials all of whose charges are on a straight line and whose equilibrium configurations are nonminimal?

4. Given a degenerate electrostatic potential V for which (32.5) holds and a ND potential U which approximates V and differs from V only in the position of one charge, what is the simplest way to characterize topologically the degenerate critical points of V in order to condition topologically, as far as possible, the ND critical points of U? (see "Neighborhood functions" by Morse [13] pp. 154–156).

§ 33

SYMMETRIC SQUARES OF MANIFOLDS AND CRITICAL CHORDS

Given a regular compact C^∞-manifold M_n in E_r, $0 < n < r$, the chords of M_n orthogonal to M_n at both endpoints are called "*critical chords*" of M_n. Such chords are represented, as we shall see, by critical points of a function h whose domain is the set of unordered pairs of distinct points x and y in M_n and whose value, given x and y, is the distance $\| x - y \| = \| y - x \|$. The domain of h is an open subset of the so-called "*symmetric square*" $| M_n^2 |$ of M_n. We shall topologize $| M_n^2 |$.

Let $| S_n^2 |$ be the symmetric square of an n-sphere. The problem of determining the connectivities of $| S_n^2 |$ mod diag $| S_n^2 |$ (to be defined) was first solved by the methods of critical point theory. These methods are here extended (see Morse [13], pp. 183–191, Richardson [1], p. 528).

The spaces χ of this section will be metric. Open subsets will be defined in the usual way in terms of the metrics, that is, each nonempty open set shall be the union of e-neighborhoods of points of χ.

The metric on M_n will be defined by taking the "distance" between two points x and y in M_n as $\| x - y \|$, where $\| x - y \|$ is the ordinary distance between x and y in E_r. The metric on the Cartesian product $| M_n \times M_n |$ will be defined by taking the "distance" between points (x, y) and (x', y') in $| M_n \times M_n |$ as

$$\| x - x' \| + \| y - y' \|.$$

The Involution ξ of $| M_n \times M_n |$. To a point $P = (x, y) \in | M_n \times M_n |$ corresponds the point $P^* = (y, x) \in | M_n \times M_n |$

termed the *mate* of P. The mapping $P \to \xi(P) = P^*$ is a homeomorphism of $| M_n \times M_n |$ onto $| M_n \times M_n |$ which is an "involution." A point $P = (x, y)$ is equal to $\xi(P)$ if and only if $x = y$, that is, if P is on the "*diagonal*" of $| M_n \times M_n |$.

We shall make important uses of the following definition:

Definition 33.0. *Prime Subsets of* $| M_n \times M_n |$. An open subset Z of $| M_n \times M_n |$ such that

$$Z \cap \xi Z = \varnothing$$

will be termed a *prime subset of* $| M_n \times M_n |$.

The Symmetric Square $| M_n^2 |$. Given M_n, we shall define a set $| M_n^2 |$, and continue by defining a metric on this set.

The Set $| M_n^2 |$. To define the set $| M_n^2 |$, let points (x, y) and (y, x) in $| M_n \times M_n |$ be identified to yield a "point in $| M_n^2 |$" denoted by $T(x, y)$ or $T(y, x)$. The set $| M_n^2 |$ is the ensemble of points $T(x, y) = T(y, x)$ for arbitrary choices of x and y in $| M_n |$. We have thus introduced a mapping

$$(x, y) \to T(x, y) : | M_n \times M_n | \to | M_n^2 | \tag{33.1}$$

onto $| M_n^2 |$. Let diag $| M_n^2 |$ denote the subset of points $T(x, x)$ of $| M_n^2 |$ for x arbitrary in $| M_n |$. The set mapping T^{-1}, if restricted to sets $\{p\}$, with $p \in \text{diag } | M_n^2 |$, maps $\{p\}$ onto $\{p\}$. Set

$$| M_n^2 | - \text{diag } | M_n^2 | = | \hat{M}_{2n} |. \tag{33.2}$$

The set mapping T^{-1} is such that for $(x, y) \in | \hat{M}_{2n} |$ and $p = T(x, y)$

$$T^{-1}(\{p\}) = \{(x, y)\} \cup \{(y, x)\}. \tag{33.3}$$

Continuity of T as yet has no meaning.

The Set $| M_n^2 |$ *Metricized*. Let p and p_0 be points in $| M_n^2 |$ with representations

$$p = T(x, y) = T(y, x), \qquad p_0 = T(x_0, y_0) = T(y_0, x_0), \tag{33.4}$$

where x, y, x_0, and y_0 are points in $| M_n |$. The distance $d(p, p_0)$ in $| M_n^2 |$ between p and p_0 will be defined by setting

$$d(p, p_0) = \min(\| x - x_0 \| + \| y - y_0 \|, \| x - y_0 \| + \| y - x_0 \|). \tag{33.5}$$

This distance function has the usual three properties. If p_0, p, and q are points in $| M_n^2 |$:

(λ_1): $\quad d(p, p_0) = d(p_0, p)$

(λ_2): $\quad d(p, p_0) = 0 \quad$ if and only if $\quad p = p_0$

(λ_3): $\quad d(p, p_0) \leqslant d(p, q) + d(q, p_0)$.

The verification of these three properties is trivial except for (λ_3). To verify (λ_3), suppose that $q = T(x_1, y_1)$. Given p, p_0 as in (33.4) and (x_1, y_1), one can choose (x, y) and (x_0, y_0) so that (33.4) holds and

$$\begin{aligned} d(p, q) &= \| x - x_1 \| + \| y - y_1 \|, \\ d(q, p_0) &= \| x_0 - x_1 \| + \| y_0 - y_1 \|. \end{aligned} \tag{33.6}$$

By virtue of the definition (33.5)

$$d(p, p_0) \leqslant \| x - x_0 \| + \| y - y_0 \|,$$

so that (λ_3) follows from (33.6).

Properties of the Mapping T. We shall verify the following lemma:

Lemma 33.1. *Let (x_0, y_0) be a point of $| M_n \times M_n |$ and $p_0 = T(x_0, y_0)$ its image in $| M_n^2 |$. If N_e is the open e-neighborhood of (x_0, y_0) in $| M_n \times M_n |$ and U_e the open e-neighborhood of p_0 in $| M_n^2 |$, then*

$$T(N_e) = U_e; \qquad T^{-1}(U_e) = N_e \cup \xi(N_e). \tag{33.7}$$

Proof. By definition of the metric on $| M_n \times M_n |$

$$N_e = \{(x, y) \in | M_n \times M_n | \mid e > \| x - x_0 \| + \| y - y_0 \|\}.$$

If u and v are points in $| M_n |$, then

$$U = \{T(u, v) \mid M_n^2 \mid \mid e > \min(\| u - x_0 \| + \| v - y_0 \|, \| v - x_0 \| + \| u - y_0 \|)\}.$$

From these representations of N_e and U_e it follows that

$$T(N_e) \subset U_e \,; \qquad T^{-1}(U_e) \subset N_e \cup \xi(N_e) \tag{33.8}$$

[using (33.3)]. On applying T^{-1} to the first of these inclusions and T to the second, we find that

$$N_e \cup \xi(N_e) \subset T^{-1}(U_e); \qquad U_e \subset T(N_e), \tag{33.9}$$

where use has been made of (33.3) and (33.1). Relations (33.7) follow trivially from (33.8) and (33.9).

Thus Lemma 33.1 is true.

It is a corollary of Lemma 33.1 that T and T^{-1} are open mappings. That is, an open set of $|M_n \times M_n|$ is mapped onto an open subset of $|M_n^2|$, and conversely an open subset ω of $|M_n^2|$ is mapped by T^{-1} onto an open subset of $|M_n^2|$.

We shall need the following lemma:

Lemma 33.2. *The compactness of* $|M_n|$ *implies the compactness of* $|M_n^2|$.

Recall that a Hausdorff space χ is compact if each collection of open subsets of χ which covers χ includes a finite subcollection which covers χ.

Let there be given a covering of $|M_n^2|$ by a collection $(\omega_i)_{i\in\alpha}$ of open subsets ω_i of $|M_n^2|$. For each $i \in \alpha$ set $z_i = T^{-1}(\omega_i)$. Then, by Lemma 33.1 z_i is open in $|M_n \times M_n|$. Thus $(z_i)_{i\in\alpha}$ is a collection of open subsets z_i of $|M_n \times M_n|$ covering $|M_n \times M_n|$. Since $|M_n \times M_n|$ is compact if $|M_n|$ is compact, there exists a finite subcollection, $w_1, w_2, \ldots, w_m$, of the sets z_i which covers $|M_n \times M_n|$. The sets $T(w_1), T(w_2), \ldots, T(w_m)$ are in the collection $(\omega_i)_{i\in\alpha}$ and have $|M_n^2|$ as union.

Since $|M_n^2|$ is a Hausdorff space, Lemma 33.2 follows.

Lemma 33.2 has the following consequence:

Lemma 33.3. *If* Z *is a "prime" subset of* $|M_n \times M_n|$ (Definition 33.0), *the restriction* $T \mid Z$ *is biunique and bicontinuous, and hence a homeomorphism of* Z *onto* $T(Z) \subset |M_n^2|$.

Taking account of the relation (33.3), with $(x, y) \in |M_n \times M_n|$ and the hypothesis $Z \cap \xi(Z) = \varnothing$, we see that the restriction $T \mid Z$

is biunique. That $T \mid Z$ is bicontinuous follows from the fact that $T(Z)$ is an open subset of $| M_n{}^2 |$, that $T \mid Z$ is biunique, and T^{-1} is an "open" mapping in accord with Lemma 33.1.

The Compact Subset diag $| M_n{}^2 |$ *of* $| M_n{}^2 |$. Diag $| M_n{}^2 |$ is by definition the subset of points $T(x, x) \in | M_n{}^2 |$. Making use of the metrics on $| M_n \times M_n |$ and $| M_n{}^2 |$, one sees that the mapping

$$(x, x) \to T(x, x) : \text{diag} \, | M_n \times M_n | \to \text{diag} \, | M_n{}^2 | \tag{33.10}$$

onto diag $| M_n{}^2 |$ is a homeomorphism. The manifold $| M_n \times M_n |$ is compact, since $| M_n |$ is compact. The subspace diag $| M_n \times M_n |$ of $| M_n \times M_n |$ is closed in $| M_n \times M_n |$ and hence compact. Its homeomorph, diag $| M_n{}^2 |$, is accordingly compact.

The following lemma greatly simplifies the problem of determining the homology characteristics of $| M_n{}^2 | -$ diag $| M_n{}^2 |$:

Lemma 33.4. *A homeomorphism* φ *of the manifold* $| M_n |$ *onto a similarly conditioned manifold* $| N_n |$ *induces a homeomorphism* Θ_φ *of* $| M_n{}^2 |$ *onto* $| N_n{}^2 |$ *that maps* diag $| M_n{}^2 |$ *homeomorphically onto* diag $| N_n{}^2 |$.

The homeomorphism φ induces the homeomorphism Φ of $| M_n \times M_n |$ onto $| N_n \times N_n |$, in which $(x, y) \in | M_n \times M_n |$ goes into the point

$$\Phi(x, y) = (\varphi(x), \varphi(y)) \in | N_n \times N_n |.$$

We shall define Θ_φ at each point $p \in | M_n{}^2 |$. To that end, let T be the mapping of $| M_n \times M_n |$ onto $| M_n{}^2 |$ defined above, and let T_* be the corresponding mapping of $| N_n \times N_n |$ onto $| N_n{}^2 |$. Introduce the point $\Theta_\varphi(p) \in | N_n{}^2 |$ by setting

$$\{\Theta_\varphi(p)\} = T_*(\Phi(T^{-1}(\{p\}))), \qquad p \in | M_n{}^2 |. \tag{33.11}$$

Recall that $T^{-1}(\{p\})$ is given by (33.3), where (x, y) and (y, x) are mated points in $| M_n \times M_n |$, coincident if $p \in$ diag $| M_n{}^2 |$. Under Φ these mated points of $| M_n \times M_n |$ go, respectively, into mated points $(\varphi(x), \varphi(y))$ and $(\varphi(y), \varphi(x))$ in $| N_n \times N_n |$. Under T_* these mated points in $| N_n \times N_n |$ go into the point $\Theta_\varphi(p)$ of $| N_n{}^2 |$, so that (33.11) is true.

That the mapping $p \to \Theta_\varphi(p)$ is biunique and onto $| N_n^2 |$ follows from the existence of the inverse of (33.11),

$$\{q\} \to T(\Phi^{-1}(T_*^{-1}(\{q\}))), \qquad q \in | N_n^2 |.$$

Θ_φ maps an open subset of $| M_n^2 |$ onto an open subset of $| N_n^2 |$ because each of the mappings T^{-1}, Φ, T_* is an *open* mapping. Since Θ_φ^{-1} also maps open sets onto open sets, Θ_φ is a homeomorphism.

One sees that Θ_φ maps diag $| M_n^2 |$ *onto* diag $| N_n^2 |$, thus completing the proof of Lemma 33.4.

It is a corollary of the preceding lemma and of Theorem 28.1 that the q-th homology groups of the spaces

$$| M_n^2 | \text{ mod diag } | M_n^2 | \qquad \text{and} \qquad | N_n^2 | \text{ mod diag } | N_n^2 | \tag{33.12}$$

(see §28) are isomorphic for each integer q provided $| M_n |$ and $| N_n |$ are topologically equivalent.

Notation. To relate the connectivities $\mathscr{R}_i$ of $| M_n^2 |$ mod diag $| M_n^2 |$ to the "critical chords" of M_n, we introduce open subspaces

$$\begin{aligned} | M_n^2 | - \text{diag } | M_n^2 | &= | \hat{M}_{2n} | \\ | M_n \times M_n | - \text{diag } | M_n \times M_n | &= | M_n \hat{\times} M_n |, \end{aligned} \tag{33.13}$$

respectively, of $| M_n^2 |$ and $| M_n \times M_n |$ [see (33.2)]. It should be understood that $| M_n \hat{\times} M_n |$ is not a symbol for a product. The notation (33.13) is permanent.

Lemma 33.5. *The open subspace $| \hat{M}_{2n} |$ of $| M_n^2 |$ is a topological $2n$-manifold.*

To verify this lemma, let p_0 be an arbitrary point of $| \hat{M}_{2n} |$. Then $p_0 = T(x_0, y_0)$ for some point (x_0, y_0) in $| M_n \hat{\times} M_n |$. Since $| M_n \hat{\times} M_n |$ is a topological $2n$-manifold, there exists an open neighborhood N of (x_0, y_0) relative to $| M_n \hat{\times} M_n |$ which is a topological $2n$-ball. Since $(x_0, y_0) \neq (y_0, x_0)$, N can be chosen as so small a neighborhood of (x_0, y_0) in $| M_n \hat{\times} M_n |$ that it is a "prime" subset of $| M_n \times M_n |$. According to Lemma 33.3 $T | N$ is a homeomorphism of N onto the open subspace $T(N)$ of $| \hat{M}_{2n} |$. Thus $T(N)$ is an open topological $2n$-ball serving as a neighborhood relative to $| \hat{M}_{2n} |$ of $p_0 = T(x_0, y_0)$.

Thus Lemma 33.5 is true.

An Extension of the Preceding. The theory of symmetric squares extends readily to symmetric squares of Hausdorff spaces χ. The preceding definitions and lemmas have their purely topological counterparts. In particular, one can define an open subset of the symmetric square χ^2 as the image of an open subset of $\chi \times \chi$ under the mapping T, where T carries a point $(x, y) \in \chi \times \chi$ into the unordered pair $T(x, y)$ in χ^2. When one comes to the study of critical chords the purely topological theory ends. It was because our principal objective was the study of critical chords of differentiable manifolds that metric rather than nonmetric Hausdorff spaces were used.

We turn now to the differential aspects of the theory, with special reference to critical chords of M_n.

$\hat{M}_{2n}$ as a Differentiable $2n$-Manifold. M_n is given as a regular compact C^∞-manifold in E_r, $r > n$. We shall show that the differentiable structure on $| M_n |$ induces a differentiable structure on $| \hat{M}_{2n} |$ in much the same way as that in which the differentiable structure on $| M_n |$ induces a product differentiable structure on $| M_n | \times | M_n |$ (see §13).

Presentations in $\mathscr{D}\hat{M}_{2n}$ *Defined.* Let p_1 be an arbitrary point in $| \hat{M}_{2n} |$ and (x_1, y_1) a point in $| M_n \hat{\times} M_n |$ such that $p_1 = T(x_1, y_1)$. By hypothesis there exist presentations

$$(F_1 : U_1, X_1) \in \mathscr{D}M_n \quad \text{and} \quad (G_1 : V_1, Y_1) \in \mathscr{D}M_n \tag{33.14}$$

of open neighborhoods X_1 and Y_1, respectively, of x_1 and y_1 so small that $X_1 \times Y_1$ is a "prime" subset of $| M_n \times M_n |$. It follows from Lemma 33.3 that the mapping

$$(u, v) \to T(F_1(u), G_1(u)) : U_1 \times V_1 \to T(X_1 \times Y_1) \tag{33.15}$$

is a homeomorphism of $U_1 \times V_1$ onto the open subset $T(X_1 \times Y_1)$ of $| \hat{M}_{2n} |$.

Definition 33.1. *Prime Presentations in* $\mathscr{D}^0\hat{M}_{2n}$. We denote the homeomorphism (33.15) by $T(F_1, G_1) = T \circ (F_1, G_1)$. This homeomorphism is a presentation [cf. (13.6)]

$$(T(F_1, G_1) : U_1 \times V_1, T(X_1 \times Y_1)) \in \mathscr{D}^0\hat{M}_{2n} \tag{33.16}$$

termed a *prime* presentation in $\mathscr{D}^0\hat{M}_{2n}$. It is characterized by the fact that $X_1 \times Y_1$ is a prime subset of $| M_n \times M_n |$.

We shall prove Lemma 33.6:

Lemma 33.6. *"Prime" presentations in $\mathscr{D}^0\hat{M}_{2n}$ are C^∞-compatible and cover $| \hat{M}_{2n} |$.*

It is clear that prime presentations in $\mathscr{D}^0\hat{M}_{2n}$ cover $| \hat{M}_{2n} |$.

To show that such presentations are C^∞-compatible, we fall back on Proposition (A_1) of §13. By virtue of Proposition (A_1) it is sufficient to verify the C^∞-compatibility of an arbitrary prime presentation $T(F_1, G_1) \in \mathscr{D}^0\hat{M}_{2n}$ with each other prime presentation $T(F_2, G_2) \in \mathscr{D}^0\hat{M}_{2n}$ for which the presentations

$$(F_2 : U_2, X) \in \mathscr{D}M_n \quad \text{and} \quad (G_2 : V_2, Y) \in \mathscr{D}M_n \tag{33.17}$$

have the same ranges X and Y on $| M_n |$, respectively, as F_1 and G_1.

The "transition" homeomorphism defined by the presentations $T(F_1, G_1)$ and $T(F_2, G_2)$, taken in that order, is a homeomorphism

$$\begin{aligned} (T(F_2, G_2))^{-1} \circ T(F_1, G_1) &= (F_2, G_2)^{-1} \circ (F_1, G_1) \qquad \text{[cf. (13.7)]} \\ &= (F_2^{-1} \circ F_1, G_2^{-1} \circ G_1) : U_1 \times V_1 \to U_2 \times V_2 \qquad \text{[cf. (13.8)]} \end{aligned} \tag{33.18}$$

of $U_1 \times V_1$ onto $U_2 \times V_2$. Since F_2 and F_1 are C^∞-compatible by hypothesis, as are G_2 and G_1, it follows from the form of (33.18) that the mapping (33.18) is a C^∞-diffeomorphism of $U_1 \times V_1$ onto $U_2 \times V_2$.

Thus Lemma 33.6 is true.

Since a suitably chosen countable subset of prime presentations in $\mathscr{D}^0\hat{M}_{2n}$ covers $| \hat{M}_{2n} |$, a C^∞-differentiable structure is thereby defined on $| \hat{M}_{2n} |$ in accord with Definition 13.1. So differentiably structured, $\hat{M}_{2n}$ will be termed the *diagonal-free symmetric square* of M_n.

In view of Lemma 33.6, prime presentations in $\mathscr{D}^0\hat{M}_{2n}$ will be termed *prime presentations in $\mathscr{D}\hat{M}_{2n}$*.

We can regard $M_n \hat{\times} M_n$ as a differentiable submanifold of $M_n \times M_n$.

Lemma 33.7. *The mapping*

$$(x, y) \to T(x, y) : | M_n \hat{\times} M_n | \to | \hat{M}_{2n} | \qquad \text{onto} \quad | \hat{M}_{2n} |$$

defines a C^∞-immersion $M_n \hat{\times} M_n \to \hat{M}_{2n}$.

This lemma will be established by the test of Definition 16.2, where φ, $F \in \mathscr{D}M_n$, and $G \in \mathscr{D}N_n$ of Definition 16.2 are replaced, respectively, [in the terms of (33.14) and (33.15)] by T and the presentations

$$((F_1, G_1) : U_1 \times V_1, X_1 \times Y_1) \in \mathscr{D}(M_n \hat{\times} M_n) \qquad \text{[cf. (13.7)]}$$

$$(T(F_1, G_1) : U_1 \times V_1, T(X_1 \times Y_1)) \in \mathscr{D}\hat{M}_{2n} \qquad \text{[cf. (33.16)].}$$

The test of Definition 16.2 is satisfied if the mapping (16.3), namely

$$G^{-1} \bar{\circ} (\varphi \bar{\circ} F) : U \to V$$

is a C^∞-diffeomorphism. In the case at hand this test is satisfied, since the mapping

$$T(F_1, G_1)^{-1} \bar{\circ} (T \bar{\circ} (F_1, G_1)) : U_1 \times V_1 \to U_1 \times V_1$$

is the identity mapping of $U_1 \times V_1$ onto $U_1 \times V_1$.

The Chord Function h on $\hat{M}_{2n}$. Let x and y be distinct points in $| M_n |$, and $p = T(x, y)$ the corresponding point in $| \hat{M}_{2n} |$. The real-valued function

$$p \to h(p) = \| x - y \|, \qquad p \in | \hat{M}_{2n} |$$

is called the *chord function* on $\hat{M}_{2n}$. Given a prime presentation $T(F_1, G_1) \in \mathscr{D}\hat{M}_{2n}$, as in (33.16), h has a representation in the local $T(F_1, G_1)$-coordinates (u_1, v_1) of the form

$$(u_1, v_1) \to (h \bar{\circ} T(F_1, G_1))(u_1, v_1) = \| F_1(u_1) - G_1(v_1) \|, \quad (u_1, v_1) \in U_1 \times V_1, \tag{33.19}$$

with $F_1(u_1) \neq G_1(v_1)$ (cf. Definition 5.5 and §13).

This representation shows that h is of class C^∞ on $\hat{M}_{2n}$ and that a point $T(x, y) \in \hat{M}_{2n}$ is a critical point of h if and only if the chord $\overline{x, y}$ in E_r with endpoints $x = F_1(u_1)$ and $y = G_1(v_1)$ in $| M_n |$ is orthogonal to M_n at x and y.

The lower bound of values of h is zero, but $h(p) > 0$ without exception for $p \in | \hat{M}_{2n} |$. The critical values of h are the lengths of the critical chords, and are bounded from zero. The set of critical points of h on $|\hat{M}_{2n}|$ is closed on $|M_n^2|$ and hence compact. It is bounded from diag $|M_n^2|$.

ND *Chord Functions h.* The chord function h on $\hat{M}_{2n}$ may be ND or degenerate. It is ND if M_n is an n-dimensional ellipsoid in E_{n+1}. It is degenerate if M_n is an n-sphere.

Morse has proved the following:

A. *An Unpublished Theorem. Corresponding to a regular, compact, connected* C^∞*-manifold* M_n *in* E_r, $n < r$, *there exists a similarly conditioned manifold* Q_n *in* E_r *with* $|Q_n|$ *homeomorphic to* $|M_n|$ *and such that the chord function* h *on* $\hat{Q}_{2n}$ *is* ND *and has just one critical point* p_a *corresponding to each critical value of* h.

The proof of the theorem shows that Q_n can be chosen so as to approximate M_n in various senses, but this fact is not used in the present study.

It follows from Lemma 33.4 that the manifolds $|\hat{M}_{2n}|$ and $|\hat{Q}_{2n}|$ of Theorem A are homeomorphic, a fact of great importance.

If the chord function h on $\hat{M}_{2n}$ is ND, its critical points are finite in number. This is because the set of critical points of h on $|\hat{M}_{2n}|$ is compact and the critical points of a ND function are isolated.

The Extended Chord Function $\mathbf{h}$. The chord function h is defined on the open subset $|\hat{M}_{2n}|$ of $|M_n^2|$. We shall extend h over all of $|M_n^2|$ by setting $\mathbf{h}(p) = 0$ when $p \in$ diag $|M_n^2|$. So defined, $\mathbf{h}$ is continuous on $|M_n^2|$.

The Principal Theorem.

Theorem 33.1. (i) *If* M_n *is a regular, compact, connected,* C^∞*-manifold in* E_r, $0 < n < r$, *the connectivities,* $\mathcal{R}_i$ *of* $|M_n^2|$ mod diag $|M_n^2|$, *are finite and vanish, excepting at most the connectivities*

$$\mathcal{R}_1, \mathcal{R}_2, \ldots, \mathcal{R}_{2n}. \tag{33.20}$$

(ii) *If the chord function* h *on* $\hat{M}_{2n}$ *is* ND, *the type numbers* $m_0, m_1, \ldots, m_{2n}$ *of the critical points of* h *and the connectivities* $\mathcal{R}_0, \mathcal{R}_1, \ldots, \mathcal{R}_{2n}$ *satisfy the relations* (30.1) *and* (30.2), *with* $2n$ *replacing* n *in these relations.*

It is not affirmed that the connectivities (33.20) are the connectivities of the domain $|\hat{M}_{2n}|$ on which h is defined. However, Theorem 30.3 will be shown to imply Theorem 33.1 once the following lemma is established:

Lemma 33.8. *If, corresponding to the manifold M_n given in Theorem 33.1, η is a positive constant less than any critical value of the chord function h on $\hat{M}_{2n}$, the isomorphisms*

$$H_q(|M_n^2|, \text{diag}\,|M_n^2|) \approx H_q(|\hat{M}_{2n}|, h_\eta) \tag{33.21}$$

are valid for each integer q.

In Theorem 33.1(i) and in this lemma it is not assumed that the chord function is ND.

To prove Lemma 33.8, it will be sufficient to verify the four following propositions. We shall refer therein to the extension $\mathbf{h}$ of h over $|M_n^2|$ defined above. The proofs depend upon a deformation D_e of $|M_n^2|$ on $|M_n^2|$ near diag $|M_n^2|$ and a second deformation $\mathbf{D}_c$ of $|M_n^2|$ which is global in character.

Proposition I. *Definition of D_e. If e is a sufficiently small positive constant, the subspace $\mathbf{h}_e$ of $|M_n^2|$ admits a deformation D_e retracting $\mathbf{h}_e$ onto* diag $|M_n^2|$ *in such a manner that for each constant c for which $0 < c < e$ $\mathbf{h}_c$ is retracted onto* diag $|M_n^2|$.

Proposition II. *Definition of $\mathbf{D}_c$. If, corresponding to e of Proposition* I, $2c = e$, $|M_n^2|$ *admits a deformation $\mathbf{D}_c$ deforming $|M_n^2|$ on itself onto itself in such a manner that $\mathbf{h}_c$ is retracted onto* diag $|M_n^2|$.

Proposition III. *If $2c = e$, as in Proposition* II, *then*

$$H_q(|M_n^2|, \text{diag}\,|M_n^2|) \approx H_q(|M_n^2|, \mathbf{h}_c), \qquad q = 0, 1, \ldots. \tag{33.22}$$

Proposition IV. *For each ordinary value $c > 0$ of h*

$$H_q(|M_n^2|, \mathbf{h}_c) \approx H_q(|\hat{M}_{2n}|, h_c), \qquad q = 0, 1, \ldots. \tag{33.23}$$

Proof of Proposition I. In establishing Proposition I we shall make use of well-known properties of short geodesic arcs on M_n. We review these properties.

Corresponding to a sufficiently small positive constant e, statements (A_1)–(A_3) are true:

(A_1) If x and y are points in $|M_n|$ such that $0 < \| x - y \| < e$, there is a unique geodesic arc $g(x, y)$, joining x to y on $|M_n|$ and absolutely minimizing length on M_n, with a length $L(x, y)$ on M_n which varies continuously with $(x, y) \in |M_n \times M_n|$ when $0 < \| x - y \| \leqslant e$. The norm $\| x - y \|$ is the length of the chord $\overline{x, y}$. The point $P(s : x, y)$ on $g(x, y)$ at a distance s from x measured along $g(x, y)$ varies continuously with s, x, y for

$$0 < \| x - y \| \leqslant e, \qquad 0 \leqslant s \leqslant L(x, y).$$

In particular, the *midpoint*,

$$z(x, y) = P(\tfrac{1}{2}L(x, y) : x, y),$$

of the arc $g(x, y)$ varies continuously with (x, y).

(A_2) A point $(x, y) \in |M_n \times M_n|$ for which $0 < \| x - y \| \leqslant e$ admits a continuous deformation on $g(x, y)$ into the point $(z(x, y), z(x, y))$ in diag $|M_n \times M_n|$. This deformation is defined by replacing x and y by points x_t and y_t which move continuously on $g(x, y)$ toward $z(x, y)$ at a velocity $L(x, y)/2$ along $g(x, y)$ as t increases from 0 to 1.

(A_3) During this deformation of (x, y) on $|M_n \times M_n|$ the chord length $\| x_t - y_t \|$ shall strictly decrease.

A Deformation D_e of the Set $\mathbf{h}_e$. Given a constant e subject to the above conditions, let p be an arbitrary point on $\mathbf{h}_e$ — diag $|M_n{}^2|$. There then exist points (x, y) and (y, x) in $|M_n \hat{\times} M_n|$ such that

$$p = T(x, y) = T(y, x). \tag{33.24}$$

The geodesic arcs $g(x, y)$ and $g(y, x)$ are identical except in sense. Under the deformation of the points (x, y) and (y, x) into $(z(x, y), z(x, y)) = (z(y, x), z(y, x))$, as defined in (A_2), (x, y) is replaced at the time t by a point (x_t, y_t) and (y, x) by the point (y_t, x_t), so that

$$T(x_t, y_t) = T(y_t, x_t), \qquad 0 \leqslant t \leqslant 1. \tag{33.25}$$

We subject the point $p \in \mathbf{h}_e$ given by (33.24) to a deformation on $| M_n{}^2 |$ in which p is replaced at the time t by the point (33.25) of $| M_n{}^2 |$.

Let $\mathbf{h}_e$ be subject to a *deformation* D_e on $| M_n{}^2 |$ in which each point $p \in \mathbf{h}_e - \operatorname{diag} | M_n{}^2 |$ is deformed as above and each point $p \in \operatorname{diag} | M_n{}^2 |$ remains fixed. That the deformation D_e thereby defined satisfies Proposition I is readily verified.

Proof of Proposition II. Let e and the deformation D_e of $\mathbf{h}_e$ be conditioned as in Proposition I. We shall show that there exists a *deformation* $\mathbf{D}_c$ of $| M_n{}^2 |$ satisfying Proposition II with $2c = e$.

The trajectory under $\mathbf{D}_c$ of a point $p \in | M_n{}^2 |$ will be defined differently depending on whether p comes under: CASE 1, $0 \leqslant \mathbf{h}(p) \leqslant c$; or CASE 2, $c < \mathbf{h}(p) \leqslant 2c$; or CASE 3, $2c < \mathbf{h}(p)$.

In case 1 the trajectory of p under $\mathbf{D}_c$ shall be the trajectory of p under D_{2c}. The set $\mathbf{h}_c$ then undergoes a deformation retracting $\mathbf{h}_c$ onto $\operatorname{diag} | M_n{}^2 |$. In case 3 the trajectory of p under $\mathbf{D}_c$ shall reduce to p.

The deformation arcs in case 2 will interpolate between the deformation arcs in cases 1 and 3 as follows: Corresponding to a point p coming under case 2, let the time t_p be determined by the condition $\mathbf{h}(p) = 2c - t_p c$. One sees that $t_p = 0$ or 1 depending on whether $\mathbf{h}(p) = 2c$ or c. In case 2 the trajectory of p under $\mathbf{D}_c$ shall be that of p under D_{2c} until the time t_p is reached. For t on the interval $t_p \leqslant t \leqslant 1$ the trajectory of p shall remain fixed as the point into which p has been deformed when $t = t_p$.

This completes the definition of a deformation $\mathbf{D}_c$ of $| M_n{}^2 |$. One sees that $\mathbf{D}_c$, with $2c = e$, satisfies Proposition II.

Proof of Proposition III. The existence of the deformation $\mathbf{D}_c$ of $| M_n{}^2 |$ implies that for each integer $q \geqslant 0$ (33.22) holds in accord with Theorem 28.4. In this application of Theorem 28.4 (χ, A), (χ', A'), and d of Theorem 28.4 are taken, respectively, as

$$(| M_n{}^2 |, \mathbf{h}_c), \qquad (| M_n{}^2 |, \operatorname{diag} | M_n{}^2 |), \qquad \mathbf{D}_c . \tag{33.26}$$

Proof of Proposition IV. To verify Proposition IV, we shall make use of the "Excision" Theorem 28.3. We wish to excise $\operatorname{diag} | M_n{}^2 |$ from $| M_n{}^2 |$ and from its modulus $\mathbf{h}_c$, understanding that c is an ordinary value of $\mathbf{h}$. We accordingly set χ, A, and A^* of Theorem 28.3

equal to $| M_n^2 |$, $\mathbf{h}_c$, and diag $| M_n^2 |$, respectively. The excision of diag $| M_n^2 |$ yields the sets

$$\chi - A^* = | \hat{M}_{2n} | \qquad \text{and} \qquad A - A^* = h_c, \tag{33.27}$$

so that once the excision hypothesis is confirmed the isomorphism (28.15)

$$H_q(\chi, A) \approx H_q(\chi - A^*, A - A^*), \qquad q = 0, 1, \ldots,$$

will yield the isomorphism (33.23).

Verification of the Excision Hypothesis (28.14). The excision hypothesis here takes the form

$$(| M_n^2 | - \mathbf{h}_c)_\epsilon \subset | M_n^2 | - \text{diag} \, | M_n^2 | \qquad \text{for some } \epsilon > 0, \tag{33.28}$$

where $(| M_n^2 | - \mathbf{h}_c)_\epsilon$ means the ϵ-neighborhood of $| M_n^2 | - \mathbf{h}_c$ in $| M_n^2 |$. Equivalently, (33.28) is the condition that for some $\epsilon > 0$ the ϵ-neighborhood in M_n^2 of the set $X = | M_n^2 | - \mathbf{h}_c$ does not meet diag $| M_n^2 |$.

To show that this condition is satisfied, let $\mathbf{h}^c$ be the subset of $| M_n^2 |$ at the $\mathbf{h}$-level c. Taking closures in $| M_n^2 |$, Cl $X = X \cup \mathbf{h}^c$. The set Cl X accordingly does not meet diag $| M_n^2 |$. Now, diag $| M_n^2 |$ is compact by Lemma 33.2. Moreover, Cl X is compact, since Cl X is closed in the compact space $| M_n^2 |$. Hence some ϵ-neighborhood of Cl X does not meet diag $| M_n^2 |$. For this ϵ (33.28) is satisfied. Excision Theorem 28.3 thus implies (33.23) and thereby Proposition IV.

Completion of Proof of Lemma 33.8. If $(0, e]$ is an interval of ordinary values of h, and if in addition $2c = e$ is conditioned as in Propositions I, II, and III, the isomorphisms of Propositions III and IV imply the isomorphisms

$$H_q(| M_n^2 |, \text{diag} \, | M_n^2 |) \approx H_q(| \hat{M}_{2n} |, h_c), \qquad q = 0, 1, \ldots. \tag{33.29}$$

The constant $\eta > 0$ of Lemma 33.8 is less than the minimum a of the critical values of h. It follows, as we shall see, that with $c < a$, as in (33.29),

$$H_q(| \hat{M}_{2n} |, h_c) \approx H_q(| \hat{M}_{2n} |, h_\eta), \qquad q = 0, 1, \ldots. \tag{33.30}$$

Verification of (33.30). That (33.30) is true will follow if we show that there exists a homeomorphism Φ of $| \hat{M}_{2n} |$ onto $| \hat{M}_{2n} |$ that maps h_c onto h_η. To this end, one can define Φ so as to leave $| \hat{M}_{2n} | - h_a$ pointwise fixed and map h_a onto h_a, leaving h^a pointwise fixed while mapping h_c onto h_η. That this is possible follows, since there are no critical points of h below the h-level a. Appropriate details can be supplied by using h-trajectories on h_a.

The isomorphism (33.21) follows from the isomorphisms (33.29) and (33.30).

This establishes Lemma 33.8.

Proof of Theorem 33.1(i). We first prove Theorem 33.1(i) under the assumption that the chord function h of $\hat{M}_{2n}$ is ND.

We shall apply Theorem 30.3(i).

We identify $\hat{M}_{2n}$ of Theorem 33.1(i) with M_n of Theorem 30.3. The chord function h on $\hat{M}_{2n}$ will be identified with the ND function f on M_n of Theorem 30.3 and the value η of h with the value η of f. The latter identification is permissible because the conditions (30.14) on the value η of f, namely, that

$$\eta \text{ be ordinary}, \qquad f_{\eta^+} \text{ be compact}, \qquad f \text{ be ND on } f_{\eta^+} \tag{33.31}$$

are satisfied by the value η of h, understanding that

$$h_{\eta^+} = \{p \in | \hat{M}_{2n} | \mid h(p) \geqslant \eta\}.$$

Theorem 30.3(i) can accordingly be applied if h, $\hat{M}_{2n}$, and $2n$ of Theorem 33.1(i) are taken, respectively, as f, M_n, and n of Theorem 30.3.

We infer from Theorem 30.3(i) that the connectivities of $| \hat{M}_{2n} | \bmod h_\eta$ are finite and vanish for $i > 2n$. The isomorphisms (33.21) imply that these connectivities are those of $| M_n^2 | \bmod \text{diag} | M_n^2 |$. Finally, $\mathscr{R}_0 = 0$, since $| M_n |$ is arcwise connected by hypothesis, so that each point of $| M_n^2 |$ is arcwise connected on $| M_n^2 |$ to the modulus diag $| M_n^2 |$.

Thus Theorem 33.1(i) is true when the chord function h of $\hat{M}_{2n}$ is ND.

Proof of Theorem 33.1(i) *when the Chord Function h is Degenerate.* It follows from the above unpublished theorem of Morse that there exists a manifold N_n, homeomorphic to M_n, conditioned as is M_n, and such that the chord function h on $\hat{N}_{2n}$ is ND. According to Lemma 33.4 the spaces $|M_n^2|$ mod diag $|M_n^2|$ and $|N_n^2|$ mod diag $|N_n^2|$ are topologically equivalent, and hence by Theorem 28.1 have isomorphic homology groups.

Theorem 33.1(i) is accordingly true without exception.

Proof of Theorem 33.1(ii). By hypothesis the chord function h on $\hat{M}_{2n}$ is ND. If $\hat{M}_{2n}$, h, η, and $2n$ of Theorem 33.1(ii) replace M_n, f, η, n, respectively, of Theorem 30.3(ii), then Theorem 33.1(ii) follows from Theorem 30.3(ii) provided the connectivities of $|\hat{M}_{2n}|$ mod h_η are understood to be the connectivities $\mathscr{R}_i$ of $|M_n^2|$ mod diag $|M_n^2|$ in accord with (33.21).

This establishes Theorem 33.1.

Chord Functions of Linking Type. A chord function h which satisfies the conditions of the following theorem will be said to be of *linking type*. These conditions are satisfied in particular by the chord function associated with an ellipsoid with unequal axes when $\mathscr{K}$ is the field of integers mod 2. (see §34).

Let h be a ND chord function on $\hat{M}_{2n}$ with disjoint critical values. Let m_q be the qth *type number* of h, that is, the number of critical points of h of index q. Let a be the minimum of the lengths of critical chords of M_n and η any positive constant $< a$. As previously, for a fixed field $\mathscr{K}$ set

$$\mathscr{R}_q = R_q(|M_n^2|, \text{diag}\ |M_n^2|), \qquad q = 0, 1, \ldots. \tag{33.32}$$

Theorem 33.2. *If under the conditions of the preceding paragraph each critical point of the chord function h is linking* mod h_η, *then*

$$\mathscr{R}_q = m_q, \qquad q = 0, 1, \ldots. \tag{33.33}$$

Proof. It follows from formula (30.15) that for each $q \geqslant 0$

$$R_q(|\hat{M}_{2n}|, h_\eta) = m_q, \tag{33.34}$$

and then from the isomorphisms (33.21) that $\mathscr{R}_q = m_q$.

Index Determination. Given a critical point p_0 of the chord function h of $\hat{M}_{2n}$, the problem of determining whether or not p_0 is ND and, if p_0 is ND, of determining the index of p_0 is equivalent to a similar problem that can be formulated in terms of the chord function

$$(x, y) \to H(x, y) = \| x - y \|, \qquad (x, y) \in | M_n \hat{\times} M_n |. \tag{33.35}$$

Recall that $| M_n \hat{\times} M_n |$ was defined in (33.13). We understand that $M_n \hat{\times} M_n$ is a differentiable manifold with a differentiable structure induced by the differentiable structure of $M_n \times M_n$, of which $M_n \hat{\times} M_n$ is an open submanifold.

We shall prove the following lemma:

Lemma 33.9. *A point (x_0, y_0) in $| M_n \hat{\times} M_n |$ is a* ND *critical point of the chord function H on $M_n \hat{\times} M_n$ of index k if and only if the point $p_0 = T(x_0, y_0)$ in $\hat{M}_{2n}$ is a* ND *critical point of index k of the chord function h on $\hat{M}_{2n}$.*

Let presentations $(F : U, X)$ and $(G : V, Y)$ be given in $\mathscr{D}M_n$ such that $x_0 \in X$, $y_0 \in Y$, and $X \cap Y = \varnothing$. Then presentations

$$((F, G) : U \times V, X \times Y) \in \mathscr{D}(M_n \hat{\times} M_n)$$

and

$$(T(F, G) : U \times V, T(X \times Y)) \in \mathscr{D}\hat{M}_{2n}$$

exist.

The critical points of $H | (X \times Y)$ and of $h | T(X \times Y)$ are determined, respectively, by the local representation

$$(u, v) \to (H \circ (F, G)(u, v) = \| F(u) - G(v) \|, \qquad (u, v) \in U \times V,$$

of H and the local representation

$$(u, v) \to (h \circ T(F, G))(u, v) = \| F(u) - G(v) \|, \qquad (u, v) \in U \times V,$$

of h. The identity of these two local representation of H and h implies Lemma 33.9.

§ 34

THE SYMMETRIC SQUARE OF AN n-SPHERE

In the terminology of §33 the connectivities $\mathscr{R}_i$ of

$$| M_n{}^2 | \bmod \operatorname{diag} | M_n{}^2 |$$

over $\mathscr{K}$ will be called the *relative connectivities* of $| M_n{}^2 |$ over $\mathscr{K}$. The connectivities $\mathscr{R}_i$ may vary with the field $\mathscr{K}$. A particular choice of the field $\mathscr{K}$ is the field $\mathbf{Z}_p$ of integers mod p, where p is a prime. When the field is $\mathbf{Z}_p$ we shall refer to the numbers $\mathscr{R}_i$ as the relative connectivities $\mathscr{R}_i(p)$ of $| M_n{}^2 | \bmod p$. Morse [13], pp. 183–191, proved the following.

Theorem 34.1. *The nonvanishing relative connectivities* mod 2 *of the symmetric product of an n-sphere are*

$$\mathscr{R}_n(2) = \mathscr{R}_{n+1}(2) = \cdots = \mathscr{R}_{2n}(2) = 1. \tag{34.1}$$

In the proof of this theorem by Morse use was made of the singular homology theory in vogue around 1930. We are here concerned with a proof of this theorem by the methods of this book.

We shall explain why the connectivities mod 2 are preferred to those mod an odd prime p. A historical reason is that the difficulties in computing the relative connectivities mod 2 in Theorem 34.1 were first surmounted by the methods of the critical point theory. A more cogent reason, which the reader is asked to take on faith, is that $\mathscr{R}_i(p) \leqslant \mathscr{R}_i(2)$, and when $p > 2$ the sum of the relative connectivities $\mathscr{R}_i(p)$ is less than their sum when $p = 2$.

These relations imply the following: Given a regular C^∞-manifold M_n in E_r, $0 < n < r$, homeomorphic to an n-sphere, the connectivities $\mathscr{R}_i(2)$ together imply more concerning the existence of critical chords of M_n than do the connectivities $\mathscr{R}_i(p)$, at least when the chord function on $\hat{M}_{2n}$ is ND.

In §34 we shall give a proof of Theorem 34.1, omitting, however, a proof of Lemma 34.2. This lemma affirms that the critical points of the chord function of the elliptical manifold $\mathscr{E}_n$ are of linking type when diag $| \mathscr{E}_n^2 |$ is a modulus. The proof of this lemma is relatively long, and it has seemed desirable to limit the pages spent on symmetric products. However, a complete proof of this lemma will be published separately, further extending the theory of symmetric products.

The Ellipsoid. Let $\mathscr{E}_n$ be the regular analytic manifold in E_{n+1} with a carrier $| \mathscr{E}_n |$ defined by the condition

$$\frac{4x_1^2}{c^2(1)} + \frac{4x_2^2}{c^2(2)} + \cdots + \frac{4x_{n+1}^2}{c^2(n+1)} = 1, \tag{34.2}$$

where

$$0 < c(1) < c(2) < \cdots < c(n+1). \tag{34.3}$$

We distinguish $\mathscr{E}_n$ from its carrier by calling $\mathscr{E}_n$ an elliptical manifold. The critical chords of $\mathscr{E}_n$ on the respective coordinate axes of E_{n+1} will be denoted by

$$g(1), g(2), \ldots, g(n+1). \tag{34.4}$$

With μ on the range $1, 2, \ldots, n+1$ the length of $g(\mu)$ is $c(\mu)$.

S_n shall denote the n-sphere structured as a regular differentiable manifold in E_{n+1}. Lemma 33.4 implies that $| \mathscr{E}_n^2 |$ is homeomorphic to $| S_n^2 |$ and $| \hat{\mathscr{E}}_{2n} |$ to $| \hat{S}_{2n} |$. The relative connectivities $\mathscr{R}_i$ of $| S_n^2 |$ are accordingly those of $| \mathscr{E}_n^2 |$ regardless of the field $\mathscr{K}$.

The following elementary proposition requires verification:

(i) *$\mathscr{E}_n$ has no critical chords other than its principal axes* $g(1), \ldots, g(n+1)$.

To prove (i), it is sufficient to verify the following two propositions: A straight line meeting $\mathscr{E}_n$ orthogonally in two points meets the origin. The *only* critical chords of $\mathscr{E}_n$ which meet the origin are the principal axes of $\mathscr{E}_n$.

Notation for Theorem 34.1. Differentiable manifolds $\mathscr{E}_n \times \mathscr{E}_n$, $\mathscr{E}_n \hat{\times} \mathscr{E}_n$, and $\hat{\mathscr{E}}_{2n}$, are understood in the senses defined in §33. For μ on its range 1, 2,..., $n + 1$ let $\alpha(\mu)$ and $-\alpha(\mu)$ be the endpoints of the chord $g(\mu)$, supposing that $x_\mu > 0$ at $\alpha(\mu)$ and $x_\mu < 0$ at $-\alpha(\mu)$. The chord function H on $\mathscr{E}_n \hat{\times} \mathscr{E}_n$ and the chord function h on $\hat{\mathscr{E}}_{2n}$ both have the critical values $c(1), c(2),..., c(\mu)$ and no other critical values. This follows from proposition (i).

These critical values are taken on by H at the critical points (x, y) in the set

$$(\alpha(1), -\alpha(1)),..., (\alpha(n + 1), -\alpha(n + 1)), \tag{34.5}$$

as well as those in the complementary set

$$(-\alpha(1), \alpha(1)),..., (-\alpha(n + 1), \alpha(n + 1)). \tag{34.5'}$$

The critical points of h are the images on $\hat{\mathscr{E}}_{2n}$ under T of the points (34.5), or, equivalently, of the points (34.5′) (for T see §33). These critical points of h will be denoted by

$$P_{c(1)}, ..., P_{c(n+1)}, \tag{34.6}$$

where the subscripts give the respective critical values.

Theorem 34.2. *The critical points* (34.5) *of H and* (34.6) *of h are* ND, *and have the respective indices*

$$n, n + 1, ..., 2n. \tag{34.7}$$

According to Lemma 33.9, to prove Theorem 34.2, it is sufficient to prove the theorem for the chord function H on $\mathscr{E}_n \hat{\times} \mathscr{E}_n$. To this end, we shall prove the following lemma:

Lemma 34.1. *For each integer μ on the range* 1,..., $n + 1$ *the critical point* $(\alpha(\mu), -\alpha(\mu))$ *of the chord function H is* ND *and has the index*

$$k(\mu) = n + \mu - 1. \tag{34.8}$$

The presentations in $\mathscr{D}\mathscr{E}_n$ to be used in proving the lemma will be taken as Monge presentations (see §5).

A Monge Presentation $(F_\mu : U, X_\mu) \in \mathscr{D}\mathscr{E}_n$. Given μ on its range $1, 2,\ldots, n+1$ and the endpoint $\alpha(\mu)$ of $g(\mu)$, we shall define a Monge presentation

$$(F_\mu : U, X_\mu) \in \mathscr{D}\mathscr{E}_n \tag{34.9}$$

of an open subset X_μ of $\mathscr{E}_n$ that contains $\alpha(\mu)$.

The domain U of F_μ will be taken in the coordinate n-plane of E_{n+1} on which $x_\mu = 0$. Let $x \to P_\mu(x)$ project E_{n+1} orthogonally onto this n-plane. For arbitrary $x \in E_{n+1}$ it is convenient to set

$$(x_1 ,\ldots, x_{\mu-1} , \check{x}_\mu , x_{\mu+1} ,\ldots, x_{n+1}) = (x)_\mu ,$$

where the sign $\check{}$ above a symbol indicates deletion of the symbol.

Supposing that $\| u \| < c(1)/2$ for $u \in U$, we shall set $(x)_\mu = (u_1 ,\ldots, u_n) = u$ for points $x \in E_{n+1}$ such that $P_\mu(x) \in U$. With U so conditioned there exists a unique open subset X_μ of $\mathscr{E}_n$ such that the endpoint $\alpha(\mu)$ is in X_μ, and X_μ has a real analytic Monge presentation of form (34.9).

Notation. It will be convenient to set

$$\left(\frac{1}{c(1)}, \ldots, \frac{1}{c(\mu-1)}, \frac{\check{1}}{c(\mu)}, \frac{1}{c(\mu+1)}, \ldots, \frac{1}{c(n+1)}\right) = (a_1 ,\ldots, a_n), \tag{34.10}$$

noting that $a_1 > a_2 > \cdots > a_n$. Products such as $a_q^2 u_q^2$ will be summed for q on the range $1,\ldots, n$. It should be noted that (34.2) and (34.10) imply that

$$4a_q^2 u_q^2 < 1 \qquad \text{when} \quad (x)_\mu = u \in U. \tag{34.11}$$

If $u \in U$ and $x = F_\mu(u)$, (34.2) implies that

$$x_\mu = \tfrac{1}{2}c(\mu)(1 - 4a_q^2 u_q^2)^{1/2}, \qquad (x)_\mu = u, \qquad \text{for} \quad u \in U. \tag{34.12}$$

The condition $\| u \| < c(1)/2$ on $u \in U$ implies (34.11), and hence that x_μ in (34.12) is given by an absolutely convergent power series in the variables $u_1 ,\ldots, u_n$, so that for $u \in U$ and $x \in F_\mu(u)$

$$x_\mu = \tfrac{1}{2}c(\mu)(1 - 2a_q^2 u_q^2 + \cdots), \qquad (x)_\mu = u, \qquad \text{for} \quad u \in U, \tag{34.13}$$

up to terms of the third order in the series for x_μ .

A Second Monge Presentation $G_\mu \in \mathscr{D}\mathscr{E}_n$. The mapping

$$x \to \rho(x) = -x : E_{n+1} \to E_{n+1} \qquad \text{onto} \quad E_{n+1}$$

is a reflection of E_{n+1} in the origin. Given the Monge presentation F_μ as in (34.9), a second Monge presentation in $\mathscr{D}\mathscr{E}_n$ is defined by setting

$$(G_\mu : V, Y_\mu) = (\rho \circ F_\mu : U, \rho(X)_\mu), \tag{34.14}$$

so that $V = U$ and $G_\mu = -F_\mu$. It follows from the preceding paragraph that for $v \in U$ and $y = G_\mu(v)$

$$y_\mu = -\tfrac{1}{2}c(\mu)(1 - 2a_q^2 v_q^2 + \cdots); \qquad (y)_\mu = v \qquad v \in U. \tag{34.15}$$

Note that $F_\mu(\mathbf{0}) = \alpha(\mu)$, $G_\mu(\mathbf{0}) = -\alpha(\mu)$.
The product presentation

$$\{(F_\mu, G_\mu) : U \times V, X_\mu \times Y_\mu\} \in \mathscr{D}(\mathscr{E}_n \hat{\times} \mathscr{E}_n) \tag{34.16}$$

is well-defined, as is the representation

$$(u, v) \to \| F_\mu(u) - G_\mu(v) \| = \| x - y \| = H(x, y)$$

of H when $x = F_\mu(u)$ and $y = G_\mu(v)$. Note that

$$\| x - y \| = (\|(x)_\mu - (y)_\mu\|^2 + (x_\mu - y_\mu)^2)^{1/2}. \tag{34.17}$$

It follows from (34.17), (34.13), and (34.15) that when $u \in U$, $v \in U$, and $x = F_\mu(u)$ and $y = G_\mu(v)$,

$$\begin{aligned} H(x, y) &= \{\| u - v \|^2 + c^2(\mu)[1 - a_q^2(u_q^2 + v_q^2) + \cdots]^2\}^{1/2} \\ &= c(\mu)\left(1 + \frac{1}{c^2(\mu)}\| u - v \|^2 - 2a_q^2(u_q^2 + v_q^2) + \cdots\right)^{1/2}, \end{aligned}$$

so that for the given μ, $H(x, y)/c(\mu)$ has the value

$$1 + \frac{1}{2}\left[\left(\frac{1}{c^2(\mu)} - a_q^2\right)(u_q - v_q)^2 - a_q^2(u_q + v_q)^2\right] + \ldots, \tag{34.18}$$

where the bracket is summed for q on its range $1, \ldots, n$.

The index of the μth critical point $(\alpha(\mu), -\alpha(\mu))$ of H is the index of the quadratic form Q in the series (34.18). One sees that the unsummed form $Q_{\mu q}$ in u_q and v_q in the series (34.18) is ND and has the index 2 or 1 depending on whether a_q is greater or less than $1/c(\mu)$. It follows that Q is a ND quadratic form in its $2n$ variables and has the index $2(\mu - 1) + (n - \mu + 1) = n + \mu - 1$, establishing Lemma 34.1.

Theorem 34.2 follows from Lemmas 34.1 and 33.9.

We state a lemma whose truth is implied by the analysis of Morse [13], pp. 186–191, and which, as stated previously, will be proved by the methods of this book in a later paper.

Lemma 34.2. *If η is a positive constant less than the minimum length of critical chords of $\mathscr{E}_n$, then each of the critical points of the chord function h is of linking type* mod h_η *over the field* $\mathbf{Z}_2$ (see §29).

Proof of Theorem 34.1, *Granting Lemma* 34.2. By virtue of Lemma 34.2 the condition of Theorem 33.2 that the critical points of the chord function h be of linking type mod h_η is satisfied. Theorem 33.2 then affirms that each connectivity $\mathscr{R}_q$ is equal to m_q, the qth type number of h. According to Theorem 34.2 the nonvanishing type numbers of h are

$$m_n = m_{n+1} = \cdots = m_{2n} = 1,$$

so that Theorem 34.1 follows, assuming, as in Theorem 34.1, that $\mathscr{K} = \mathbf{Z}_2$.

§ 35

THE COMPLEX PROJECTIVE n-SPACE $\mathbf{CP}_n$

Introduction. The complex projective n-space $\mathbf{CP}_n$, as defined below, has a complex dimension n. With the definition of a suitable structure $\mathbf{CP}_n$ becomes a compact differentiable manifold M_{2n} of class C^∞. Milnor ([2], pp. 25–27) has defined a real-valued ND function g on M_{2n} with the following properties. The critical points of g are $n + 1$ in number with indices

$$0, 2, \ldots, 2(n - 1), 2n \tag{35.1}$$

and critical values which increase with the indices of the critical point. We shall term such a function a *Milnor function g*.

Granting the existence of a Milnor function g on M_{2n}, the "lacunary theorem" of Morse ([7], p. 151) or Corollary 30.2 implies the following:

Theorem 35.1. *The connectivities R_q, $q \geqslant 0$, over $\mathscr{K}$ of M_{2n} are independent of the choice of the field, and in order of their dimensions are alternatingly* 1 *or* 0 *for $q \leqslant 2n$, with $R_q = 0$ for $q > 2n$.*

To establish Theorem 35.1, we shall define and topologize $\mathbf{CP}_n$. The topologized $\mathbf{CP}_n$ will be shown to be a compact, connected, topological manifold $|\mathbf{CP}_n|$ which admits a differentiable structure as a C^∞-manifold M_{2n}. On M_{2n} a C^∞-function g will then be defined and shown to be a Milnor function.

$\mathbf{CP}_n$ *Defined.* Let $\mathbf{CE}_{n+1}$ be the Cartesian product of $n + 1$ complex planes of points $z_0, z_1, \ldots, z_n$ respectively. Let $\mathbf{CE}^0_{n+1}$ denote $\mathbf{CE}_{n+1}$ with the origin deleted from $\mathbf{CE}_{n+1}$. Let $\mathbf{z} = (z_0, z_1, \ldots, z_n)$ denote an arbitrary point of $\mathbf{CE}^0_{n+1}$ (cf. Chern [1], pp. 1, 2).

Two points $\mathbf{z}'$ and $\mathbf{z}''$ in $\mathbf{CE}^0_{n+1}$ are termed *projectively equivalent* if there exist nonvanishing complex numbers ρ' and ρ'' such that $\rho'\mathbf{z}' = \rho''\mathbf{z}''$. This relation of equivalence is reflexive, symmetric, and transitive.

Let $\varphi(\mathbf{z})$ denote the projective equivalence class of an arbitrary $\mathbf{z} \in \mathbf{CE}^0_{n+1}$. The set $\mathbf{CP}_n$ of equivalence classes $\varphi(\mathbf{z})$ is defined by the mapping

$$\mathbf{z} \to \varphi(\mathbf{z}) : \mathbf{CE}^0_{n+1} \to \mathbf{CP}_n , \tag{35.2}$$

regarded as onto. The space $\mathbf{CE}_{n+1}$ has a well-defined product topology. Taken as a subspace of $\mathbf{CE}_{n+1}$, $\mathbf{CE}^0_{n+1}$ has an induced topology. One topologizes $\mathbf{CP}_n$ by requiring that a subset X of $\mathbf{CP}_n$ be *open* if and only if it is the image under φ of an open subset Y of $\mathbf{CE}^0_{n+1}$. The set $\varphi^{-1}(X)$ will then be the ensemble of points in $\mathbf{CE}^0_{n+1}$ projectively equivalent to points in Y, and will be open in $\mathbf{CE}^0_{n+1}$. It follows that φ is continuous. $\mathbf{CP}_n$, so topologized, is a Hausdorff space $| \mathbf{CP}_n |$.

Lemma 35.1. *The space $| \mathbf{CP}_n |$ is a connected, compact, topological $2n$-manifold which can be differentiably structured so as to be a C^∞-manifold M_{2n}.*

$| \mathbf{CP}_n |$ is connected, since it is the continuous image under φ of the connected space $\mathbf{CE}^0_{n+1}$.

The Compactness of $| \mathbf{CP}_n |$. Given $\mathbf{z} \in \mathbf{CE}^0_{n+1}$ set

$$\| \mathbf{z} \| = (| z_0 |^2 + \cdots + | z_n |^2)^{1/2}, \qquad K = \{\mathbf{z} \in \mathbf{CE}^0_{n+1} \mid \| \mathbf{z} \| = 1\}. \tag{35.3}$$

Topologized by $\mathbf{CE}^0_{n+1}$, K is compact. The mapping $\varphi \mid K$ of K is onto $| \mathbf{CP}_n |$ and continuous. Hence $| \mathbf{CP}_n |$ is compact.

$| \mathbf{CP}_n |$ *as a Topological Manifold.* Let $U_n \times V_n$ be the real $2n$-plane of points

$$(u, v) = (u_1 , v_1 ; u_2 , v_2 ; \cdots ; u_n , v_n). \tag{35.4}$$

We shall define $n + 1$ presentations

$$(F^k : U_n \times V_n , X^k), \qquad k = 0, 1,..., n, \tag{35.5}$$

of open subspaces $X^0, X^1,..., X^n$ of $| \mathbf{CP}_n |$ whose union is $| \mathbf{CP}_n |$.

For each k on the range 0, 1,..., n and each point $(u, v) \in U_n \times V_n$ a point $\mathbf{z} = (z_0, z_1, ..., z_n)$ in $\mathbf{CE}^0_{n+1}$ of the form

$$\mathbf{z}(u, v; k) = (u_1 + iv_1, ..., u_k + iv_k, 1, u_{k+1} + iv_{k+1}, ..., u_n + iv_n) \tag{35.6}$$

is introduced with $z_k = 1$. We then set

$$F^k(u, v) = \varphi(\mathbf{z}(u, v; k)) \in \mathbf{CP}_n, \qquad (u, v) \in U_n \times V_n \tag{35.7}$$

and $F^k(U_n \times V_n) = X^k$. The set

$$G^k = \{\mathbf{z} \in \mathbf{CE}^0_{n+1} \mid z_k \neq 0\}$$

is open in $\mathbf{CE}^0_{n+1}$ and $X^k = \varphi(G^k)$. Hence X^k is *open* in $|\mathbf{CP}_n|$ by definition of open sets in $|\mathbf{CP}_n|$.

The mapping F^k is *biunique*, since distinct points (u, v) and (u', v') in $U_n \times V_n$ yield points $\mathbf{z}(u, v; k)$ and $\mathbf{z}(u', v'; k)$ in $\mathbf{CE}^0_{n+1}$ which are projectively nonequivalent.

The mapping F^k of $U_n \times V_n$ into the Hausdorff space $|\mathbf{CP}_n|$ is continuous, since φ is continuous and $F^k(u, v)$ defined by (35.7). The mapping F^k is bicontinuous, since the restriction of F^k to a compact $2n$-ball in $U_n \times V_n$ is both biunique and continuous. Hence F^k is a homeomorphism onto X_k. Moreover,

$$|\mathbf{CP}_n| = X^0 \cup X^1 \cup \cdots \cup X^n. \tag{35.8}$$

Thus $|\mathbf{CP}_n|$ is a topological $2n$-manifold covered by the presentations

$$F^0, F^1, ..., F^n. \tag{35.9}$$

The C^∞-Compatibility of the Presentations F^k. Let h and k be distinct integers on the range 0, 1,..., n with $h > k$. We shall establish the C^∞-compatibility of F^k and F^h.

To that end, let $(u, v) \in U_n \times V_n$ be F^k-parameters on X^k and let $(u', v') \in U_n \times V_n$ be F^h-parameters on X^h. Set

$$Z_{2n} = (F^k)^{-1}(X^k \cap X^h); \qquad Z'_{2n} = (F^h)^{-1}(X^k \cap X^h). \tag{35.10}$$

We wish to show that the transition homeomorphism

$$\lambda_{kh} = (F^h \mid Z'_{2n})^{-1} \circ (F^k \mid Z_{2n}) : Z_{2n} \to Z'_{2n} \qquad \text{(onto)} \tag{35.11}$$

associated with F^k and F^h is a C^∞-diffeomorphism.

Points $(u, v) \in Z_{2n}$ and $(u', v') \in Z'_{2n}$ correspond under the transition homeomorphism λ_{kh} if and only if the corresponding points $\mathbf{z}(u, v; k)$

and $\mathbf{z}(u', v'; h)$ in $\mathbf{CE}^0_{n+1}$ are projectively equivalent, that is, if and only if

$$(u_h + iv_h)(u'_{k+1} + iv'_{k+1}) = 1 \tag{35.12}$$

and

$$\mathbf{z}(u', v'; h) = \frac{\mathbf{z}(u, v; k)}{u_h + iv_h}, \qquad (u, v) \in Z_{2n}, \tag{35.13a}$$

$$\mathbf{z}(u, v; k) = \frac{\mathbf{z}(u', v'; h)}{u'_{k+1} + iv'_{k+1}}, \qquad (u', v') \in Z'_{2n}. \tag{35.13b}$$

One could omit (35.13a) or (35.13b) and the preceding statement would remain true. One sees that λ_{kh} is a real analytic diffeomorphism of Z_{2n} onto Z'_{2n} uniquely determined by the relations (35.13) between (u, v) and (u', v').

It follows that the $n + 1$ presentations F^k of open subsets of $| \mathbf{CP}_n |$ are C^∞-compatible.

Hence $| \mathbf{CP}_n |$ is the carrier of a C^∞-manifold M_{2n}, in accord with Definition 13.1.

This completes the proof of Lemma 35.1.

Milnor's Function g on $\mathbf{CP}_n$. Corresponding to an arbitrary point $\mathbf{z} \in \mathbf{CE}^0_{n+1}$ and to any set of real constants $c_0 < c_1 < \cdots < c_n$, a real-valued function g on $\mathbf{CP}_n$, or, equivalently, on M_{2n}, will be defined by setting

$$g(\varphi(\mathbf{z})) = c_i \mid z_i \mid^2 / \| \mathbf{z} \|^2. \tag{35.14}$$

(summing with respect to i on the range $0, 1, \ldots, n$). We shall prove that g is a Milnor function on M_{2n} with critical values $c_0, c_1, \ldots, c_n$.

Note first that g is real-valued. It is uniquely defined at each point of $\mathbf{CP}_n$, since the right side of (35.14) is the same for any two points $\mathbf{z} \in \mathbf{CE}^0_{n+1}$ which are projectively equivalent.

The Critical Points of g on M_{2n}. The critical points of g on X^k (if any exist) are the images under F^k of the critical points $(u_0, v_0) \in U_n \times V_n$ of the function

$$(u, v) \to (g \circ F^k)(u, v) = g(\varphi(\mathbf{z}(u, v; k))) = g_k(u, v), \quad (u, v) \in U_n \times V_n, \tag{35.15}$$

introducing g_k (see Definition 5.5).

Note that

$$\| \mathbf{z}(u, v; k)\|^2 = 1 + \omega^2(u, v), \qquad (u, v) \in U_n \times V_n ,$$

where $\omega^2(u, v)$ is the quadratic form

$$\omega^2(u, v) = \| u \|^2 + \| v \|^2, \qquad u \in U_n ; \quad v \in V_n .$$

From (35.14) and (35.15) it follows that for $(u, v) \in U_n \times V_n$

$$\begin{aligned}(1 + \omega^2(u, v))\, g_k(u, v) - c_k \\ = c_0(u_1{}^2 + v_1{}^2) + \cdots + c_{k-1}(u_k{}^2 + v_k{}^2) \\ + c_{k+1}(u_{k+1}^2 + v_{k+1}^2) + \cdots + c_n(u_n{}^2 + v_n{}^2).\end{aligned} \tag{35.16}$$

Subtracting $c_k\omega^2(u, v)$ from both sides of (35.16), we find that

$$\begin{aligned}(1 + \omega^2(u, v))(g_k(u, v) - c_k) \\ = (c_0 - c_k)(u_1{}^2 + v_1{}^2) + \cdots + (c_{k-1} - c_k)(u_k{}^2 + v_k{}^2) \\ + (c_{k+1} - c_k)(u_{k+1}^2 + v_{k+1}^2) + \cdots + (c_n - c_k)(u_n{}^2 + v_n{}^2).\end{aligned} \tag{35.17}$$

Thus g_k is real-analytic on $U_n \times V_n$.

A straightforward computation, using both (35.14) and (35.17), shows that the only critical point of g_k on $U_n \times V_n$ occurs at the origin in $U_n \times V_n$ and that the corresponding critical value is c_k . The quadratic terms in the Taylor's expansion of g_k about the origin are the terms on the right of (35.17). Taking into account the ordering $c_0 < c_1 < \cdots < c_n$, one sees that the origin is a ND critical point of g_k of index $2k$.

Thus g is a Milnor function on M_{2n} with critical values $c_0 < c_1 < \cdots < c_n$, assumed respectively at unique ND critical points with indices 0, 2, 4,..., $2n$.

Theorem 35.1 *follows from the "lacunary principle" enunciated in Corollary* 30.2.

The critical point of g on $\mathbf{CP}_n$ with index $2k$ is the projective equivalence class of the point $\mathbf{z}(0, 0; k) \in \mathbf{CE}_{n+1}^0$, that is, the point in $\mathbf{CE}_{n+1}^0$ whose ith complex coordinate z_i is $\delta_i{}^k$.

From Appendix III and Theorem 35.1 we infer that the Betti numbers of $\mathbf{CP}_n$ are equal to the corresponding connectivities, and there are no torsion coefficients. The homology groups $H_i(\mathbf{CP}_n , \mathbf{Z})$

over the ring $\mathbf{Z}$ of integers are thus trivial for i odd, and for i on the range $0, 2, 4, \ldots, 2n$ are cyclic, isomorphic with $\mathbf{Z}$.

Remark. The problem of determining whether or not a critical point of a Milnor function g is of *linking* type has not entered because the "lacunary" nature of the sequence of indices of the critical points of g implies that each critical point is of linking type. In problems of more general type the set of indices of the critical points will not in general be of lacunary type. In these more general problems it is necessary to associate a definite k-cap z^k with each critical point of positive index k in order to determine whether or not this k-cap is linkable (see §29).

The next section is concerned with this problem.

§ 36

CAPS AND SADDLES

Objective of §36. Let a ND $f \in C^\infty(M_n)$ be given such that f_c is compact for each value c of f. As affirmed in Theorem 23.5, such an f exists on a prescribed C^∞-manifold M_n.

Given a field $\mathscr{K}$, the critical points of f, with their indices and their linking characteristics, determine, up to an isomorphism, the homology groups over $\mathscr{K}$ of each sublevel set f_c of M_n, in accord with Theorem 29.3. The importance of simple criteria as to whether or not a critical point p_a of f is of linking type is clear.

A critical point p_a at the f-level a, of positive index k, is of linking type by definition in §29 if and only if given a k-cap z^k associated with p_a, $\partial z^k \sim 0$ on $\dot{f}_a$. A problem of first importance is accordingly to find or recognize a k-cap belonging to a prescribed critical point p_a. The characterization of a k-cap z^k belonging to p_a should be independent of the particular presentations in $\mathscr{D}M_n$ used to find or recognize a k-cap.

The two main theorems of this section meet these objectives. The first theorem, termed the *Saddle Theorem*, applies in case $0 < k < n$ and reduces the problem of characterizing a k-cap of p_a to the problem when $k = n$. The second theorem, termed the *Carrier Theorem*, gives an effective characterization of a k-cap of p_a when $k = n$. The case $k = 0$ is trivial.

Notation. Corresponding to a prescribed positive dimension and a number $\rho > 0$, let B_j^ρ be an origin-centered open j-ball of radius ρ in any Euclidean space of dimension j, and let $\mathbf{B}_j^\rho$ be the C^∞-manifold with carrier B_j^ρ and with a Euclidean differentiable structure.

Our first lemma shows the essentially *local* character of a k-cap of a critical point p_a of f.

Lemma 36.0. (α) *Given a presentation $(F : U, X) \in \mathscr{D}M_n$ such that $p_a \in X$, then p_a is a critical point of $f \mid X$ and has the same index relative to $f \mid X$ as to f.*

(β) *Moreover, a k-cap of p_a relative to $f \mid X$ is a k-cap of p_a relative to f.*

The truth of (α) is clear.

Proof of (β). We shall make use of a coset-contracting isomorphism

$$H_k(f_a, \dot{f}_a) \approx H_k(f_a \cap X, \dot{f}_a \cap X). \tag{36.1}$$

This isomorphism is a consequence of the Excision Theorem 28.3 on setting $\chi = f_a$, $A = \dot{f}_a$, $A^* = f_a \cap (M_n - X)$, and noting that $p_a = \chi - A$.

By virtue of (29.7) there is just one element in a "prebase" for the homology group of either member of (36.1). A k-cap z^k of p_a relative to $f \mid X$ is such an element for the group on the right of (36.1). Since (36.1) is a coset-contracting isomorphism, z^k is a k-cap of p_a relative to f.

Thus Lemma 36.0 is true.

Canonical Coordinates near P_a. Before defining saddle manifolds and stating the Saddle Theorem we shall define a special presentation in $\mathscr{D}M_n$, termed *f-canonical*, of a neighborhood of the critical point p_a of f. Such a presentation is defined with the aid of Theorem 22.2. We suppose $0 < k < n$.

Let E_n be regarded as a Cartesian product, $U_k \times V_{n-k}$, of Euclidian spaces U_k and V_{n-k} of dimensions k and $n - k$, respectively. Let $u = (u_1, ..., u_k)$ be an arbitrary point in U_k and $v = (v_1, ..., v_{n-k})$ an arbitrary point in V_{n-k}. Let O_u and O_v denote the origin in U_k and V_{n-k}, respectively.

It follows from Theorem 22.2 that if σ is a sufficiently small positive constant, there exists a presentation

$$(G : B_k^\sigma \times B_{n-k}^\sigma, Y^\sigma) \in \mathscr{D}M_n \tag{36.2}$$

such that $G(O_u, O_v) = p_a$ and

$$(f \circ G)(u, v) = a - \| u \|^2 + \| v \|^2, \qquad (u, v) \in B_k^\sigma \times B_{n-k}^\sigma. \tag{36.3}$$

Definition 36.1. *G and Y^σ.* We term G an *f-canonical presentation* of a canonical neighborhood Y^σ of p_a relative to M_n and term the G-coordinates (u, v) *canonical f-coordinates* on Y^σ.

Having shown in Lemma 36.0 that the determination of a k-cap of p_a is a local matter, we shall narrow the determination of a k-cap z^k of p_a, $0 < k < n$, still further by the use of k-dimensional "saddles" on M_n associated with p_a. As previously, p_a is a critical point of f at the f-level a.

Definition 36.2. An f-Saddle L_k of M_n at p_a. A C^∞-manifold L_k, $0 < k < n$, which is the C^∞-diffeomorph of an open Euclidean k-ball B_k^e and which is C^∞-embedded in M_n so as to meet p_a will be termed an *f-saddle* of M_n *at* p_a if, together with $|\dot{L}_k| = |L_k - p_a|$, it has the properties:

(i) *The point p_a is a* ND *critical point of $f \mid L_k$ of index k.*

(ii) $|\dot{L}_k| \subset \dot{f}_a$.

We can suppose that B_k^e is in a k-plane of Cartesian coordinates $\alpha_1, \ldots, \alpha_k$ and that there is a homeomorphism onto $|L_k|$ of form

$$\alpha \to F(\alpha) : B_k^e \to |L_k|, \qquad F(\mathbf{0}) = p_a, \tag{36.4}$$

which C^∞-embeds $\mathbf{B}_k^e$ in M_n. Then F is in $\mathscr{D}L_k$ and determines $\mathscr{D}L_k$. If an f-saddle L_k exists at p_a, there exists a submanifold L_k^* of L_k with carrier included in a *prescribed* neighborhood of p_a relative to M_n and with L_k^* again an f-saddle of M_n at p_a. We term L_k^* a *subsaddle* of L_k.

That f-saddles exist is shown by the following lemma:

Lemma 36.1. *The restriction, both of domain and range, of form*

$$(u, O) \to G(u, O) : B_k^\sigma \to \hat{L}_k \qquad \text{onto } \hat{L}_k \tag{36.4$'$}$$

of the canonical presentation $G \in \mathscr{D}M_n$, given by (36.2), *defines an f-saddle $\hat{L}_k$ of M_n at p_a.*

Let ψ be the inclusion mapping of $\hat{L}_k$ into M_n. The inclusion ψ defines a C^∞-embedding of $\hat{L}_k$ in M_n. The test (16.3) for such an embedding is satisfied on taking φ, G, and F of (16.3), respectively, as ψ, G of (36.2), and the presentation (36.4$'$). That conditions (i) and (ii) of Definition 36.2 are satisfied by $\dot{L}_k$ follows from (36.3) and the relation $G(O_u, O_v) = p_a$.

Most of the remainder of this section will be spent in proving the following fundamental theorem:

Saddle Theorem 36.1. *If L_k is an f-saddle of M_n at a critical point p_a of f of positive index $k < n$, then a k-cap z^k of p_a, relative to $f \mid L_k$, is a k-cap of p_a on M_n relative to f.*

The following lemma both motivates and simplifies the proof of Theorem 36.1:

Lemma 36.2. *If the conclusion of Theorem* 36.1, *modified by replacing L_k by some subsaddle L_k^* of L_k, is true, then Theorem* 36.1 *is true as stated.*

Proof. Given the k-cap z^k of p_a relative to $f \mid L_k$, as in Theorem 36.1, we must prove under the hypothesis of Lemma 36.2 that

$$z^k \not\sim 0 \qquad \text{on} \quad f_a \bmod \dot{f}_a. \tag{36.5}$$

Let u^k be a k-cap of p_a relative to $f \mid L_k^*$. By virtue of Lemma 36.0 u^k is a k-cap relative to $f \mid L_k$. Since u^k and z^k are both k-caps relative to $f \mid L_k$ it follows from (29.7) that on $\mid L_k \mid \bmod \mid \dot{L}_k \mid$, for some non-null r_1 and r_2 in $\mathscr{K}$,

$$r_1 u^k \sim r_2 z^k, \tag{36.6}$$

and hence that (36.6) holds on $f_a \bmod \dot{f}_a$. By hypothesis of Lemma 36.2 $u^k \not\sim 0$ on $f_a \bmod \dot{f}_a$, so that (36.5) follows from the validity of (36.6) on $f_a \bmod \dot{f}_a$.

This establishes Lemma 36.2.

Presentations F of f-Saddles L_k. Of the presentations F of f-saddles of form (36.4) there are some essentially simpler than others. If $\mid L_k \mid$ is in a sufficiently small neighborhood of p_a relative to $\mid M_n \mid$, L_k admits a special presentation which we shall term *simple* and presently define. It will be seen that for f-saddles L_k which are "simply" presented, the Saddle Theorem can be readily proved. We shall also see that each f-saddle has a "simply" presented f-subsaddle, so that by Lemma 36.2 the Saddle Theorem is true for arbitrary f-saddles.

In order to define *simple* presentations of f-saddles, "canonical" coordinates (u, v) near p_a are needed. Recall that Y^σ is the open neighborhood of p_a on M_n represented by canonical coordinates (u, v).

f-Saddles L_k Carried by Y^σ. The C^∞-embedding by F in (36.4) of $\mathbf{B}_k^e$ in M_n as L_k satisfies the embedding condition of Definition 16.2. If $| L_k |$ is included in the range Y^σ of the f-canonical presentation G given by (36.2), the mapping

$$G^{-1} \circ F : B_k^e \to B_k^\sigma \times B_{n-k}^\sigma \tag{36.7}$$

is well-defined, and by the embedding condition on F is a biunique C^∞-mapping into the domain of G of the form

$$\alpha \to (u(\alpha), v(\alpha)) : B_k^e \to B_k^\sigma \times B_{n-k}^\sigma . \tag{36.8}$$

It implies a C^∞-representation

$$F(\alpha) = G(u(\alpha), v(\alpha)), \qquad \alpha \in B_k^e , \tag{36.9}$$

of the presentation F of L_k. In particular, it can happen that the local F-coordinates $(\alpha_1 ,..., \alpha_k)$ have been so chosen that (36.9) takes the form

$$F(\alpha) = G(\alpha, \mathbf{v}(\alpha)), \qquad \alpha \in B_k^e, \tag{36.10}$$

where $\alpha \to \mathbf{v}(\alpha)$ is a C^∞-mapping of B_k^e into B_{n-k}^σ. This leads to a definition.

Definition 36.3. *Simply Presented f-Saddles L_k.* If an f-saddle L_k at p_a is so small in diameter that $| L_k |$ is included in the open neighborhood Y^σ of p_a and if a presentation F of L_k of form (36.4) admits a representation of form (36.10), then L_k will be said to be *simply* presented by F.

The following lemma, taken with Lemma 36.2, enables us to prove the Saddle Theorem 36.1:

Lemma 36.3. *Each f-saddle L_k of M_n at p_a has a subsaddle L_k^* at p_a which admits a "simple" presentation F^*.*

Proof. According to (36.3) and (36.9) if F is the presentation (36.4) of L_k, then

$$(f \circ F)(\alpha) = a - \| u(\alpha)\|^2 + \| v(\alpha)\|^2, \qquad \alpha \in B_k^e. \tag{36.11}$$

By Condition (i) on L_k of Definition 36.2 p_a is a critical point of $f \mid L_k$ of index k. This implies that the quadratic form in the Taylor's development about the origin of the right side of (36.11) in terms of the variables $\alpha_1, \ldots, \alpha_k$ is a negative definite quadratic form in $\alpha_1, \ldots, \alpha_k$. This is possible only if the quadratic form in a similar Taylor's development of $-\| u(\alpha)\|^2$ is a negative-definite quadratic form. From this we infer that

$$\frac{D(u_1(\alpha), \ldots, u_k(\alpha))}{D(\alpha_1, \ldots, \alpha_k)} (O_\alpha) \neq 0. \tag{36.12}$$

When (36.12) holds the mapping $\alpha \to u(\alpha)$ of B_k^e into E_k, if restricted in domain to a suitable open neighborhood N of the origin in B_k^e, can be restricted in range so as to have an inverse

$$u \to \theta(u) : B_k^\rho \to N \subset E_k, \tag{36.13}$$

which, for suitable positive ρ, is a C^∞-diff of B_k^ρ onto N. From the presentation F of L_k given in (36.4) one obtains a presentation

$$u \to F^*(u) = F(\theta(u)) : B_k^\rho \to L_k^* \tag{36.14}$$

in $\mathscr{D}L_k$ of an f-subsaddle L_k^* of M_n at p_a with carrier $F^*(B_k^\rho)$. By virtue of (36.14) and (36.9)

$$F^*(u) = G(u(\theta(u)), v(\theta(u))) = G(u, \mathbf{v}(u)), \qquad u \in B_k^\rho, \tag{36.15}$$

where we have set $v(\theta(u)) = \mathbf{v}(u)$. Thus the subsaddle L_k^* of L_k is "simply" presented by F^*.

Proof of the Saddle Theorem 36.1. According to Lemmas 36.2 and 36.3 we can assume without loss of generality in proving Theorem 36.1 that the f-saddle L_k given in Theorem 36.1 is "simply" presented in the sense of Definition 36.3. We accordingly assume that L_k is an f-saddle on M_n at p_a with a presentation $\in \mathscr{D}L_k$

$$u \to F(u) : B_k^e \to |L_k|, \qquad F(0) = p_a, \tag{36.16}$$

such that

$$F(u) = G(u, \mathbf{v}(u)), \qquad u \in B_k^e \subset B_k^\sigma, \tag{36.17}$$

where $u \to \mathbf{v}(u)$ is a C^∞-mapping of B_k^e into B_{n-k}^σ.

We shall make use of the subset

$$G(B_k^{\ e} \times B_{n-k}^{\sigma}) = Y \tag{36.18}$$

of Y^σ [cf. (36.2)]. To prove Theorem 36.1, we shall show that there exist coset-contracting isomorphisms

$$H_q(f_a, \dot{f}_a) \approx H_q(|L_k|, |\dot{L}_k|), \qquad q = 0, 1,..., \tag{36.19}$$

since the validity of (36.19) when $q = k$ implies that a k-cap of p_a relative to $f \mid L_k$ is also a k-cap of p_a relative to f.

Method of Proof of (36.19). The isomorphism (36.19) will be verified by establishing the coset-contracting isomorphisms [Y from (36.18)]

$$H_q(f_a, \dot{f}_a) \approx H_q(f_a \cap Y, \dot{f}_a \cap Y) \tag{36.20}$$

and

$$H_q(f_a \cap Y, \dot{f}_a \cap Y) \approx H_q(|L_k|, |\dot{L}_k|) \tag{36.21}$$

and then composing these isomorphisms to obtain (36.19).

Verification of (36.20). This isomorphism follows from the Excision Theorem 28.3 in the usual way on taking (χ, A) of Theorem 28.3 as $(f_a, \dot{f}_a)$ and noting that $\chi - A = p_a$.

Verification of (36.21). This isomorphism follows from Theorem 28.4 once we have established Propositions 36.1 and 36.2.

Proposition 36.1. *There exists a deformation d retracting $f_a \cap Y$ onto $|L_k|$ and $\dot{f}_a \cap Y$ onto $|\dot{L}_k|$.*

The deformation d, to be defined, is a deformation of a subset of Y on itself. We can define d by suitably defining a deformation δ in the domain of the canonical coordinates (u, v) representing Y.

The image of $f_a \cap Y \subset Y$ under G^{-1} is the set

$$W = \{(u, v) \in B_k^{\ e} \times B_{n-k}^{\sigma} \mid -\|u\|^2 + \|v\|^2 \leqslant 0\}, \tag{36.22}$$

in accord with (36.3) and (36.18). The image of $|L_k| \subset Y$ under G^{-1} is the set,

$$w = \{(u, v) \in B_k^{\ e} \times B_{n-k}^{\sigma} \mid v = \mathrm{v}(u),\ u \in B_k^{\ e}\}, \tag{36.22'}$$

in accord with (36.16) and (36.17). Let W_0 and w_0 denote the sets W and w, respectively, with the origin in E_n deleted.

Proposition 36.1 will follow from proposition 36.2.

Proposition 36.2. *The deformation*

$$(u, v, t) \to \delta(u, v, t) = (u, (1-t)v + t\mathbf{v}(u)), \qquad (u, v) \in W,\ 0 \leqslant t \leqslant 1, \tag{36.23}$$

continuously retracts W onto w and W_0 onto w_0.

To establish Proposition 36.2, it will be sufficient to verify the following relations in accord with Definition 23.1 of a retracting deformation:

$$\begin{array}{lll}
(\mathrm{A}_1): & \delta(u, v, 0) = (u, v), & (u, v) \in W, \\
(\mathrm{A}_2): & \delta(u, v, 1) = (u, \mathbf{v}(u)), & (u, v) \in W, \\
(\mathrm{A}_3): & \delta(u, \mathbf{v}(u), t) = (u, \mathbf{v}(u)), & u \in B_k^{\,e};\ 0 \leqslant t \leqslant 1, \\
(\mathrm{A}_4): & \delta(O_u, O_v, t) = (O_u, O_v), & 0 \leqslant t \leqslant 1, \\
(\mathrm{A}_5): & \delta(u, v, t) \in W, & (u, v) \in W;\ 0 \leqslant t \leqslant 1, \\
(\mathrm{A}_6): & \delta(u, v, t) \in W_0, & (u, v) \in W_0;\ 0 \leqslant t \leqslant 1.
\end{array}$$

That relations (A_1)–(A_4) are true follows immediately from (36.23).

Proof of (A_5). Since the points $\delta(u, v, 0) = (u, v)$ and $\delta(u, v, 1) = (u, \mathbf{v}(u))$ are in W and W is convex in E_n, (A_5) follows.

Proof of (A_6). We must prove that if $(u_0, v_0) \neq (O_u, O_v)$, then $\delta(u_0, v_0, t) \neq (O_u, O_v)$.

Note first that when $(u_0, v_0) \neq (O_u, O_v)$ then $u_0 \neq O_u$; otherwise, by (36.22)′, $(u_0, v_0) = (O_u, O_v)$. In case $u_0 \neq O_u$ (36.23) implies that

$$\delta(u_0, v_0, t) \neq (O_u, O_v), \qquad 0 \leqslant t \leqslant 1.$$

Proposition 36.2 follows from relations (A_1)–(A_6).

Verification of Proposition 36.1. Proposition 36.2 implies the following: Corresponding to an arbitrary point $p \in f_a \cap Y$, let $(\bar{u}(p), \bar{v}(p))$ be canonical coordinates of p; then $p = G(\bar{u}(p), \bar{v}(p))$ and the mapping

$$(p, t) \to d(p, t) = G(\delta(\bar{u}(p), \bar{v}(p), t)), \qquad p \in f_a \cap Y,\ 0 \leqslant t \leqslant 1, \tag{36.24}$$

defines a deformation satisfying Proposition 36.1. This follows from the relations (A_1)–(A_6) and the character of G as a homeomorphism.

Conclusion of Proof of Theorem 36.1. Proposition 36.2 implies Proposition 36.1, while Proposition 36.1 implies the isomorphism (36.21). Isomorphisms (36.20) and (36.21) compose to yield the coset-contracting isomorphism (36.19). Theorem 36.1 follows from the isomorphism (36.19) when $q = k$.

Thus Saddle Theorem 36.1 *is true.*

A k-cycle u^k on $f_a \bmod \mathring{f}_a$ will be a k-cap of p_a relative to f if and only if the barycentric subdivisions of u^k are k-caps of p_a relative to f (cf. Lemma 28.2). From this we infer the following useful corollary of the Saddle Theorem:

Corollary 36.1. *A k-cycle u^k on f_a* mod $\mathring{f}_a$ *which admits a subdivision $z^k + w^k$ such that $| w^k | \subset \mathring{f}_a$ and z^k is a k-cap of p_a relative to $f \mid L_k$ is a k-cap of p_a on M_n.*

Proof. By Theorem 36.1 z^k is a k-cap relative to f, so that $z^k + w^k$ is a k-cap relative to f.

The Carrier Theorem. When the index k of the given critical point p_a of the ND f on M_n is such that $0 < k < n$ the Saddle Theorem reduces the problem of associating a k-cap with p_a to the problem when $k = n$. The Carrier Theorem gives an effective sufficient condition that an n-cycle y^n on $f_a \bmod \mathring{f}_a$ be an n-cap of p_a by conditioning the way $| y^n |$ is carried on $| M_n |$ arbitrarily near p_a.

Barycentric Subdivision. Recall that if y^n is an n-cycle on $f_a \bmod \mathring{f}_a$, the first barycentric subdivision πy^n of y^n is also an n-cycle on $f_a \bmod \mathring{f}_a$ because $\partial\pi = \pi\partial$. Moreover, Lemma 28.2 implies that $\pi y^n \sim y^n$ on $f_a \bmod \mathring{f}_a$, so that y^n is an n-cap associated with p_a if and only if πy^n is an n-cap associated with p_a.

Before coming to the Carrier Theorem two definitions are needed:

Definition 36.4. *Subchains Meeting a Point.* Let χ be a Hausdorff space, q a point of χ, and y^n a singular n-chain $n > 0$ on χ such that

$q \in | y^n |$. If y^n has a reduced form $r_i \sigma_i^n$ over $\mathscr{K}$, the deletion of the terms of this sum for which $| \sigma_i^n |$ does not meet q will leave a subchain of y^n, termed *the subchain of y^n meeting q.*

Definition 36.5. *n-Chains on M_n Simply-carried* **at** *q.* Let M_n be a C^∞-manifold and q a point of M_n. An n-chain y^n on M_n whose carrier meets q will be said to be *simply-carried* by M_n **at** q if the following two conditions are satisfied:

Condition (m_1). The subchain u^n of y^n which "meets" q has for carrier a closed topological n-disk Δ_n on M_n in which q is an *interior* point.

Condition (m_2). ∂u^n is carried by $\beta\Delta_n$, the geometric boundary of Δ_n, and in the sense of Definition 30.3 has at least one *simply-carried separate* $(n - 1)$-cell.

The second principal theorem of this section follows:

Carrier Theorem 36.2 (i) *Let there be given a C^∞-manifold M_n, a* ND $f \in C^\infty(M_n)$, *a critical point p_a of f of index n at the f-level a, and a cycle y^n on f_a* mod $\dot{f}_a$ *such that*

$$| y^n | - p_a \subset \dot{f}_a . \tag{36.25}$$

(ii) *A sufficient condition that y^n be an n-cap associated with p_a is that for some integer $\mu \geqslant 0$ the μ-fold barycentric subdivision of y^n be simply-carried by M_n* **at** p_a.

We shall prove the theorem in case (ii) holds for $\mu = 0$. It will then follow in case (ii) holds for some $\mu > 0$.

Let u^n be the "subchain of y^n meeting p_a." We identify p_a with the point q of Definition 36.5 and introduce the topological n-disk Δ_n as in Condition (m_1). Under Condition (m_2) Lemma 30.2 implies that $\partial u^n \not\sim 0$ on $\beta\Delta_n$. If we understand that $\dot{\Delta}_n = \Delta_n - p_a$, it follows from Lemma 29.0 that $u^n \not\sim 0$ on Δ_n mod $\dot{\Delta}_n$. We see from (36.25) that $\dot{\Delta}_n \subset \dot{f}_a$. The Excision Theorem 28.3 then implies that there exists a coset-contracting isomorphism

$$H_n(f_a , \dot{f}_a) \approx H_n(\Delta_n , \dot{\Delta}_n),$$

so that the condition $u^n \not\sim 0$ on Δ_n mod $\dot{\Delta}_n$ implies that $u^n \not\sim 0$ on f_a mod $\dot{f}_a$. Hence $y^n \not\sim 0$ on f_a mod $\dot{f}_a$.

This establishes the Carrier Theorem.

Remark. The Carrier Theorem is false if the condition on y^n of the second paragraph of the theorem is deleted.

§ 37

THE REAL PROJECTIVE n-SPACE P_n

The homology groups of P_n are well known. However, the derivation of these homology groups of P_n by the study of a suitably chosen ND function f on P_n will reveal much concerning critical point theory, and, incidentally, disclose the nature of the homology groups in question.

Notation. Let E_{n+1} be the Euclidean space of points $x = (x_0, x_1, \ldots, x_n)$ and E^0_{n+1} the space E_{n+1} with the origin deleted. Two points x' and x'' in E^0_{n+1} are termed *projectively equivalent* if there exist real nonzero numbers r' and r'' such that $r'x' = r''x''$. Let $\psi(x)$ denote the projective equivalence class of an arbitrary point $x \in E^0_{n+1}$. Let P_n denote the set of equivalence classes $\psi(x)$ into which E^0_{n+1} is partitioned. One has the mapping

$$x \to \psi(x) : E^0_{n+1} \to P_n \qquad \text{onto} \quad P_n . \tag{37.1}$$

The space E^0_{n+1} has a standard topology. One topologizes P_n by requiring that a subset X of P_n be open if and only if X is the image under ψ of an open subset Y of E^0_{n+1}. With P_n so topologized, ψ is continuous and P_n becomes a Hausdorff space $| P_n |$.

That $| P_n |$ is *compact* and connected is proved when $n > 0$, as in the case of $\mathbf{CP}_n$, and is trivial when $n = 0$.

$| P_n |$ as a Topological Manifold. To give $| P_n |$ a differentiable structure,[†] let U_n be a Cartesian n-plane of points $u = (u_1, \ldots, u_n)$. We shall define presentations

$$(F^k : U_n, Y^k), \qquad k = 0, 1, \ldots, n, \tag{37.2}$$

[†] P_0 is defined above, but no differentiable structure is to be defined on P_0 nor any function f. Similarly, $| S_0 |$ is a pair of points, but no differentiable structure is to be defined on $| S_0 |$ nor any function similar to f.

of open subsets Y^k of $| P_n |$ whose union is $| P_n |$.

For each k on the range 0, 1,..., n and each point $u \in U_n$ a point

$$\mathbf{x}(u : k) = (u_1, u_2, ..., u_k, 1, u_{k+1}, ..., u_n) \in E^0_{n+1} \tag{37.3}$$

is introduced. We then set

$$F^k(u) = \psi(\mathbf{x}(u : k)) \in | P_n |, \qquad u \in U_n, \quad n > 0, \tag{37.4}$$

and $F^k(U_n) = Y^k$. The lemma that Y^k is open in $| P_n |$ and that F^k is biunique and continuous follows as in the case of $\mathbf{CP}_n$. Thus $| P_n |$ is a topological manifold covered by the presentations $F^0, F^1, ..., F^n$.

The C^∞-Compatibility of the Presentations F^k. Let h and k be distinct integers on the range 0, 1,..., n, with $h > k$. The proof of the C^∞-compability of F^h and F^k is essentially as in the case of $\mathbf{CP}_n$, u replacing (u, v) and u' replacing (u', v'). In particular, the compatibility condition (35.12) here takes the form $u_h u'_{k+1} = 1$, and real analytic transition homeomorphisms are defined by analytic relations similar to (35.13a) and (35.13b).

One concludes that the $n + 1$ presentations $F^0, F^1, ..., F^n$ are pairwise C^∞-compatible, so that when $n > 0$ $| P_n |$ becomes the carrier of a C^∞-manifold M_n whose differentiable structure is determined by these $n + 1$ presentations.

A ND Function f on P_n, $n > 0$. Corresponding to an arbitrary point $x = (x_0, x_1, ..., x_n) \in E^0_{n+1}$ and a set of real constants $0 < c_0 < c_1 < \cdots < c_n$, a real-valued function f on P_n, or, equivalently, on M_n, can be defined by setting

$$f(\psi(x)) = c_i x_i^2 / \| x \|^2 \tag{37.5}$$

(summing with respect to i for i on the range 0, 1,..., n). The function f is thereby uniquely defined at each point of P_n.

The critical points of f on Y^k are the images under F^k of the critical points (if any exist) of the mapping

$$u \to (f \circ F^k)(u) = f(\psi(\mathbf{x}(u : k))) = f_k(u), \qquad u \in U_n, \quad n > 0, \tag{37.6}$$

introducing f_k. It follows from (37.5) and (37.6) that $(1+\|u\|^2)f_k(u)-c_k$ is equal to

$$c_0u_1^2+\cdots+c_{k-1}u_k^2+c_{k+1}u_{k+1}^2+\cdots+c_nu_n^2. \tag{37.7}$$

One sees that f_k is real and analytic on U_n. It follows, as in the case of $\mathbf{CP}_n$, that the only critical point of f_k on U_n is a ND critical point $u=\mathbf{0}$. The corresponding point $p_k=F^k(\mathbf{0})$ is a critical point of f on Y^k of index k. We draw the following conclusion:

Theorem 37.0. *The critical points of f are the $n+1$ points p_k, $k=0,\ldots,n$, of which p_k is in the projective equivalence class on E_{n+1}^0 of the point x_k whose real coordinates x_i are δ_i^k, $i=0,1,\ldots,n$, whose critical value is c_k and whose index is k.*

The indices of the critical points of f on M_n do not form a "lacunary sequence." As a consequence, the linking characteristics of these critical points are *a priori* dependent upon the field $\mathscr{K}$. To clarify this dependence we shall make use of the classical representation of P_n in which diametrically opposite points of an n-sphere S_n represent the same point of P_n.

The Sphere S_n. In the Cartesian plane E_{n+1} of points $x=(x_0, x_1,\ldots, x_n)$ let S_n be the regularly structured C^∞-manifold whose carrier is the origin-centered n-sphere $|S_n|$ of unit radius. Among the presentations in $\mathscr{D}S_n$ are $2n+2$ Monge presentations

$$(G_k^\epsilon : B_n, W_k^\epsilon), \qquad k=0,1,\ldots,n;\ \epsilon=\pm 1, \tag{37.8}$$

of open subsets of S_n which we now define.

W_k^ϵ shall be the open hemisphere of S_n on which sign $x_k=$ sign ϵ. The domain B_n shall be an origin-centered open unit n-ball of coordinates

$$(x_0, x_1,\ldots, x_{k-1}, \check{x}_k, x_{k+1},\ldots, x_n)=(\alpha_1,\ldots,\alpha_n), \tag{37.9a}$$

deleting x_k. We complete the definition of the Monge presentation (37.8) of W_k^ϵ by setting

$$x_k=\epsilon(1-\alpha_1^2-\cdots-\alpha_n^2)^{1/2}, \qquad \alpha\in B_n. \tag{37.9b}$$

The $2n+2$ presentations in $\mathscr{D}S_n$ thereby defined cover $|S_n|$, and are C^∞-compatible, in accord with Theorem 5.1.

A Function ζ on S_n. A real-valued C^∞-function $x \to \zeta(x)$ is defined on S_n by setting

$$\zeta(x) = c_i x_i^2, \qquad x \in S_n, \tag{37.10}$$

(summing with respect to i), where $c_0 < c_1 < \cdots < c_n$ are the constants introduced in (37.5). Note that $\zeta(x) = f(\psi(x))$ for $x \in S_n$.

The critical points, if any, of ζ on the hemisphere W_k^ϵ are the images under G_k^ϵ of the critical points of the mapping

$$\alpha \to (\zeta \circ G_k^\epsilon)(\alpha) = \zeta_k(\alpha), \qquad \alpha \in B_n, \tag{37.11}$$

introducing ζ_k. One finds that $\zeta_k(\alpha) - c_k$ is equal to

$$(c_0 - c_k)\,\alpha_1^{\ 2} + \cdots + (c_{k-1} - c_k)\,\alpha_k^{\ 2} + (c_{k+1} - c_k)\,\alpha_{k+1}^2 + \cdots + (c_n - c_k)\,\alpha_n^{\ 2}$$

for $\alpha \in B_n$. It follows that the only critical point of ζ in the open hemisphere W_k^ϵ of $|\, S_n \,|$ is the point q_k^ϵ at the intersection of W_k^ϵ with the x_k-axis. One sees that q_k^ϵ is ND and has the critical value c_k and index k.

The Chain Transformation $\hat{\psi}$ Induced by ψ. As defined, ψ maps E_{n+1}^0 onto $|\, P_n \,|$. According to Definition 26.5 ψ induces the chain transformation $\hat{\psi}$ of the singular complex $S(E_{n+1}^0)$ into the singular complex $S(|\, P_n \,|)$.

On $|\, S_n \,|$ the most useful chains are the unitary chains, which we shall now define.

Unitary k-Chains on $|\, S_n \,|$. The point

$$q_k^\epsilon, \qquad k = 0, 1, \ldots, n, \quad \epsilon = \pm 1, \tag{37.12}$$

is that point of intersection of the x_k-axis with $|\, S_n \,|$ at which $x_k = \epsilon$. The points q_k^ϵ will be called the *unitary* points of $|\, S_n \,|$. An ordered subset of $k + 1$ distinct unitary points, no two of which are antipodal on $|\, S_n \,|$, determine an ordered Euclidean k-simplex a^k in E_{n+1}^0. The projection τ from the origin of a^k into $|\, S_n \,|$ defines a singular k-simplex on $|\, S_n \,|$. The corresponding singular k-cell σ^k on $|\, S_n \,|$ will be called *unitary*. The unitary singular n-cells on $|\, S_n \,|$ are "simply-carried" by $|\, S_n \,|$ (Definition 30.2).

A "reduced" k-chain (Definition 24.0)

$$u^k = e_1\sigma_1{}^k + \cdots + e_r\sigma_r{}^k, \qquad e_i = \pm 1, \tag{37.13}$$

in which each k-cell $\sigma_i{}^k$ is unitary and no two cells $\sigma_i{}^k$ have the same carrier, will be called a *unitary k-chain* on $| S_n |$. We say that $| u^k |$ is *simply-covered.*

The Operator Θ. Given a point $q \in | S_n |$, let θq denote the point on $| S_n |$ antipodal to q. If $q_0 q_1 \cdots q_k$ is an ordered unitary simplex, let $(q_0, q_1, \ldots, q_k)$ denote the corresponding singular unitary k-cell σ^k on $| S_n |$. We term the singular unitary k-cell $(\theta q_0, \ldots, \theta q_k)$ the k-cell $\Theta\sigma^k$ *opposite* σ^k. If $u^k = e_i\sigma_i{}^k$, with $e_i = \pm 1$, is a unitary k-chain on $| S_n |$, we introduce the unitary k-chain $\Theta u^k = e_i\Theta\sigma_i{}^k$ and term Θu^k *opposite* u^k. Note that

$$\Theta(\Theta u^k) = u^k, \qquad \partial(\Theta u^k) = \Theta\partial u^k. \tag{37.14}$$

Some Special Unitary k-Cycles on $| S_n |$. Given S_n, a sequence of spheres

$$S_1, S_2, \ldots, S_{n-1}, S_n \tag{37.15}$$

carried by $| S_n |$ is introduced, where

$$| S_k | = \{x \in | S_n | \mid x_n = x_{n-1} = \cdots = x_{k+1} = 0\}$$

for each k on the range $1, 2, \ldots, n-1$.

We refer to a corresponding sequence of closed hemispheres

$$H_1, H_2, \ldots, H_{n-1}, H_n \tag{37.16}$$

of which

$$H_k = \{x \in | S_k | \mid x_k \geqslant 0\}, \qquad k = 1, 2, \ldots, n. \tag{37.17}$$

It is convenient to denote by $E_2{}^0, \ldots, E_n{}^0$ the respective subspaces

$$E^0_{\mu+1} = \{x \in E^0_{n+1} \mid x_n = x_{n-1} = \cdots = x_{\mu+1} = 0\}$$

of E^0_{n+1} for μ on the range $1, 2, \ldots, n-1$, and to denote by $P_1, P_2, \ldots, P_{n-1}$ the real projective subspaces P_μ of P_n of dimension μ defined in terms of $E^0_{\mu+1}$ as P_n is defined in terms of E^0_{n+1}, and correspondingly structured both topologically and differentiably. Theorem 37.0 holds if one replaces f by $f \mid P_\mu$ and n by μ.

In preparation for Theorem 37.1 we introduce the unitary 1-chain,

$$\omega^1 = (q_0^{-1}, q_1^{1}) + (q_1^{1}, q_0^{1}) \tag{37.18}$$

as a sum of two unitary 1-cells. On $| S_1 |$ there exists a unitary singular 1-cycle

$$\gamma^1 = \omega^1 + \Theta\omega^1. \tag{37.19}$$

Taking ω^0 as the singular 0-cell (q_0^{1}), we find that

$$\partial\omega^1 = (q_0^{1}) - (q_0^{-1}) = \omega^0 - \Theta\omega^0 = \gamma^0, \tag{37.20}$$

introducing the 0-cycle γ^0.

The definitions of ω^1, γ^1, and γ^0 are the initial steps in an induction from which the following basic theorem results:

Theorem 37.1 (α) *There exists a sequence $\omega^1, \omega^2, \ldots, \omega^n$ of unitary chains on $| S_n |$ whose carriers are the respective simply-covered hemispheres $H_1, \ldots, H_n$, and a sequence $\gamma^1, \ldots, \gamma^n$ of unitary cycles whose carriers are the respective simply-covered spheres $| S_1 |, \ldots, | S_n |$, where the chains ω^i and cycles γ^i are such that the following is true:*

(β) *If the chains ω^1, γ^1, and γ^0 are defined as above, then for k on the range* 1, 2,..., n

$$\gamma^k = \omega^k + \Theta\omega^k \qquad \text{when } k \text{ is odd,} \tag{37.21}$$

$$\gamma^k = \omega^k - \Theta\omega^k \qquad \text{when } k \text{ is even,} \tag{37.22}$$

$$\partial\omega^k = \gamma^{k-1}, \qquad k = 1, 2, \ldots, n. \tag{37.23}$$

The theorem is true for $n = 1$. We assume that the theorem is true when n is replaced by a positive integer $m < n$, and prove it true when n is replaced by $m + 1$.

By hypothesis γ^m is a unitary m-cycle whose carrier simply covers $| S_m |$. If

$$\gamma^m = e_1\sigma_1^{m} + \cdots + e_r\sigma_r^{m}, \qquad e_1 = \pm 1, \quad r = 2^{m+1}, \tag{37.24}$$

is the reduced form of γ^m, one defines ω^{m+1} by setting

$$\omega^{m+1} = e_1\sigma_1^{m+1} + \cdots + e_r\sigma_r^{m+1}, \tag{37.25}$$

where σ_i^{m+1} is the unitary $(m + 1)$-cell on $| S_n |$ whose defining vertices are the vertices of σ_i^m in their given order, preceded by q^1_{m+1}. One sees that $| \omega^{m+1} |$ simply covers H_{m+1}. Setting $k = m + 1$, one then defines a unitary chain γ^{m+1} by (37.21) or (37.22), depending on whether $(m + 1)$ is odd or even. One sees that $| \gamma^{m+1} |$ simply covers $| S_{m+1} |$. For $m + 1$ odd

$$\partial\gamma^{m+1} = \partial\omega^{m+1} + \partial\Theta\omega^{m+1} = \gamma^m + \Theta\gamma^m.$$

By our inductive hypothesis (37.21) and (37.22) are valid when $k = m$, so that

$$\partial\gamma^{m+1} = \omega^m - \Theta\omega^m + \Theta(\omega^m - \Theta\omega^m) = 0. \tag{37.26}$$

Thus γ^{m+1} is a cycle. The proof that γ^{m+1} is a cycle when $m + 1$ is even is similar.

It follows from the definition (37.25) of ω^{m+1} that (37.23) holds.

The k-Caps of f on $| P_n |$. Theorem 37.1 leads to Theorem 37.2 below. In Theorem 37.2 we shall refer to the unitary chains ω^k of Theorem 37.1.

Theorem 37.2. *For k on the range $1, 2,..., n$, $\widehat{\psi}\omega^k$ is a k-cap of p_k relative to $f | P_k$ as well as to f. Moreover,*

$$\partial\widehat{\psi}\omega^k = 2\widehat{\psi}\omega^{k-1} \qquad \text{when } k \text{ is even,} \tag{37.27}$$

$$\partial\widehat{\psi}\omega^k = 0 \qquad \text{when } k \text{ is odd,} \tag{37.28}$$

$$\widehat{\psi}\omega^k \not\sim 0 \quad \text{on} \quad | P_k | \qquad \text{when } k \text{ is odd.} \tag{37.29}$$

That $\widehat{\psi}\omega^k$ is a k-cap of p_k relative to $f | P_k$ follows from Theorem 36.2; in fact, the first barycentric subdivision of $\widehat{\psi}\omega^k$ is the image under $\widehat{\psi}$ of the first barycentric subdivision of ω^k, and, as such, is seen to be "simply-carried **at** p_k" in the sense of Definition 36.5. By Theorem 36.2 $\widehat{\psi}\omega^k$ is then a k-cap of p_k relative to $f | P_k$.

One shows as follows that $\widehat{\psi}\omega^k$ is a k-cap of p_k, relative to f:

There exists an open neighborhood N of p_k relative to $| P_k |$ which is an f-saddle on $| P_n |$ of p_k. Among the barycentric subdivisions of $\widehat{\psi}\omega^k$ of sufficiently high order there exists one of the form $z^k + w^k$, where $|z^k| \subset N$ and $|w^k| \subset f_a$. Since z^k is "simply carried at p_k," z^k is a k-cap of p_k relative to $f \mid L_k$, in accord with Carrier Theorem 36.2. By Saddle Theorem 36.1 z^k is a k-cap of p_k relative to f. It follows that $z^k + w^k$, and hence $\widehat{\psi}\omega^k$, is a k-cap of p_k relative to f.

Verification of (37.27) *and* (37.28). Note that

$$\partial\omega^k = \omega^{k-1} + \Theta\omega^{k-1} \qquad \text{when } k \text{ is even} \tag{37.30}$$

$$\partial\omega^k = \omega^{k-1} - \Theta\omega^{k-1} \qquad \text{when } k \text{ is odd} \tag{37.31}$$

by virtue of Theorem 37.1. Relations (37.27) and (37.28) follow on applying $\widehat{\psi}$ to the terms of (37.30) and (37.31), recalling that $\partial\widehat{\psi} = \widehat{\psi}\partial$ and $\widehat{\psi}\Theta u^j = \widehat{\psi}u^j$ for any unitary j-chain u^j, $j = 0, 1, \ldots, n$.

Verification of (37.29). Suppose k odd. Then $\widehat{\psi}\omega^k$ is a k-cycle by (37.28). The first conclusion of Theorem 37.2 is that $\widehat{\psi}\omega^k$ is a k-cap of p_k relative to $f \mid P_k$. By definition of a linking k-cycle $\widehat{\psi}\omega^k$ is then a linking k-cycle associated with the critical point p_k of $f \mid P_k$. Theorem 29.3(ii) applied to $f \mid P_k$ gives (37.29).

Thus Theorem 37.2 is true.

Retracting Deformations on $|P_n|$. To determine the linking characteristics of the critical points p_k of f on P_n, a set of retracting deformations will be needed. These deformations will be the *products* of deformations, taking "products" in a sense which we shall now define.

A Deformation D. A deformation D of a subset ξ of a Hausdorff space χ is defined by a continuous mapping

$$(p, t) \to D(p, t) : \xi \times [0, 1] \to \chi \qquad \text{into } \chi.$$

For each point $p \in \xi$, $D(p, 0) = p$ by hypothesis, and the partial mapping $t \to D(p, t)$ of $[0, 1]$ into χ is called the *D-trajectory* of p. We term ξ the *initial* set of D and $D(\xi, 1)$ the *terminal* set of D. A deformation D is termed *continuable* by a deformation D' if the terminal set of D is included in the initial set of D'. If $X \subset \xi$, by the *restriction* of D to X is meant the deformation defined by the mapping $D \mid (X \times [0, 1])$.

In defining products of deformations the following terminology will be used. Let $[a, b]$ and $[c, d]$ be intervals of the real axis and $t \to \theta(t)$ the sense-preserving linear mapping of $[c, d]$ onto $[a, b]$. If an arc h in χ of form $t \to h(t) : [a, b] \to \chi$ is given, then the arc $t \to h(\theta(t)) : [c, d] \to \chi$ is termed *the arc h retaken over* $[c, d]$. Let the terminal point of h be denoted by Ter h. It is not assumed that a mapping $t \to h(t)$ is biunique.

Definition 37.1. *Product Deformations.* For some integer $r > 0$ let $D^0, D^1, D^2, \ldots, D^r$ be a sequence of $r + 1$ deformations of subsets of χ such that each deformation of the sequence except the last is "continuable" by its successor. Let the interval $[0, 1]$ be partitioned into $r + 1$ successive intervals of equal length whose closures are denoted by $I_0, I_1, I_2, \ldots, I_r$. The product deformation

$$D = D^r \cdots D^2 D^1 D^0 \tag{37.32}$$

of the initial set ξ of D is inductively defined as follows: If p is a prescribed point of ξ, the D-traj of p is an arc $t \to g(t) : [0, 1] \to \chi$ such that

$$\begin{aligned} g \mid I_0 &= \text{the } D^0\text{-traj of } p, \text{ retaken over } I_0, \\ g \mid I_1 &= \text{the } D^1\text{-traj of Ter}(g \mid I_0), \text{ retaken over } I_1, \\ &\vdots \\ g \mid I_r &= \text{the } D^r\text{-traj of Ter}(g \mid I_{r-1}), \text{ retaken over } I_r. \end{aligned} \tag{37.33}$$

One readily verifies that the resultant mapping $(p, t) \to D(p, t)$ of $\xi \times [0, 1]$ into χ is continuous.

Our deformation lemma will refer to the subset $\dot{f}_{c_k} = f_{c_k} - p_k$ of $| P_n |$, and in the proof to the subsets $| \dot{P}_j | = | P_j - p_j |$ of $| P_n |$ for j on the range $1, \ldots, n$.

Lemma 37.1. *For* $k = 1, 2, \ldots, n$ *there exists a deformation* d_k *retracting* $\dot{f}_{c_k}$ *onto* $| P_{k-1} |$.

Before starting the formal proof of Lemma 37.1 we shall describe the subset f_{c_k} of $| P_n |$ in terms of the subset ζ_{c_k} of $| S_n |$.

For k on the range $1, \ldots, n$, ζ_{c_k} is the intersection with $| S_n |$ of the solid $(n + 1)$-dimensional ellipsoid in E_{n+1} on which

$$c_0 y_0^2 + c_1 y_1^2 + \cdots + c_n y_n^2 \leqslant c_k. \tag{37.34}$$

Moreover, $f_{c_k} = \psi(\zeta_{c_k})$. Of the unitary points q_i^ϵ of $| S_n |$ those for which $i > k$, $i = k$, and $i < k$ are on the exterior, boundary, and interior, respectively, of ζ_{c_k} relative to $| S_n |$, provided $k < n$. The corresponding critical points p_i of f on $| P_n |$ are, respectively, on the exterior, boundary, and interior of f_{c_k} relative to $| P_n |$.

We continue with a definition:

The Deformation δ^k *Retracting* $| \dot{P}_k |$ *onto* $| P_{k-1} |$. Let q be prescribed in $\dot{H}_k$, and thereby $\psi(q)$ in $\dot{P}_k$. Corresponding to q, let λ_q be the unique quarter circle on H_k which issues from $q_k{}^1$, meets q, and terminates at a point q' on the boundary $| S_{k-1} |$ of H_k. The point q may coincide with q'. Under δ^k the point $\psi(q)$ shall be deformed on $| \dot{P}_k |$ into the point $\psi(q')$ on $| P_{k-1} |$ by moving q along λ_q, and thereby $\psi(q)$ along $\psi(\lambda_q)$, at a velocity (possibly zero) equal to the length of the subarc qq' of λ_q.

Remark. A simple calculation shows that the value $\zeta(p)$, and hence the value $f(\psi(p)) = \zeta(p)$, decreases whenever q moves along λ_q so that y_k decreases.

The Deformation d_k *defined*, $k = 1, 2, \ldots, n$. Set $k = n - r$. The restrictions of the deformations

$$\delta^n, \delta^{n-1}, \delta^{n-2}, \ldots, \delta^{n-r} \tag{37.35}$$

to the respective subsets

$$\dot{f}_{c_k} \cap | \dot{P}_n |, \dot{f}_{c_k} \cap | \dot{P}_{n-1} |, \ldots, \dot{f}_{c_k} \cap | \dot{P}_{n-r} | \tag{37.36}$$

of $\dot{f}_{c_k}$ are well-defined deformations

$$D^0, D^1, \ldots, D^r. \tag{37.37}$$

Let the sets (37.36) be denoted, respectively, by

$$\xi_0, \xi_1, \ldots, \xi_r.$$

We shall verify statement (A):

(A) *For* $0 \leqslant i < r$ *the deformation* D^i *retracts* ξ_i *onto* ξ_{i+1}.

Recall that $f(p_{n-i-1}) = c_{n-i-1} \geqslant c_k$.

The deformation D^i either leaves a point $p \in \xi_i$ fixed or deforms p so that on the D^i-traj of p, f strictly decreases as t increases. It follows that D^i deforms ξ_i on ξ_i. The least upper bound of f on the sets ξ_i is c_k. Hence there is no D^i-trajectory terminating at the point p_{n-i-1} of $| P_{n-i-1} |$. The set ξ_{i+1} is a proper subset of ξ_i and is fixed under the deformation D^i. These facts imply (A).

A product deformation

$$d_k = D^r \cdots D^1 D^0, \qquad r = n - k, \tag{37.38}$$

of $\dot{f}_{c_k}$ is well-defined. The set $\xi_r = | \dot{P}_k |$. The deformation D^r is then δ^r, and retracts $| \dot{P}_k |$ onto $| P_{k-1} |$. The deformation d_k accordingly retracts $\dot{f}_{c_k}$ onto $| P_{k-1} |$.

Thus Lemma 37.1 is true.

Linking Characteristics of Critical Points of f. The following theorem is a consequence of Theorem 37.2 and Deformation Lemma 37.1. Let $\mathbf{Q}$ be the field of rational numbers.

Theorem 37.3. *For a choice of the field $\mathscr{K}$ as $\mathbf{Q}$ or as $\mathbf{Z}_p$, with p a prime, the classification of the critical points p_k of f on P_n for positive index k as of linking or nonlinking types is in accord with the Table* III.

TABLE III

Field	Index k	Classification
$\mathbf{Z}_p$, $p = 2$	$1, 2, \ldots, n$	Linking
$\mathbf{Z}_p$, p arbitrary	Odd	Linking
$\mathbf{Z}_p$, $p \neq 2$	Even	Nonlinking
$\mathbf{Q}$	Odd	Linking
$\mathbf{Q}$	Even	Nonlinking

According to Theorem 37.2, for any field $\mathscr{K}$ and any critical point p_k of positive index, $\widehat{\psi}\omega^k$ is a k-cap of p_k relative to f. The point p_k will be of linking type over $\mathscr{K}$ if

$$\partial \widehat{\psi}\omega^k \sim 0, \qquad \text{on } \dot{f}_{c_k} \quad \text{over } \mathscr{K}. \tag{37.39}$$

The first line of the table follows from (37.27) and (37.28), the second and fourth from (37.28). The third and fifth lines of the table are valid by (37.27) if

$$\widehat{\psi\omega}^{k-1} \not\sim 0 \qquad \text{on} \quad \dot{f}_{c_k} \quad \text{over } \mathscr{K} \tag{37.40}$$

for an even index k and an arbitrary field $\mathscr{K}$. We shall verify (37.40) for k even.

By virtue of the Deformation Lemma 37.1 and Theorem 28.4 there exists a coset-contracting isomorphism

$$H_{k-1}(\dot{f}_{c_k}) \approx H_{k-1}(P_{k-1}) \qquad \text{over} \quad \mathscr{K}. \tag{37.41}$$

Moreover, (37.29) implies that over any field $\mathscr{K}$

$$\widehat{\psi\omega}^{k-1} \not\sim 0 \qquad \text{on} \quad | P_{k-1} | \tag{37.42}$$

for $k - 1$ even, or, equivalently, for k odd. From (37.41) and (37.42) we infer the truth of (37.40), so that the third and fifth lines of the table are valid.

This establishes Theorem 37.3.

It is an immediate consequence of Theorems 37.3 and 29.3 that when the field is $\mathbf{Q}$ or $\mathbf{Z}_p$, with p a prime, the nonvanishing connectivities of $| P_n |$ are given by the Table IV.

TABLE IV

Field $\mathscr{K}$	Dimension n	Nonvanishing connectivities
$\mathbf{Z}_p$, $p = 2$	$n > 0$	$R_0 = 1, R_1 = 1, R_2 = 1, \ldots, R_n = 1$
$\mathbf{Z}_p$, $p \neq 2$	Odd n	$R_0 = 1, R_n = 1$
$\mathbf{Z}_p$, $p \neq 2$	Even $n > 0$	$R_0 = 1$
$\mathbf{Q}$	Odd n	$R_0 = 1, R_n = 1$
$\mathbf{Q}$	Even $n > 0$	$R_0 = 1$

The results of this section yield all of the homology characteristics of P_n over the ring $\mathbf{Z}$ of integers, in particular, the Betti numbers of P_n and the torsion coefficients of the different dimensions. This will be shown in detail in the paper which presents the results abstracted in

Appendix III. The theorem that the qth Betti number of P_n is equal to the qth connectivity of P_n over **Q**, taken with the above table, gives the Betti numbers of P_n.

The space P_n, provided with the ND function f of this section, is easily seen to come under the *prime-simple* case of Appendix III. The determination of the torsion subgroups of the respective homology groups is particularly simple in the case of prime-simple spaces.

A Unitary n-Cycle on S_n. In §30 a singular n-chain which is "simply-carried" on a topological n-manifold has been defined and Lemma 30.3 established. Theorem 37.1(α) has (i) of the following theorem as by-product. Statement (ii) follows from Lemma 30.3.

Theorem 37.4. (i) *On an origin-centered n-sphere* S_n *in* E_{n+1} *there exists a simply-carried n-cycle* γ^n *the n-cells of whose reduced form are unitary.*

(ii) *Such an n-cycle on* S_n *is nonbounding in* S_n *with* $|\gamma^n| = |S_n|$.

§ 38

STEIN MANIFOLDS. A THEOREM OF LEFSCHETZ

Andreotti and Frankel (AF) have given a proof [1] of what is known as the Lefschetz Theorem on "hyperplane sections" of a nonsingular projective algebraic variety of complex dimension n. They refer to an unpublished proof by Thom as "the first to use Morse's theory of critical points" in proving the Lefschetz Theorem. See AF for a formulation of the Lefschetz Theorem (see also Lefschetz [1]).

The proof as given by AF is based on a theorem on "Stein manifolds" stated and proved by AF. This theorem on Stein manifolds is more general than what is required to prove the theorem of Lefschetz. We restrict ourselves to this theorem because Stein manifolds, in the sense of AF, can be readily defined, and because the application to algebraic geometry is relatively simple for one familiar with the terminology of algebraic geometry.

As we shall see, the theorem of AF on Stein manifolds of complex dimension n is a theorem on a ND function, defined on a real differentiable manifold M_{2n} of dimension $2n$, where f is endowed with certain special properties because of a special complex analytic origin.

The Spaces $\mathbf{CE}_m$ *and* E_{2m} . $\mathbf{CE}_m$ is a space of m complex variables, the Cartesian product of m complex planes of complex variables

$$w_1 ,\ldots, w_m . \tag{38.1}$$

Let E_{2m} be a real Euclidean space of $2m$ variables, the Cartesian product of m real 2-planes with coordinates

$$x_1 , y_1 ; x_2 , y_2 ;\ldots; x_m , y_m , \tag{38.2}$$

respectively.

There is an analytic homeomorphism ξ of $\mathbf{CE}_m$ onto E_{2m} defined by setting

$$x_\mu + iy_\mu = w_\mu, \qquad \mu = 1, \ldots, m. \tag{38.3}$$

One can review the early abstract definition of a Stein manifold as found, for example, in Séminaire Henri Cartan, 1951–52, Chapters VII–IX, containing lectures by Cartan.

Remmert [1] has shown that a Stein n-manifold, as originally defined, can be biholomorphically embedded as a Stein manifold Σ_n in a space $\mathbf{CE}_m$ of sufficiently high dimension m. With AF we are thus concerned with a Stein manifold Σ_n in $\mathbf{CE}_m$, as now to be defined.

Definition 38.1. **A Stein Manifold Σ_n in $\mathbf{CE}_m$.** Here $0 < n < m$, and Σ_n shall have for carrier a closed subset of points in $\mathbf{CE}_m$ whose real image in E_{2m} under ξ is a topological manifold in E_{2m}, and shall satisfy the following conditions.

Properties of Σ_n in $\mathbf{CE}_m$. Corresponding to a prescribed point $w^0 \in \Sigma_n$ the complex coordinates of $\mathbf{CE}_m$ shall admit a reordering such that the following is true. Set

$$z_1 = w_1 - w_1^0, \ldots, z_n = w_n - w_n^0, \tag{38.4}$$

$$z = (z_1, \ldots, z_n), \tag{38.5}$$

$$\| z \| = (| z_1 |^2 + \cdots + | z_n |^2)^{1/2}. \tag{38.6}$$

Let e be a prescribed positive constant. We shall restrict z, by the condition $\| z \| < e$, to an open neighborhood $\mathbf{N}_e$ of the origin in the space $\mathbf{CE}_n$ of the points z. If $e > 0$ is sufficiently small and $\| z \| < e$, there shall exist absolutely convergent power series

$$P_\rho(z_1, \ldots, z_n), \qquad \rho = 1, \ldots, m - n, \tag{38.7}$$

in the complex variables $(z_1, \ldots, z_n)$ with complex coefficients such that for each ρ, $P_\rho(\mathbf{0}) = w_{n+\rho}^0$ and the mapping

$$(z_1, \ldots, z_n) \rightarrow (w_1, \ldots, w_m) : \mathbf{N}_e \rightarrow \mathbf{CE}_m \tag{38.8}$$

in which

$$\begin{aligned} w_j^i &= z_j, & j &= 1, 2, \ldots, n, \\ w_{n+\rho} &= P_\rho(z_1, \ldots, z_n), & \rho &= 1, \ldots, m - n, \end{aligned} \tag{38.9}$$

defines an analytic homeomorphism of $\mathbf{N}_e$ onto an open neighborhood $\mathbf{X}$ of w^0 relative to Σ_n.

We term the above mapping of $\mathbf{N}_e$ *onto* $\mathbf{X}$ *a complex Monge presentation of* $\mathbf{X} \subset \Sigma_n$ *based on* $\mathbf{N}_e$.

Each complex Monge presentation such as (38.9) of an open neighborhood $\mathbf{X}$ relative to Σ_n of a point $w^0 \in \Sigma_n$ gives rise to a unique real analytic Monge presentation of the real neighborhood $X = \xi(\mathbf{X})$ relative to $M_{2n} = \xi(\Sigma_n)$ of the real point $p^0 = \xi(w^0)$ of M_{2n}.

To verify this affirmation, let the complex conjugate of a complex number c be denoted by conj c. We can suppose that w^0 is the origin, and in accord with (38.3) and (38.4) set

$$z_j = x_j + iy_j, \qquad j = 1, \ldots, n, \tag{38.10}$$

where x_j and y_j are real. For convenience we set

$$(x_1, \ldots, x_n; y_1, \ldots, y_n) = (v_1, \ldots, v_{2n}) = v. \tag{38.11}$$

Corresponding to the complex Monge presentation (38.9) of $\mathbf{X}$, a real analytic Monge presentation of $X = \xi(\mathbf{X})$ is given as follows.

The coordinates of a point in E_{2m} are given by (38.2). Of these coordinates the $2n$ coordinates in (38.11) can be supposed given, as in (38.11), by the parameters $v_1, \ldots, v_{2n}$, subject to the condition $\| v \| < e$. One obtains the remaining $2m - 2n$ coordinates of a point on M_{2n} by setting

$$\begin{aligned} 2x_{n+\rho} &= P_\rho(z) + \operatorname{conj} P_\rho(z) = \varphi_\rho(v), \\ 2y_{n+\rho} &= i \operatorname{conj} P_\rho(z) - iP_\rho(z) = \psi_\rho(v), \end{aligned} \qquad \rho = 1, \ldots, m-n, \tag{38.12}$$

introducing φ_ρ and ψ_ρ. One observes that φ_ρ and ψ_ρ so defined for $\| v \| < e$ are real and analytic. The equations (38.11) and (38.12) give a real analytic Monge presentation of the real neighborhood $X = \xi(\mathbf{X})$ of the origin in M_{2n}.

Focal Points of M_{2n}. It follows from Theorem 6.1 that the points $q \in E_{2m}$ which are neither on M_{2n} nor focal points of M_{2n} are everywhere dense in E_{2m}. With this fact as a starting point the following theorem can be proved:

Theorem 38.1. *Let Σ_n in $\mathbf{CE}_m$, $n < m$, be a Stein manifold with a real image $\xi(\Sigma_n) = M_{2n}$ in E_{2m}. If q is a fixed point of E_{2m} which is neither a point of M_{2n} nor a focal point of M_{2n}, the real-valued distance function*

$$p \to \| p - q \| = f_q(p) : M_{2n} \to R \qquad (q \text{ fixed}) \tag{38.13}$$

is ND *on M_{2n} and has no critical point with an index $k > n$.*

The Index Forms. The first part of the proof makes no essential use of the fact that the real image M_{2n} of the Stein manifold Σ_n is anything more than an arbitrary differentiable manifold which is regular and analytic in E_{2m}. Corresponding to a critical point p^0 of f_q on M_{2n} one obtains a representation such as (31.7) of a quadratic form determining the index of p^0.

We term the point q of Theorem 38.1 the *pole* of the distance function f_q.

As in the proof of Theorem 31.3, suppose that p^0 is the origin of coordinates in E_{2m} and that the $2n$-plane, say V_{2n}, of coordinates (38.11) is tangent to M_{2n} at the origin p^0. Suppose further that the pole q is on the x_{n+1}-axis and that $x_{n+1} = c > 0$ at q.

If $\| v \| < e$ and if e is sufficiently small, the points

$$(x_1, \ldots, x_m; y_1, \ldots, y_m) \in E_{2m} \tag{38.14}$$

in the neighborhood X of p^0 relative to M_{2n}, as presented by (38.11) and (38.12), are such that the following is true: For $\| v \| < e$ (38.12) holds and can be given the form

$$\begin{aligned} x_{n+\rho} &= \tfrac{1}{2} b^{\rho}_{ij}\, v_i v_j ++ \\ & \qquad\qquad\qquad \rho = 1, \ldots, m - n, \\ y_{n+\rho} &= \tfrac{1}{2} c^{\rho}_{ij}\, v_i v_j ++ \end{aligned} \tag{38.15}$$

where the quadratic forms on the right of (38.15) are real and symmetric, and the remainders (indicated by $++$) are real and analytic in the variables $v_1, \ldots, v_{2n}$ for $\| v \| < e$ and vanish with their first and second partial derivatives at the origin $v = \mathbf{0}$.

Let the coordinates $v_1, \ldots, v_{2n}$ be subjected to an orthogonal transformation $\mathbf{T}$ into coordinates $u_1, \ldots, u_{2n}$ such that under $\mathbf{T}$

$$b^1_{ij} v_i v_j = \sigma_1 u_1^{\,2} + \cdots + \sigma_{2n} u^2_{2n}. \tag{38.16}$$

As in the proof of Theorem 31.2, it follows that if $p(v) \in X$ is a point with the local coordinates $v = (v_1, ..., v_{2n})$, the square $f_q^2(p(v))$ of the distance from q to $p(v)$ is equal to c^2 plus

$$\| v \|^2 - cb_{ij}^1 v_i v_j ++ \qquad \text{for} \quad \| v \| < e \tag{38.17}$$

[cf. (31.6)] and in terms of the coordinates $u_1, ..., u_{2n}$ is equal to c^2 plus

$$(1 - c\sigma_1)u_1^2 + \cdots + (1 - c\sigma_{2n})u_{2n}^2, \tag{38.18}$$

omitting terms of order higher than the second. The index k of the quadratic form (38.18) is the index of the critical point p^0 of the distance function f_q. This form is ND by virtue of our choice of q.

We shall now make our first use of the hypothesis that M_{2n} is the real image of a Stein manifold Σ_n. The basic lemma follows.

Lemma 38.1. *Let $\sigma_1, ..., \sigma_{2n}$ be the characteristic roots of the matrix $\| b_{ij}^1 \|$. If $\sigma_1', ..., \sigma_{2n}'$ is a suitable reordering of these roots, then*

$$(\sigma_1', ..., \sigma_{2n}') = (-\sigma_1, ..., -\sigma_{2n}). \tag{38.19}$$

Notation. We suppose that (38.9) gives a complex Monge presentation of a neighborhood $\mathbf{X}$ relative to Σ_n of the origin w^0 and that (38.12) and (38.11) give the corresponding real Monge presentation of $X = \xi(\mathbf{X})$, with the $2n$-plane of coordinates $(x_1, ..., x_n; y_1, ..., y_n) = (x, y)$ tangent to M_{2n} at the origin p^0. From the first equation in (38.12) we infer that

$$P_1(z) + \text{conj } P_1(z) = \varphi_1(x_1, ..., x_n; y_1, ..., y_n), \tag{38.20}$$

subject to the conditions

$$z_j = x_j + iy_j, \qquad j = 1, ..., n. \tag{38.21}$$

Suppose that $\mathbf{Q}(z_1, ..., z_n)$ is the quadratic form that gives the second order terms in the series for $2P_1(z)$, and that $Q(x_1, ..., x_n; y_1, ..., y_n)$ is the real, symmetric, quadratic form that gives the second order terms in the power series for $2\varphi_1$. Then (38.20) implies that

$$\mathbf{Q}(z) + \text{conj } \mathbf{Q}(z) = Q(x_1, ..., x_n; y_1, ..., y_n), \tag{38.22}$$

subject to (38.21).

The relation (38.22) is an identity in the variables $(x_1,...,x_n; y_1,..., y_n)$ subject to the conditions $z_j = x_j + iy_j$ of (38.21). If in (38.22) one substitutes $(-y_j , x_j)$ for (x_j , y_j) and in accord with (38.21) substitutes iz_j for z_j and $-i$ conj z_j for conj z_j , then the left side of (38.22), so evaluated, is equal to

$$-(\mathbf{Q}(z) + \text{conj}\ \mathbf{Q}(z)),$$

while the right side of (38.22) is equal to

$$Q(-y_1 ,..., -y_n ; x_1 ,..., x_n).$$

This evaluation accordingly gives the identity

$$-Q(x_1 ,..., x_n ; y_1 ,..., y_n) = Q(-y_1 ,..., -y_n ; x_1 ,..., x_n). \tag{38.23}$$

The right side of (38.23) is a quadratic form $Q'(x_1 ,..., x_n ; y_1 ,..., y_n)$, which reduces to $Q(x_1 ,..., x_n ; y_1 ,..., y_n)$ after a suitable orthogonal transformation. Hence the forms Q' and Q have the same set of characteristic roots, in different orders. On the other hand, the quadratic form $Q''(x_1 ,..., x_n ; y_1 ,..., y_n)$ given by the left side of (38.23) has a set of characteristic roots which are the negatives of the characteristic roots of Q. The roots of Q' are equal to those of Q'' because of the identity (38.23), and in some order are equal to the roots of Q.

The function $v \to \varphi_1(v)$ originated in (38.12). Subject to (38.11) the terms of second order in $2\varphi_1$ in an expansion about the origin are the forms

$$b_{ij}^1 v_i v_j = Q(x_1 ,..., x_n ; y_1 ,..., y_n). \tag{38.24}$$

Lemma 38.1 accordingly follows from the above properties of the characteristic roots of Q.

The Proof of Theorem 38.1 *Concluded.* To verify that the index k of the critical point p^0 of the distance function $p \to f_q(p)$ satisfies the condition $k \leqslant n$, we make use of the fact that k is the number of coefficients in the quadratic form (38.18) which are negative.

If a coefficient $1 - c\sigma_j$ in the form (38.18) is negative, σ_j must be positive, and hence by Lemma 38.1 some other characteristic root σ_j' of Q must be equal to $-\sigma_j$, and hence yield a positive coefficient of the quadratic form (38.18). Lemma 38.1 thus implies that $k \leqslant n$.

Theorem 38.1 follows.

We state a first corollary of Theorem 38.1:

Corollary 38.1. *No component of the Stein manifold Σ_n is compact.* Were a component of Σ_n compact, the real image under ξ of this component would be a compact, regular, analytic, real manifold N_{2n}, a submanifold of M_{2n}. The distance function $p \to f_q(p)$ would then have a critical point of index $2n$ on N_{2n}, contrary to Theorem 38.1.

Corollary 38.2. *The singular rth homology groups of the Stein manifold Σ_n are trivial over each field for $r > n$.*

This is a consequence of Theorem 38.1 and Theorem 29.3.

Corollary 38.3. *The homology groups $H_r(\Sigma_n, \mathbf{Z})$ of Σ_n over the ring $\mathbf{Z}$ of integers are trivial for $r > n$, and their torsion groups vanish for $r > n - 1$.*

Propositions concerning the homology theory over $\mathbf{Z}$ are summarized in Appendix III. The homology theory over $\mathbf{Z}$ presented in Appendix III will be given a full exposition in a paper supplementary to this book without making any use of a *triangulation* of the underlying differentiable manifold M_n. The proposition implying Corollary 38.3 is as follows:

Proposition 38.1. *If M_n is a C^∞-manifold on which there exists a* ND *function $f \in C^\infty(M_n)$ such that f_c is compact for each value of f and the indices k of critical points of f are at most m, then the homology groups $H_r(M_n, \mathbf{Z})$ vanish for $r > m$ and the torsion subgroups vanish for $r > m - 1$.*

§ 39

SUPPLEMENTARY CONCEPTS AND THEOREMS

A. Bowls and Special Homotopies. We shall supplement the homotopy theorems of §23. Singular chain-homotopies were introduced in §27 with the aid of continuous deformations (27.1) of a Hausdorff space χ on itself. We shall define continuous singular chain families by first defining continuous deformations of singular *cells* on χ rather than of points of χ.

Continuous Families of Singular q-Simplices. Singular q-simplices are defined, following Eilenberg, as in §26. In the sense of Definition 26.1, a singular q-simplex on χ is a continuous mapping

$$x \to \tau(x) : s \to \chi \tag{39.1}$$

into χ of a vertex-ordered Euclidean simplex $s = p_0 p_1 \cdots p_q$. Let I be the unit interval on the t-axis. We understand that a continuous family of singular simplices with the *initial* singular simplex τ is defined by a continuous map

$$(x, t) \to \mathscr{T}(x, t) : s \times I \to \chi, \qquad \text{with} \quad \mathscr{T}(x, 0) = \tau(x). \tag{39.2}$$

The partial maps

$$x \to \mathscr{T}(x, t) = \tau^t(x), \qquad 0 \leqslant t \leqslant 1, \tag{39.3}$$

introducing $\tau^t(x)$, are the *singular simplices* τ^t defined, as we shall say, by $\mathscr{T}$.

Continuous Families of Singular q-cells. Let "*equivalent*" singular q-simplices

$$\tau' : s' \to \chi \tag{39.4a}$$

and

$$\tau'' : s'' \to \chi \tag{39.4b}$$

be defined as in §26. We understand, as in §26, that s' and s'' are two vertex-ordered Euclidean simplices barycentrically mapped one onto the other with preservation of vertex order. Let continuous mappings

$$(x, t) \to \mathscr{T}'(x, t) : s' \times I \to \chi \tag{39.5a}$$

and

$$(x, t) \to \mathscr{T}''(x, t) : s'' \times I \to \chi \tag{39.5b}$$

be given such that

$$\mathscr{T}'(x, 0) = \tau'(x), \qquad x \in s', \tag{39.6a}$$

and

$$\mathscr{T}''(x, 0) = \tau''(x), \qquad x \in s'', \tag{39.6b}$$

and such that for each t the partial mappings

$$x \to \mathscr{T}'(x, t) : s' \to \chi \tag{39.7a}$$

and

$$x \to \mathscr{T}''(x, t) : s'' \to \chi \tag{39.7b}$$

define "equivalent" singular simplices τ'^t and τ''^t on χ. The "initial" equivalent singular simplices of $\mathscr{T}'$ and $\mathscr{T}''$ are given, respectively, by (39.4a) and (39.4b).

The singular q-cells $\sigma^q(t)$ which are the equivalence classes of the respective singular simplices τ^t, $0 \leqslant t < 1$, of the continuous family $\mathscr{T}$ will be said to define *a continuous family* $\mathring{\mathscr{T}}$ *of singular q-cells* $\sigma^q(t)$.

Families of Singular Boundary Cells. We refer to the vertex-ordered $(q-1)$-simplex, $q > 0$,

$$s(i) = p_0 \cdots \hat{p}_i \cdots p_q\,, \qquad p_i \text{ deleted,} \tag{39.8}$$

of (26.4) and to $\tau_i = \tau \mid s(i)$ of (26.5), where τ is given by (39.1). The mapping $\mathscr{T}$ of (39.2), if restricted to $s(i) \times I$, defines a continuous family of singular $(q-1)$-simplices $\tau_i{}^t$, with τ_i as an initial singular $(q-1)$-simplex. For each $t \in [0, 1]$ the relation (26.6) has the formal extension

$$\partial\sigma^q(t) = (-1)^i\, \sigma_i^{q-1}(t), \tag{39.9}$$

introducing the singular $(q-1)$-cells $\sigma_i^{q-1}(t) = \mathring{\tau}_i{}^t$, where $0 \leqslant t \leqslant 1$ and i is on the range $0, 1, \ldots, q$.

Continuous Families of Singular q-Chains. Let

$$z^q = r_1\sigma_1{}^q + \cdots + r_m\sigma_m{}^q, \qquad 0 \neq r_\mu \in \mathscr{K}, \quad m > 0, \tag{39.10}$$

be a "reduced" representation on χ of the q-chain z^q (Definition 24.0). Suppose that each q-cell $\sigma_\mu{}^q$ is the initial q-cell in a continuous family $\sigma_\mu{}^q(t)$ of q-cells subject to the following *compatibility conditions*. These conditions are satisfied only exceptionally. We shall give an example in the critical point theory.

Compatibility Conditions. For μ and ν on the range $1,\ldots, m$ with $\mu \neq \nu$ let σ^{q-1} be a singular $(q-1)$-cell which appears formally in a representation (39.9) of $\partial\sigma_\mu{}^q(0)$ and of $\partial\sigma_\nu{}^q(0)$, with coefficients on the right of (39.9) which are e_μ and e_ν, respectively, on the range ± 1. The compatibility condition ***requires*** that the two continuous families $\sigma^{q-1}(t)$, $0 \leqslant t \leqslant 1$, of singular boundary $(q-1)$-cells with initial $(q-1)$-cell σ^{q-1} associated as above with the families $\sigma_\mu{}^q(t)$ and $\sigma_\nu{}^q(t)$, $0 \leqslant t \leqslant 1$ be identical. For each t the cell $\sigma^{q-1}(t)$ will then appear in the formulas (39.9) for $\partial\sigma_\mu{}^q(t)$ and $\partial\sigma_\nu{}^q(t)$ with coefficients e_μ and e_ν, respectively.

Let the reduced q-chain (39.10) be given as an initial chain, and let mutually compatible continuous families of cells $\sigma_\mu{}^q(t)$ be defined with initial q-chains $\sigma_\mu{}^q(0) = \sigma_\mu{}^q$, $\mu = 1, 2,\ldots, m$. The family of chains defined by setting

$$z^q(t) = r_1\sigma_1{}^q(t) + \cdots + r_m\sigma_m{}^q(t), \qquad 0 \leqslant t \leqslant 1, \tag{39.11}$$

will be termed an *admissible continuous family of singular q-chains $z^q(t)$*, with initial chain $z^q = z^q(0)$ and terminal chain $z^q(1)$. It is clear that if z^q is a q-cycle on χ, then $z^q(t)$ is a q-cycle on χ for each t.

The following theorem connects the preceding special homotopy theory with homology theory:

Theorem 39.1. *If the chain* (39.10) *is a cycle, so that the family* (39.11) *is a family of cycles, then for each value t_1 of t on the interval $0 < t_1 \leqslant 1$ the initial cycle $z^q(0)$ is homologous to $z^q(t_1)$ on*

$$\operatorname*{Union}_{0 \leqslant t \leqslant t_1} | z^q(t)|. \tag{39.12}$$

We prove no theorems in this section.

Deformations through a Critical Level. Let M_n be a C^∞-manifold and f a ND function in $C^\infty(M_n)$. We suppose f_c compact for each value c of f. Let a be a critical value of f, assumed at a single critical point p_a, and set $\mathring{f}_a = f_a - p_a$.

If a reduced chain (39.10) is given, the maximum diameter on M_n of carriers $| \sigma_\mu{}^q |$ will be called the *mesh* of the chain z^q. We are supposing that M_n has been metricized, so that such a mesh is well-defined.

The following theorems, taken with the homotopy theorems of §23, give a basis for the homotopical aspects of the theory of critical points on M_n.

Theorem 39.2. *Corresponding to the critical point p_a of f there exists a positive constant e so small that the following is true: Suppose that the index k of the critical point p_a is such that $0 < k \leqslant n$ and let r be an integer such that $0 \leqslant r < k$. Corresponding to any r-cycle z^r on f_a with mesh $< e$ there exists an admissible continuous family of r-cycles on f_a whose initial cycle is z^r and whose terminal cycle is on $\mathring{f}_a$.*

Comments on Theorem 39.2 The condition in Theorem 39.2 that the mesh of z^r be less than e is not restrictive as far as homology theory is concerned, since a suitable subdivision of z^r on f_a will be homologous to z^r on f_a and satisfy this mesh condition.

Three aids are invoked in proving Theorem 39.2. The theorems on deformation retractions into f_a are useful. The concept of "simply carried" q-cells of §30 is employed. When $k < n$ the fact that the bowl "ascending from p_a" has the dimension $n - k$ is what makes the condition $r < k$ effective.

Extensions. The preceding theorems can be extended to the case in which the ring $\mathbf{Z}$ of integers replaces a field. One can also include the case in which in Theorem 39.2 there is given a relative r-cycle z^r on $f_a \bmod \mathring{f}_a$ with mesh $< e$. One can then prove the existence of an admissible continuous family of relative r-cycles $z^r(t)$ on $f_a \bmod \mathring{f}_a$ whose initial relative cycle is z^r, whose boundaries $\partial z^r(t)$ are independent of t, and whose terminal cycle vanishes $\bmod \mathring{f}_a$.

B. Global Alteration of Critical Values. Given a ND $f \in C^\infty(M_n)$, we seek modifications of f as a ND function in $C^\infty(M_n)$ which leave the critical points of f invariant, together with their indices. That f can be modified so that its replacement $\hat{f}$ has distinct critical values at distinct critical points has been seen in Lemma 22.4. We seek alterations of critical values which are more than the infinitesimal modifications of Lemma 22.4.

Let a ND f be given, and a particular critical point z of f. A ND function $\hat{f} \in C^\infty(M_n)$ replacing f will be termed *admissible* relative to z if the critical points of f and $\hat{f}$ are identical and have the same indices and if $\hat{f}$ is identical with f in some neighborhood of each critical point of f other than z.

Definition 39.1. *A Replacement Interval for f and z.* Given a ND $f \in C^\infty(M_n)$ and a critical point z of f, an open interval I_z^f of real values will be termed a *replacement interval for f and z* if for each value $a \in I_z^f$ there exists a ND $\hat{f} \in C^\infty(M_n)$ which replaces f admissibly relative to z and has the critical value $\hat{f}(z) = a$ at z.

We shall show how the bowls ascending and descending from z determine a replacement interval for f and z. To that end, two definitions are needed. For simplicity we assume that M_n is connected and compact.

Definition 39.2. *The k-Dome $B^-(z)$.* If $k > 0$, each f-arc of the descending bowl $B_-(z, k)$ of Definition 22.2 has a critical point z' as lower limiting endpoint. Given z, these critical points z' are finite in number, and so have a maximum f-value $M(z)$. The differentiable k-manifold $B^-(z)$ with carrier

$$|B^-(z)| = \{q \in B_-(z, k) \mid f(q) > M(z)\} \tag{39.13}$$

C^∞-embedded in M_n will be called the *k-dome with zenith z*.

Definition 39.3. *The Inverted $(n - k)$-Dome $B^+(z)$.* If $k < n$, each f-arc of the ascending bowl $B_+(z, n - k)$ has a critical point z'' as upper limiting endpoint. Given z, these critical points z'' are finite in number, and so have a minimum f-value $m(z)$. The differentiable $(n - k)$-manifold $B^+(z)$ with carrier

$$|B^+(z)| = \{q \in B_+(z, n - k) \mid f(q) < m(z)\} \tag{39.14}$$

C^∞-embedded in M_n will be called the *inverted* $(n - k)$*-dome* with *nadir* z.

The following theorem is a consequence of Theorems 4.1 and 4.2 of Morse [16] and Morse [12], pp. 249–259:

Theorem 39.3. *Let* M_n *be a compact connected* C^∞*-manifold and* z *a prescribed critical point of* f. *The open interval of values of* f *assumed by* f *on the union of the domes* $B^\pm(z)$ *is a replacement interval* I_z^f *for* f *and* z.

On referring to Definitions 39.2 and 39.3 we see that the replacement interval affirmed to exist in Theorem 39.3 has the form

$$I_z^f = (f(z'), f(z'')); \qquad f(z') < f(z) < f(z''), \tag{39.15}$$

where z' and z'' are critical points of f.

Modifying a Riemannian Structure **S** *on* M_n. We admit modifications of the Riemannian structure **S** on M_n that leave M_n and f invariant. It is desirable to preserve the character of **S** as "*f-preferred*" by leaving invariant the Riemannian structure in sufficiently small open neighborhoods of each critical point of f. To this end, let c be an ordinary value of f and N_c an open neighborhood of f^c relative to M_n such that $\mathrm{Cl}\, N_c$ contains no critical point of f. A replacement of an f-preferred Riemannian structure **S** on M_n by a Riemannian structure $\hat{\mathbf{S}}$ on M_n identical with that of **S** on the submanifold $\hat{M}_n$ with carrier $| M_n - \mathrm{Cl}\, N_c |$ will be called a *sectional modification* of **S**. The new Riemannian structure $\hat{\mathbf{S}}$ will be f-preferred because **S** is f-preferred.

Each "sectional" modification of a Riemannian structure **S** on M_n presumably will modify the k- or $(n - k)$-domes whose f-trajectories meet f^c. On exploiting this fact we are led by a simple proof to the following theorem:

Theorem 39.4. *Let* M_n, f, *and a critical point* z *of* f *be given as in Theorem* 39.3, *together with an f-preferred Riemannian structure* **S** *on* M_n.

It is then possible to make a finite number of successive sectional modifications of **S** *such that the replacement interval* (39.15) *defined by the resultant domes* $B^\pm(z)$ *is such that*

$$\text{index } z' < \text{index } z < \text{index } z''. \tag{39.16}$$

The reader can readily verify the theorem for $n = 2$ and $k = 1$. The general case will then become transparent by virtue of the elementary nature of trajectories orthogonal to the level manifolds of a ND quadratic form.

One can illuminate the nature of the proof still further.

Suppose that some of the f-trajectories emanating from the zenith of the original k-dome $B^{-}(z)$ were obstructed in their downward continuation by a critical point w with index $h \geqslant k$. Let c be a value on the open interval $(f(w), f(z))$. The original k-dome $B^{-}(z)$ meets f^c in a differentiable $(k - 1)$-sphere. The inverted $(n - h)$-dome of $B^{+}(w)$ meets f^c in a differentiable $(n - h - 1)$-sphere. Since

$$(n - h - 1) + (k - 1) < n - 1 \qquad \text{when} \quad h \geqslant k,$$

it is clear that after a suitable sectional modification of the original Riemannian structure the f-trajectories of $B^{-}(z)$ will *bypass* w. Similar arguments apply to the ascending f-trajectories of $B^{+}(z)$.

Theorem 39.4 follows.

Theorems 39.3 and 39.4 have the following corollary:

Corollary 39.1. *Corresponding to a compact connected C^∞-manifold there exists a* ND $F \in C^\infty(M_n)$ *such that at each critical point z of index k* $F(z) = k$.

By virtue of Morse [9], p. 383, there exists a ND $f \in C^\infty(M_n)$ with just one point p of index 0 and just one point q of index n.

It follows from Theorems 39.3 and 39.4 that a finite sequence of "sectional modifications" of preferred Riemannian structures $\mathbf{S}$, $\mathbf{S}_1, \ldots, \mathbf{S}_r$ on M_n and suitable modifications of critical values of f and of its replacements will lead to a ND F satisfying Corollary 39.1.

Smale. Corollary 39.1 is Theorem B of Smale [1] and is used by Smale in deriving his solution of the Poincaré problem when $n \geqslant 5$.

C. Orientability without Triangulation. Let M_n be a compact connected C^∞-manifold. In the absence of a triangulation of M_n the homological characterization of the "orientability" of M_n requires study. We shall give a definition of orientability of M_n very similar to the classical definition of the orientability of a surface and supplement this definition by two equivalent characterizations belonging to homology theory:

Sense-Compatible Presentations. Two overlapping presentations F and G in $\mathscr{D}M_n$ will be said to be *sense-compatible* if the corresponding transition diff λ given by (5.6) has a positive Jacobian.

Test Sequences of Presentations. A finite sequence of presentations

$$(F_i : U_i, X_i) \in \mathscr{D}M_n, \qquad i = 1,..., m > 2, \tag{39.17}$$

will be termed a *test sequence* if it has the following properties: (1) Each U_i is an open Euclidean n-ball; (2) $X_i \cap X_{i+1} \neq \varnothing$, $i = 1,..., m - 1$; (3) $X_1 \cap X_m \neq \varnothing$; (4) for $i = 1,..., m - 1$ the presentations F_i and F_{i+1} are sense-compatible.

Definition 39.4. *Orientability.* A connected C^∞-manifold will be said to be geometrically *orientable* if the first and last presentations of each test sequence of presentations in $\mathscr{D}M_n$ are sense-compatible.

Corresponding to an arbitrary, simple, regular arc γ joining a point $q \in M_n$ to itself it is easily shown that there exists a partition of γ into a finite sequence of arcs γ_i, $i = 1,..., m$, corresponding to which there exists a test sequence (39.17) of presentations F_i such that $|\gamma_i| \subset X_i$ for each i.

The fundamental theorem follows:

Theorem 39.4. *A compact connected C^∞-manifold is orientable in the geometric sense if and only if either one of the following two equivalent conditions is satisfied:*

Condition 1. *Corresponding to a* ND f *for which p is the sole point of maximum of f on M_n, Condition* 1 *requires that p be of linking type* (Definition 29.2).

Condition 2. *The* nth *connectivity of M_n over the field* Q *of rational numbers is* 1.

The existence of a ND f for which p is the sole point of maximum of f follows from the work of Morse [9]. The equivalence of Conditions 1 and 2 follows from Theorem 29.3. A proof that geometric orientability of M_n is equivalent to Condition 2 will be published separately.

Geometric orientability is well-defined for C^∞-manifolds which are not compact. Equivalent homological conditions in the noncompact case will be studied separately.

APPENDIX I

PRELIMINARY DEFINITIONS

This book presupposes mathematical knowledge at the level of a good first-year graduate student. The object of Appendix I is to recall, or clarify, a number of terms which are used without definition in the text, or are used in different senses by current mathematical writers.

In general our use of terms is in accord with that of Bourbaki.

A *Hausdorff space* is a topological space X in which distinct points have disjoint neighborhoods.

A *topological n-manifold* is a Hausdorff space in which each point has an open neighborhood homeomorphic to Euclidean n-space.

A *subspace* of a topological space X is a subset of X whose topology is "induced" by that of X. The terms "a subspace of X" and "a subset of X" are by no means synonymous.

A *relatively compact subset* U of a topological space X is a subset U of X whose closure in X is compact.

Mappings into E_n *of class* C^r. Let E_n, $n > 0$, be a Euclidean space of points $y = (y_1, \ldots, y_n)$. Let X be a nonempty open subset of E_m, $0 < m$, of points $x = (x_1, \ldots, x_m)$. A mapping

$$x \to (\varphi_1(x), \ldots, \varphi_n(x)) : X \to E_n \tag{I.1}$$

will be said to be of *class* C^r, $0 \leqslant r$, on X if for i on the range $1, \ldots, n$, the partial derivatives of φ_i of orders at least r exist and, together with φ_i, are continuous. A mapping φ of class C^r for each $r > 0$ is said to be of class C^∞.

Grad g. If a mapping

$$x \to g(x) = g(x_1, \ldots, x_m) : X \to R \tag{I.2}$$

of a nonempty open subset X of E_m into R is of at least class C^1, the vector

$$\partial g(x)/\partial x_1, \ldots, \partial g(x)/\partial x_m \tag{I.3}$$

will be termed the *gradient of g at x* and will be denoted by (grad g)(x).

Domain, Range, Image Set. If $F : U \to X$ is a mapping of a set U into a set X, then U is termed the *domain* of F and X the *range* of F. The subset $F(U)$ of X is termed the *image* set of U under F.

Composition $g \circ f$. Let two functions

$$f : U \to V \qquad \text{and} \qquad g : X \to Y \tag{I.4}$$

be given such that $V = X$. A *composite* function

$$g \circ f : U \to Y \tag{I.5}$$

is then defined with values

$$(g \circ f)(u) = g(f(u)), \qquad u \in U. \tag{I.6}$$

Such composition is *associative*; if three functions

$$f : U \to V, \qquad g : V \to W, \qquad h : W \to Z \tag{I.7}$$

are given, the composite functions

$$h \circ (g \circ f) \qquad \text{and} \qquad (h \circ g) \circ f \tag{I.8}$$

mapping U into Z are defined and equal.

Extended Compositions $g \bar{\circ} f$. Let functions f and g be given as in (I.4) subject to the conditions $f(U) \subset X$. An *extended composite* function

$$g \bar{\circ} f : U \to Y \tag{I.9}$$

is then defined with values

$$(g \bar{\circ} f)(u) = g(f(u)), \qquad u \in U. \tag{I.10}$$

Extended composition of functions is not in general associative, as simple examples show. If $f \circ g$ is well-defined, then it is always true that

$$f \circ g = f \mathbin{\bar{\circ}} g. \tag{I.11}$$

Note. An extended composition $g \mathbin{\bar{\circ}} f$ of functions f and g is equal to a composition

$$g \circ i \circ f', \tag{I.12}$$

where f' is a mapping of U onto $f(U)$ with the same values as f and i is the inclusion mapping of $f(U)$ into X. It seems notationally simpler for our purposes to use $g \mathbin{\bar{\circ}} f$ rather than the composition (I.12).

A *field* $\mathscr{K}$ is a nontrivial commutative ring in which each nonzero element has a multiplicative inverse.

APPENDIX II

ON ELEVATING MANIFOLD DIFFERENTIABILITY

We shall make use of the following theorem (see Morse [10]):

Theorem A. *Let M_n be a regular, compact, differentiable n-manifold in E_{n+1} of class C^μ, $\mu > 0$.*

There then exists a sense-preserving C^μ-diff,

$$x \to (\varphi_1(x),\ldots, \varphi_{n+1}(x)) : E_{n+1} \to E_{n+1} \qquad \text{onto} \quad E_{n+1} \tag{II.1}$$

such that $\varphi(M_n)$ is a regular n-manifold of class C^∞ in E_{n+1}.

Corresponding to an arbitrary neighborhood N of M_n relative to E_{n+1} and to an arbitrary positive constant e the diff φ can be chosen so that

$$| \varphi_j(x) - x_j | < e, \qquad j = 1,\ldots, n + 1, \tag{II.2}$$

and

$$\left| \frac{\partial \varphi_j(x)}{\partial x_i} - \delta_i^j \right| < e, \qquad i = 1,\ldots, n + 1, \tag{II.3}$$

for each $x \in E_{n+1}$, and φ reduces to the identity for $x \in E_{n+1} - N$, so that $\varphi(N) = N$.

Proof of Theorem 9.1′. We take M_n of Theorem A as Σ of Theorem 9.1′, so that $\mu = 2$ in Theorem A. By hypothesis $| \Sigma |$ bounds the compact set Z. In terms of the diff φ of Theorem A, set $\varphi(\Sigma) = \Sigma^*$ and $\varphi(Z) = Z^*$. Then Σ^* is a regular manifold of class C^∞ and $| \Sigma^* |$ bounds Z^*.

Choice of N in Theorem A. Let N be so small a neighborhood of $|\,\Sigma\,| = |\,M_n\,|$ that Z^* is included in the open domain of f of Theorem 9.1′ and the critical points of $f \mid \mathring{Z}$ are those of $f \mid \mathring{Z}^*$.

Choice of e of Theorem A. Let e be so small that grad f is emergent on Σ^*. This is possible, since (2) and (3) hold.

With e and N so chosen f is admissible relative to the regular C^∞-domain Z^* bounded by Σ^*, and grad f is emergent on Σ^*.

Theorem 9.1 implies that the type numbers of $f \mid Z^*$ and the connectivities of Z^* satisfy the relations (9.4). Since the type numbers of $f \mid Z^*$ are equal to those of $f \mid Z$ and the connectivities of Z are equal to those of Z^* (see Part III), Theorem 9.1′ follows.

Proof of Theorem 10.2′. Let M_n of Theorem A be taken as Σ of Theorem 10.2′, so that $\mu = 2$ in Theorem A. By hypothesis of Theorem 10.2′ Σ bounds Z of Theorem 10.2′. If φ is the diff of Theorem A, set $\varphi(\Sigma) = \Sigma^*$ and $\varphi(Z) = Z^*$. Then Σ^* is a regular manifold of class C^∞ and $|\,\Sigma^*\,|$ bounds Z^*.

The Functions f and f^*. Let D be an open neighborhood of Z on which f is of class C^2. Set $\varphi(D) = D^*$ and let f^* be defined on D^* by setting $f^*(y) = f(x)$ for $x \in D$ subject to the condition $y = \varphi(x)$. Then f^* is of class C^2 on D^*, ND on Z^*, and ordinary on Σ^*, with the critical points of $f \mid \mathring{Z}$ and of $f^* \mid \mathring{Z}^*$ corresponding biuniquely with preservation of indices. Thus the type numbers of $f \mid \mathring{Z}$ will be equal to the corresponding numbers of $f^* \mid \mathring{Z}^*$.

The restrictions $f \mid \Sigma$ and $f^* \mid \Sigma^*$. These functions have equal values at points $p \in \Sigma$ and $q \in \Sigma^*$ when $q = \varphi(p)$. It follows from Theorem 5.5 that the critical points of $f \mid \Sigma$ and $f \mid \Sigma^*$ correspond biuniquely under φ with preservation of indices. If the constant e of Theorem A is sufficiently small, the critical points of $f \mid \Sigma_-$ and $f \mid \Sigma^*_-$ (Definition 10.2) will in particular correspond under φ.

One concludes that the augmented type numbers of $f \mid Z$ are equal to those of $f^* \mid Z^*$. But the augmented type numbers of $f^* \mid Z^*$ satisfy the relations (9.4) of Theorem 9.1 by virtue of Theorem 10.2. Since the connectivities of Z are equal to those of Z^*, the augmented type numbers of $f \mid Z$ likewise satisfy the relations (9.4).

This establishes Theorem 10.2′.

APPENDIX III

SINGULAR HOMOLOGY THEORY ON M_n OVER $\mathbf{Z}$

Such a homology theory has been developed by the authors without making any use of a global "triangulation" of M_n and will be presented in the near future. Appendix III is an introduction to this theory.

Hypotheses. For simplicity let the manifold M_n be of class C^∞ and connected. As has been seen in Theorem 23.5 there then exists a ND $f \in C^\infty(M_n)$ such that for each value c of f, f_c is compact and has just one critical point of index 0. One can suppose, in accordance with Lemma 22.4, that each critical value a of f is assumed at just one critical point p_a of f. The principal theorems concern the singular homology groups of f_c over $\mathbf{Z}$. Our theorems are proved by an induction with respect to the increasing values of a in $\{a\}_c$, where $\{a\}_c$ is the set of critical values $a \leqslant c$ of f other than the absolute minimum a_0 of f.

We shall state a number of principal theorems, omitting proofs. Unless otherwise stated all chains, cycles, and homology groups will be singular chains, cycles, and homology groups over $\mathbf{Z}$. A first major theorem follows.

Theorem 1. *For each value c of f and each integer $q \geqslant 0$ the qth homology group of f_c over $\mathbf{Z}$ is "finitely generated."*

Theorem 1 is implied by the following lemma. If $a > a_0$ and if the qth homology group $\dot{H}_q^a$ of $\dot{f}_a$ over $\mathbf{Z}$ is finitely generated, then the qth homology group H_q^a of f^a is finitely generated. A proof of this lemma and of Theorem 1 without any global triangulation of $\dot{f}_a$, f_a, or f_c, is a major departure from classical methods.

A Finitely Generated Abelian Group A. For Abelian group theory see Ledermann [1] and Mac Lane and Birkhoff [2], pp. 344–355.

The group A is a direct sum $B \oplus T$ of its torsion subgroup T and a free Abelian subgroup B. We term B a *Betti subgroup* of A and term $\dim B$ the *Betti number* β of A. The torsion subgroup T of A is uniquely determined by A. This is not true of Betti subgroups of A if T is nontrivial. Given a base

$$u_1, \ldots, u_\beta \tag{III.1}$$

of B, one obtains a base of a second Betti subgroup B' of A by adding an arbitrary element t_i of T to u_i in (III.1) for each i. Now, $B' \neq B$ unless each $t_i = 0$. However, $A = B' \oplus T$ and $\dim B = \dim B'$.

One can obtain an arbitrary base for B by subjecting the base (III.1) of B to a "unimodular" transformation.

Torsion Subgroups T. The torsion subgroup of a finitely generated Abelian group A has a finite number of elements. It may be regarded as a **Z**-module. The group T, if nontrivial, admits a "cyclic primary decomposition" (CPD) (see MacLane and Birkoff [2], pp. 353–354). Such a CPD of T is by definition a direct sum

$$g_1 \oplus g_2 \oplus \cdots \oplus g_r \qquad (\text{no } g_i = 0) \tag{III.2}$$

of "primary subgroups" of T which are cyclic.

The order of a summand g_i of a CPD of T of form (III.2) is a power $p_i^{e_i}$ of a prime p_i with exponent $e_i \geqslant 1$. One thus has a list

$$p_1^{e_1}, \ldots, p_r^{e_r} \tag{III.3}$$

of "prime powers," the orders of the respective summands in (III.2). Moreover, a second CPD of T is isomorphic to the first CPD of T and so (after a suitable reordering of its direct summands) yields the same list (III.3) of prime power orders. The prime powers listed in (III.3) are termed *elementary divisors* of T, or, if one pleases, of the Abelian group A of which T is the torsion subgroup. Elementary divisors in the list (III.3) are regarded as distinct if they have distinct indices i even if they are numerically equal.

The indexed elementary divisors of T uniquely determine (by elementary algebraic processes) the classical ordered set of "torsion coefficients" of T with their respective multiplicities. Conversely, the "torsion coefficients" of T, if given with their multiplicities, determine (except for order) a list of indexed elementary divisors of T.

The prime-simple case. An important and, in homology theory, a very general case, occurs if T is nontrivial and if the exponents in (III.3) are equal to 1. Then T and A are termed *prime-simple*. In the case of projective spaces each homology group with a nontrivial torsion subgroup comes under the prime-simple case.

We add a definition to Abelian group theory.

Definition 1. *The Free Index of* $w \in A$. With an arbitrary element w in a finitely generated Abelian group A we shall associate a unique integer $\mathbf{s} \geqslant 0$ termed the *free index* of w. We define $\mathbf{s}$ as follows.

The free index $\mathbf{s}$ of w shall be zero if w is of finite order.

If w has an infinite order the free index $\mathbf{s}$ of w shall be a positive integer $\mathbf{s}$ with the following property: Corresponding to each Betti subgroup B of A there is a base for B containing an element u_B such that

$$w = \mathbf{s}u_B + \tau_B, \qquad \tau_B \in T. \tag{III.4}$$

Such a positive integer $\mathbf{s}$ exists and is independent of the choice of the Betti group B of A. One terms $\mathbf{s}$ the free index of w.

Program. We turn now to the problem of determining the qth homology group H_q^c, $q \geqslant 0$, of f_c, up to an isomorphism.

A group H_q^c, as we have shown, is a finitely generated Abelian group, and as such is determined, up to an isomorphism, by the dimension of a Betti subgroup B_q^c and the elementary divisors of the unique torsion subgroup T_q^c of H_q^c. It follows from the retracting deformations of §23 that these homology characteristics of H_q^c are equal to the corresponding characteristics of H_q^α, where α is the largest value of a in $\{a\}_c$. These characteristics of H_q^α are the terminal result of an inductive determination of the corresponding characteristics of H_q^a from those of $\dot{H}_q^a$ as a takes on successive increasing values in $\{a\}_c$.

Our principal problem is then to determine H_q^a from $\dot{H}_q^a$, up to an isomorphism, for each $a \in \{a\}_c$, making use of the index k_a of the critical point p_a at the f-level a and the homological characteristics of ∂-boundaries of *universal* k*-caps* κ_a^k associated with p_a as follows.

The Role of k-Caps. In §29 we have seen that the homology groups of f_c over a prescribed field $\mathscr{K}$ are determined, up to an isomorphism, by the indices k_a of the critical points p_a at the respective critical levels $a \in \{a\}_c$ and the characterization of each critical point p_a as being of "linking" or "non-linking" type. This characterization was in terms of k-caps associated in §29 with p_a. This association depended upon the choice of the field $\mathscr{K}$, so that the k-caps of p_a defined in §29 should in the present context be termed k-caps of p_a *over* $\mathscr{K}$.

Universal k-Caps. The k-caps of p_a over $\mathbf{Z}$, as we shall define them, will be called *universal* k-caps because they satisfy the definition in §29 of a k-cap of p_a over every field $\mathscr{K}$.

An "f-saddle L_k of M_n at p_a," as introduced in Definition 36.2, will satisfy the isomorphisms (36.19) (here understood as between homology groups over $\mathbf{Z}$) provided the f-saddle L_k is a sufficiently small "subsaddle" of a prescribed f-saddle of M_n at p_a.

Definition 2. *A Universal k-Cap of p_a. Given an f-saddle L_k, restricted as in the preceding paragraph, a universal k-cap κ_a^k of p_a shall be any singular k-cell which is "simply carried" by $|L_k|$ and such that $p_a \in |\kappa_a^k|$ and $|\partial\kappa_a^k| \subset \dot{f}_a$* (see Definition 30.2 of simply carried).

The universal k-caps κ_a^k associated with the critical point p_a enter our homology theory by way of the following fundamental lemma:

A Lemma on $\mathbf{w}_a^{k-1}$. *If $\mathbf{w}_a^{k-1}$ denotes the homology class on $\dot{f}_a$ of a preferred universal k-cap κ_a^p, then the homology class on $\dot{f}_a$ of the ∂-boundary of any other universal k-cap of p_a is $e\mathbf{w}_a^{k-1}$, where $e = \pm 1$.*

Definition 3. *The Free Index $\mathbf{s}_a$ of $\mathbf{w}_a^{k-1}$.* The first of the essential homological characteristics of $\mathbf{w}_a^{k-1}$ is its invariant *free index* $\mathbf{s}_a$ as an element in the Abelian group $\dot{H}_{k-1}^a$.

The distinction between the cases $\mathbf{s}_a = 0$, $\mathbf{s}_a = 1$, and $\mathbf{s}_a > 1$ is fundamental in our program of determination (up to an isomorphism) of H_q^a from $\dot{H}_q^a$. The determination of the Betti numbers of f_a from those of $\dot{f}_a$ is as follows:

Theorem 2. *For an arbitrary integer $r > 0$ and value $a \in \{a\}_c$ the following relations hold*:

$$\dim B_r^{\,a} - \dim \dot{B}_r^{\,a} = 1, \qquad \dim B_{r-1}^{a} - \dim \dot{B}_{r-1}^{a} = 0, \quad (r = k_a\,;\ \mathbf{s}_a = 0),$$

$$\dim B_r^{\,a} - \dim \dot{B}_r^{\,a} = 0, \qquad \dim B_{r-1}^{a} - \dim \dot{B}_{r-1}^{a} = -1, \ (r = k_a\,;\ \mathbf{s}_a > 0),$$

while $\dim B_q^{\,a} = \dim \dot{B}_q^{\,a}$ *when* $q \neq k_a$ or $k_a - 1$.

Elementary Divisors of $H_q^{\,a}$. Our program includes the determination of the elementary divisors of the torsion subgroup $T_q^{\,a}$ of $H_q^{\,a}$ given the Betti numbers of $\dot{f}_a$, the elementary divisors of $\dot{T}_q^{\,a}$, the index k_a of p_a, and homological characteristics of $\mathbf{w}_a^{k-1}$ on $\dot{f}_a$, such as the free index $\mathbf{s}_a$ and *linking index* $\mathbf{t}_a$ of p_a, presently to be defined.

The simplest cases occur when $q \neq k_a - 1$. *In each such case the elementary divisors of $T_q^{\,a}$ are equal to those of $\dot{T}_q^{\,a}$. The same is true if $q = k_a - 1$ and $\mathbf{s}_a = 1$.*

We introduce another homological characteristic of $\mathbf{w}_a^{k-1}$:

The linking index $\mathbf{t}_a$ of p_a. By virtue of relation (III.4) applied to $\mathbf{w}_a^{k-1}$ as an element in $\dot{H}_{k-1}^{a}$

$$\mathbf{w}_a^{k-1} = \mathbf{s}_a u_B^{\;a} + \tau_B^{\;a}, \qquad \tau_B^{\;a} \in \dot{T}_{k-1}^{a} \tag{III.5}$$

where we have added the superscript a to u_B and τ_B of (III.4). Set

$$\mathbf{t}_a = \min_B(\text{order } \tau_B^{\,a}), \tag{III.6}$$

where B ranges over all Betti subgroups of $\dot{H}_{k-1}^{a}$. The integer $\mathbf{t}_a$ is at least 1. We term $\mathbf{t}_a$ the *linking index* of p_a relative to f.

The above determination of the elementary divisors of $H_q^{\,a}$ is complete except in the case in which $q = k_a - 1$ and $s_a \neq 1$.

The following theorem is of interest:

Theorem 3. *A necessary and sufficient condition that each nontrivial torsion group $T_q^{\,a}$ for which $a \in \{a\}_c$ be prime-simple is that each linking index $\mathbf{t}_a = 1$ and each free index $\mathbf{s}_a > 0$ be the product of distinct primes.*

In the paper to which Appendix III is an introduction the determination of the elementary divisors will be made more explicit in accord with the program outlined above.

It should be noted that the carriers of the universal k-cap ∂-boundaries are *topological* $(k - 1)$*-spheres*, so that the determination (up to an isomorphism) of the homology groups of f_c has been reduced to the determination of the indices of the critical points p_a of f and homological properties on $\mathring{f}_a$ of the spherically carried $(k - 1)$-cycles $\partial\kappa_a^{\,k}$.

We shall close Appendix III by stating two theorems useful in interpreting results on homology groups over the field $\mathbf{Q}$ in terms of homology groups over $\mathbf{Z}$.

Given the above sublevel set f_c of M_n , let N_c be the set of all indices of critical points of f on f_c . We say that f_c is of *lacunary type* if there are no two positive integers in N_c which differ by 1. If f is a Milnor function of a complex projective space, each f_c is of lacunary type.

Theorem 4. *The homology groups of a sublevel set f_c of lacunary type are torsion free.*

Another theorem of general use follows.

Theorem 5. *Let χ be a Hausdorff space for which finitely generated singular homology groups over $\mathbf{Z}$ and over $\mathbf{Q}$ exist. Then*

$$R_q(\chi, \mathbf{Q}) = \beta_q(\chi), \qquad q = 0, 1, \ldots,$$

where $\beta_q(\chi)$ is the qth Betti number of χ.

FUNDAMENTAL SYMBOLS

Symbol	Meaning	Page
	PART I	
	SECTION 1	
E_n	a Euclidean n-space	7
C^r, $r \geqslant 0$	types of differentiability, see Appendix I	7
$\varphi \circ \psi$	composition of φ and ψ	8
$\delta_\nu{}^\mu$	Kronecker delta	8
	SECTION 2	
$\Vert z \Vert$	norm of vector z	10
$\dfrac{D(F_1, \ldots, F_n)}{D(x_1, \ldots, x_n)}$	Jacobian	13
	SECTION 3	
$\Vert a_{ij} \Vert$	matrix	15
$\vert a_{ij} \vert$	determinant	15
	SECTION 4	
grad f	the gradient vector of f	21
	SECTION 5	
$(F : V, X)$	a presentation F of X	29
Γ_n	a topological n-manifold	30
M_n	a differentiable manifold	30
$\vert M_n \vert$	the carrier of M_n	30
$\mathscr{D}M_n$	the set of presentations of M_n	30
f	a differentiable function on M_n	35
$f \mathbin{\bar{\circ}} F$	composition in the larger sense, see Appendix I	35

Symbol	Meaning	Page
	SECTION 6	
$(x, a) \to U(x, a)$	a family of mappings U^a into R	44
	SECTION 7	
Ω	the regular manifold defined by (7.2); in Theorem 6.3 the set of x-critical points of U	50
Ω^*	the set of x-singular points in Ω; in §6 the set of degenerate x-critical points of U	50
	SECTION 9	
$\mathscr{K}$	a commutative field, see §24	65
Z	a regular domain in E_{n+1}	64
$\mathring{Z}$	interior of Z	64
m_k	the kth type number of $f \mid Z$	65
$R_k(Z, \mathscr{K})$	the kth connectivity of Z	65
	SECTION 10	
Σ_-	the entrant portal of Z	70
Σ_+	the emergent portal of Z	70
m_k'	the kth augmented type number of $f \mid Z$	70
	PART II	
	SECTION 13	
$C^m(M_n)$	class of C^m-functions on M_n	95
$M_n \times N_r$	product of manifolds M_n and N_r	95
(F, G)	product of presentations F and G	95
$\mathscr{D}^0\Gamma_n$	set of homeomorphic presentations of open subsets of Γ_n	96
	SECTION 14	
$(p, a) \to \mathbf{G}(p, a)$	a family of differentiable mappings into R	98
$(v, a) \to U(v, a)$	a local representation of $\mathbf{G}$	99
$\Gamma(U)$	the set of v-critical points of U	99
$\Gamma^{\text{deg}}(U)$	the set of degenerate v-critical points of U	99
$\mathbf{H}(U)$	the v-critical matrix of U	100
	SECTION 16	
$\varphi(M_n)$	a C^m-manifold defined in Theorem 16.3b	117
$((\Gamma_n, Q_r, m))$	a Q_r-embedded C^m-manifold	119
	SECTION 17	
$(\eta^1(u),\ldots, \eta^n(u))$	a vector at $u \in E_n$ transformed contravariantly	127
$(c_1(u),\ldots, c_n(u))$	a vector at $u \in E_n$ transformed covariantly	127

Symbol	Meaning	Page
$\mathbf{V}_u$	the vector space over R at u in E_n	128
$(F : U, X)$	a presentation in $\mathscr{D}M_n$	128
$(G : V, Y)$	a presentation in $\mathscr{D}M_n$	128
$u \to \mathbf{v}(u)$	transition diffeomorphism defined by F and G	129
$v \to \mathbf{u}(v)$	transition diffeomorphism defined by G and F	129
$\mathbf{J}(\mathbf{v}, u)$	Jacobian matrix of $\mathbf{v}$ at u	129
$\mathbf{J}(\mathbf{u}, v)$	Jacobian matrix of $\mathbf{u}$ at v	129
$\{\mathbf{d}/\varphi_H\}$	a dynamical system on M_n	132
	SECTION 18	
Π_q	an infinite set product at $q \in M_n$	138
T_q	the vector space tangent to M_n at q	138
T_q^*	the vector space cotangent to M_n at q	140
	SECTION 19	
Q_H	a Riemannian form indexed by a presentation $H \in \mathscr{D}M_n$	143
γ	a regular arc on M_n	145
$L(\gamma)$	length of γ on M_n	145
Q^H	a Riemannian coform dual to Q_H	148
$M_n{}^f$	the submanifold of M_n with the critical points of f deleted	153
	SECTION 20	
f^c	the c-level subset of M_n	158
$\mathscr{F}$	an f-presentation in $\mathscr{D}M_n{}^f$	158
$\mathscr{F}^c$	a partial presentation in $\mathscr{D}^0 f^c$ induced by F for c ordinary	160
$\mathbf{f}^c$	an M_n-embedded C^∞-manifold with $\lvert \mathbf{f}^c \rvert = f^c$, c ordinary	160
	SECTION 21	
$f_{(a,b)}$	an open subset of M_n defined by (21.1)	166
$f_{[a,b]}$	a closed subset of M_n defined by (21.2)	166
$\mathbf{f}_{(a,b)}$	the "submanifold" of M_n with carrier $f_{(a,b)}$	166
	SECTION 22	
$\mathbf{S}$	a designation of a Riemannian structure on M_n	172
$\mathbf{S}^f$	an f-preferred Riemannian structure	175
$B_-(z, k)$	a k-bowl descending from a critical point z	178
$B_+(z, n-k)$	an $(n-k)$-bowl ascending from a critical point z	178
	SECTION 23	
k_a	index of the critical point z_a	184
k_b	index of the critical point z_b	184
$T^c_{z_a}$	trace of $B_+(z_a, n-k_a)$ on f^c	184
$T^c_{z_b}$	trace of $B_-(z_b, k_b)$ on f^c	184
Ω	a mapping of Theorem 23.1	184

Symbol	Meaning	Page
$\mathbf{D}$	a retracting deformation	187
f_α	a sublevel set defined in (23.17)	191
$f_{\alpha+}$	a superlevel set defined in (23.18) and (30.13)	191
$\hat{\Omega}$	a modification of Ω defined in (23.21)	192
$\simeq$	is of the same homotopy type	196

PART III

Section 24

Symbol	Meaning	Page
G	a vector space over $\mathscr{K}$	205
L	a linear subspace of G	205
G/L	a quotient group	205
α	the linear homomorphism of Lemma 24.1	206
α_*	the linear homomorphism induced by α (Lemma 24.1)	206
$G_1 \oplus G_2$	the direct sum of vector spaces G_1 and G_2	207
K	a ∂-structured complex over $\mathscr{K}$	208
$C_q(K)$	the vector space over $\mathscr{K}$ of q-chains of K	208
∂	the boundary operator of K	208
$\sim$	is homologous to	209
$H_q(K)$	the qth homology group of K over $\mathscr{K}$	209
$R_q(K)$	the qth connectivity of K over $\mathscr{K}$	209
$\mathbf{b}_q(K)$	homology prebase for K of q-cycles	209
a^q	a geometric simplex	211
$s = s^q$	an ordered q-simplex $p_0p_1 \cdots p_q$	212
$\vert s \vert$	the carrier of s	212
$s(i)$	the ith face of s	212

Section 25

Symbol	Meaning	Page
P	a simplicial n-polyhedron in E_m	215
$\mathbf{P}$	a ∂-structured complex based on P	215
$\vert u^q \vert$	carrier of a chain u^q on $\mathbf{P}$	216
$b(a^n)$	an n-polyhedron, the barycentric subdivision of a geometric simplex a^n	217
B	the barycentric operator on $C_q(\mathbf{a}^n)$	218
$w \times I$	the product of a geometric simplex and an interval $I = [0, 1]$	219
$P'(w)$	a simplicial polyhedron subdividing the prism $w \times I$	220
$P''(w)$	another simplicial polyhedron subdividing $w \times I$	220

Section 26

Symbol	Meaning	Page
χ	a Hausdorff space	224
τ	a singular q-simplex, a mapping $s = s^q \to \chi$	224
$\mathring{\tau}$	a singular q-cell, the class of singular q-simplices equivalent to τ	224
$\vert \mathring{\tau} \vert = \vert \tau \vert$	the carrier of τ and $\mathring{\tau}$	225

Symbol	Meaning	Page
$S(\chi)$	the complex of singular cells on χ	225
$\partial\mathring{\tau}$	the algebraic boundary of $\mathring{\tau}$	225
$\mid w^q \mid$	the carrier of a singular q-chain w^q	226
$\hat{g}$	a chain transformation $\hat{g} : \mathbf{P} \to S(\chi)$ induced by a mapping $g : P \to \chi$	227
$\widehat{\varphi}$	a chain transformation $\widehat{\varphi} : S(\chi) \to S(\chi')$ induced by a mapping $\varphi : \chi \to \chi'$	227

Section 27

Symbol	Meaning	Page
d	a continuous deformation $(p, t) \to d(p, t)$	234
d_1	the terminal mapping $p \to d(p, 1)$ of d	234
$\widehat{d}_1$	the terminal chain-transformation induced by d_1	234
$\mathbf{d}$	the operator of the chain-homotopy (27.8) induced by d	235
d-traj η	the trajectory under d of a subset η of χ	237
πz	the singular barycentric subdivision of the chain z	238
Π	the linear operator of the chain homotopy (27.20)	238
mesh z	the mesh of the singular chain z	242
$\pi^{(n)}z$	the n-fold iterate of πz	241

Section 28

Symbol	Meaning	Page
$Z_q(S(\chi), S(A))$	the vector space over $\mathscr{K}$ of q-cycles on χ mod A	245
$B_q(S(\chi), S(A))$	the vector space over $\mathscr{K}$ of bounding q-cycles on χ mod A	245
$H_q(S(\chi), S(A))$	the singular qth homology group on χ mod A over $\mathscr{K}$	245
$R_q(\chi, A)$	the qth connectivity of χ mod A over $\mathscr{K}$	246
$\mathbf{b}_q(\chi, A)$	a homology prebase on χ mod A	246
$H_q(\chi, A)$	a notational abbreviation for $H_q(S(\chi), S(A))$	246
p_a	a critical point at the f-level a	252
D^σ	origin-centered open n-ball of radius σ	252
$\dot{Y}$	$Y - p_a$, where Y is a subset of M_n containing p_a	252
Δ_k	an origin-centered k-disk (closed)	253
$\dot{\Delta}_k$	Δ_k with the origin deleted	253

PART IV

Section 33

Symbol	Meaning	Page
ξ	an involution of $M_n \times M_n$	295
diag $\mid M_n \times M_n \mid$	the diagonal of $\mid M_n \times M_n \mid$	296
$M_n{}^2$	the symmetric square of M_n	296
T	a mapping of $\mid M_n \times M_n \mid$ onto $\mid M_n{}^2 \mid$	296
diag $\mid M_n{}^2 \mid$	the diagonal of $\mid M_n{}^2 \mid$	296
$\hat{M}_{2n}$	$M_n{}^2$ with diag $\mid M_n{}^2 \mid$ deleted	296
$M_n \hat{\times} M_n$	$M_n \times M_n$ with diagonal $\mid M_n \times M_n \mid$ deleted	300
h	the chord function on $\hat{M}_{2n}$	303
$\mathbf{h}$	the extended chord function on $M_n{}^2$	304

Symbol	Meaning	Page
H	the chord function on $M_n \hat{\times} M_n$, (33.35)	311
$\mathscr{R}_i$	the ith connectivity of $\mid M_n{}^2 \mid$ mod(diag $\mid M_n{}^2 \mid$)	304
	SECTION 34	
$\mathbf{Z}_p$	the field of integers mod p, p a prime	312
$\mathscr{E}_n$	an n-dimensional ellipsoid in E_{n+1}	313
	SECTION 35	
$\mathbf{CP}_n$	the complex projective n-space	318
	SECTION 36	
$B_j{}^\rho$	an origin-centered open j-ball of radius ρ in E_j	324
Y^σ	the range of a presentation G of canonical coordinates	326
L_k	an f-saddle of M_n at a critical point of index k	326
	SECTION 37	
P_n	the real projective n-space	335
	SECTION 38	
w_μ	a complex variable $x_\mu + iy_\mu$	349
$\mathbf{CE}_m$	the Cartesian product of m complex planes	349
$\xi : \mathbf{CE}_m \to E_{2m}$	a homeomorphism defined by setting $w_\mu = x_\mu + iy_\mu$	349
Σ_n	a Stein manifold in $\mathbf{CE}_m$, $n < m$	349

BIBLIOGRAPHY

ALEXANDROFF, P. and HOPF, H.

1. "Topologie." Chelsea, New York, 1965.

ANDREOTTI, A., and FRANKEL, T.

1. The Lefschetz theorem on hyperplane sections. *Ann. of Math.* **69** (1959), 713–717.

ARNOLD, V.

1. Sur la courbure de Riemann des groupes de difféomorphismes. *Compt. Rend.* **260** (1965), 5668–5671.

BAIADA, E., and MORSE, M.

1. Homotopy and homology related to the Schoenflies problem. *Ann. of Math.* **58** (1953), 142–165.

BIRKHOFF, G., and MAC LANE, S.

1. "A Survey of Modern Algebra." Rev. ed. Macmillan, New York, 1953.
2. "Algebra." See Mac Lane, S., and Birkhoff, G.

BIRKHOFF, G. D.

1. Dynamical systems with two degrees of freedom. *Trans. Am. Math. Soc.* **18** (1917), 199–300.
2. "Dynamical systems," vol. **9**. *Am. Math. Soc. Colloq. Publ.* Rev. ed. Providence, Rhode Island, 1966.

BISHOP, R. L., and CRITTENDEN, R. J.

1. "Geometry of Manifolds." Academic Press, New York, 1964.

BÔCHER, M.

1. "Introduction To Higher Algebra." Macmillan, New York, 1938.

BORSUK, K.

1. Sur les rétractes. *Fund. Math.* **17** (1931), 152–170.

BOTT, R.

1. An application of the Morse theory to the topology of Lie groups. *Bull. Soc. Math. France* **84** (1956), 251–281.
2. The stable homotopy of the classical groups. *Ann. of Math.* **70** (1959), 313–337.
3. "Morse Theory and Its Application to Homotopy Theory." Univ. of Bonn, Bonn, 1960.

Bott, R., and Samelson, H.
1. Applications of the theory of Morse to symmetric spaces. *Am. J. Math.* **80** (1958), 964–1029.

Bourbaki, N.
1. "Éléments de Mathématique : Livre I, Théorie des Ensembles." Chap. I, 3rd éd., Hermann, Paris, 1958.
2. "Éléments de Mathématique : Livre II, Algèbre." Chap. II, 3rd éd., Hermann, Paris, 1962.
3. "Éléments de Mathématique : Livre III, Topologie Générale." Chap. I, 3rd éd., Hermann, Paris, 1961.

Brown, A. B.
1. Critical sets of an arbitrary real analytic function of n variables. *Ann. of Math.* **32** (1931), 512–520.

Cairns, S. S.
1. Homeomorphisms between topological manifolds and analytic manifolds. *Ann. of Math.* **41** (1940), 796–808.
2. The manifold smoothing problem. *Bull. Am. Math. Soc.* **67** (1961), 237–238.

Cantwell, J., and Morse, M.
1. See Morse, M., and Cantwell, J. [1].

Chern, S. S.
1. "Complex Manifolds Without Potential Theory." Van Nostrand, Princeton, New Jersey, 1967.

Crowell, R. H., and Fox, R. H.
1. "Introduction to Knot Theory." Ginn, New York, 1963.

Deheuvels, R.
1. Topologie d'une fonctionnelle. *Ann. of Math.* **61** (1955), 13–72.

de Rham, G.
1. "Variétés Différentiables." Hermann, Paris, 1960.

Dubovickii, A. Ya.
1. On differentiable mappings of an n-dimensional cube into a k-dimensional cube (Russian). *Mat. Sb.* **32** (74) (1953), 443–464.

Eells, J., and Kuiper, N. H.
1. Manifolds which are like projective planes. *Publ. Inst. Hautes Études Sci.* **14** (1962), 181–222.

Eilenberg, S.
1. Singular homology theory. *Ann. of Math.* **45** (1944), 407–447.

Eilenberg, S., and Steenrod, N.
1. "Foundations of Algebraic Topology." Princeton Univ. Press, Princeton, New Jersey, 1952.

Èl'sgol'c, L. È.
1. Zur Theorie der Invarianten, die zur Bestimmung der unteren Grenze der Anzahl der kritischen Punkte einer stetigen Funktion, die auf einer Mannigfaltigkeit definiert ist, dienen können. *Mat. Sb.* **5** (47) (1939), 551–558.

Frankel, T.
1. Critical submanifolds of the classical groups and Stiefel manifolds. "Differential and Combinatorial Topology," pp. 37–53. Princeton Univ. Press, Princeton, New Jersey, 1965.

FROLOV, S., and ÈL'SGOL'C, L.

1. Limite inférieure pour le nombre des valeurs critiques d'une fonction, donnée sur une variété. *Mat. Sb.* **42** (1935), 637–643.

GOURSAT, E.

1. "Cours d'Analyse Mathématique," Vol. I. 5th ed. Gauthier-Villars, Paris, 1927.

HADAMARD, J.

1. Les surfaces à courbures opposées et leur lignes géodésiques. *J. Math. Pures Appl.* [5], **4** (1898), 27–73.
2. Sur quelques applications de l'indice de Kronecker. See Tannery, J. "Introduction a la Théorie des Fonctions d'une Variable," Vol. 2, pp. 437–477. 2nd ed. Hermann, Paris, 1910.

HELGASON, S.

1. "Differential Geometry and Symmetric Spaces." Academic Press, New York, 1962.

HILTON, P. J.

1. "An Introduction to Homotopy Theory." Cambridge Univ. Press, London and New York, 1953.

HUREWICZ, W., and WALLMAN, H.

1. "Dimension Theory." 3rd printing. Princeton Univ. Press, Princeton, New Jersey, 1952.

JOHN, F.

1. Über die Vollständigkeit der Relationen von Morse für die Anzahlen kritischer Punkte. *Math. Ann.* **109** (1934), 381–394.

JORDAN, C.

1. "Cours d'Analyse," Vol. I. 3rd ed. Gauthier-Villars, Paris, 1909.

KALMBACH, G.

1. Über niederdimensionale CW-Komplexe in nichtkompakten Mannigfaltigkeiten. Dissertation, Univ. of Göttingen, Göttingen, 1966.

KELLOGG, O. D.

1. "Foundations of Potential Theory." Springer, Berlin, 1929.

KERVAIRE, M.

1. A manifold which does not admit any differentiable structure. *Comment. Math. Helv.* **34** (1961), 257–270.

KIANG, T.

1. On the critical points of non-degenerate Newtonian potentials. *Am. J. Math.* **54** (1932), 92–109.

KNOPP, K., AND SCHMIDT, R.

1. Funktionaldeterminanten und Abhängigkeit von Funktionen. *Math. Z.* **25** (1926), 373–381.

KUIPER, N. H.

1. C^r-functions near non-degenerate critical points. *Math. Inst., Univ. Amsterdam.*

LEDERMANN, W.

1. "Introduction to the Theory of Finite Groups." 3rd rev. ed. Oliver & Boyd, Edinburgh and London and Wiley (Interscience), New York, 1957.

LEFSCHETZ, S.

1. "L'analysis Situs et la Géométrie Algébrique." Gauthier-Villars, Paris, 1950.

LJUSTERNIK, L. A.

1. The topology of calculus of variations in the large. Translations of "Mathematical Monographs," Vol. 16. Am. Math. Soc., Providence, Rhode Island, 1966.

LJUSTERNIK, L., AND SNIREL'MAN, L.

1. Topological methods in variational problems. *Tr. Sci. Invest. Inst. Math. Mech.* **11** (1930).

MAC LANE, S., and BIRKHOFF, G.

1. "A Survey of Modern Algebra." See Birkhoff, G., and Mac Lane, S.
2. "Algebra." Macmillan, New York, 1967.

MAZUR, B.

1. Morse theory. "Differential and Combinatorial Topology," pp. 145–165. Princeton Univ. Press, Princeton, New Jersey, 1965.

MÉTIVIER, M.

1. Valeurs critiques des applications différentiables. *Anais Acad. Brasil. Cienc.* **36** (1964), 383–397.

MILNOR, J.

1. On manifolds homeomorphic to the 7-sphere. *Ann. of Math.* **64** (1956), 399–405.
2. "Morse Theory." Princeton Univ. Press, Princeton, New Jersey, 1963.
3. "Lectures on the *h*-Cobordism Theorem." Princeton Univ. Press, Princeton, New Jersey, 1965.

MORSE, A. P.

1. The behaviour of a function on its critical set. *Ann. of Math.* **40** (1939), 62–70.

MORSE, M.

1. Relations between the critical points of a real function of n independent variables. *Trans. Am. Math. Soc.* **27** (1925), 345–396.
2. The analysis and analysis situs of regular n-spreads in $(n + r)$-space. *Proc. Natl. Acad. Sci. U.S.* **13** (1927), 813–817.
3. The foundations of a theory of the calculus of variations in the large in m-space (second paper). *Trans. Am. Math. Soc.* **32** (1930), 599–631.
4. "Introduction to Analysis in the Large." 1947 Lectures. Univ. Microfilms, Inc., Ann Arbor, Michigan, 1947.
5. Functional topology and abstract variational theory. *Mémor. Sci. Math.* **92** (1939), 1–79.
6. "Topological Methods in the Theory of Functions of a Complex Variable." Princeton Univ. Press, Princeton, New Jersey, 1947.
7. Recent advances in variational theory in the large. *Proc. Intern. Congr. Math.*, 1950, 2.
8. Topologically non-degenerate functions on a compact n-manifold M. *J. Analyse Math.* **7** (1959), 189–208.
9. The existence of polar non-degenerate functions on differentiable manifolds. *Ann. of Math.* **71** (1960), 352–383.
10. On elevating manifold differentiability. *J. Indian Math. Soc.* **24** (1960), 379–400.
11. The elimination of critical points of a non-degenerate function on a differentiable manifold. *J. Analyse Math.* **13** (1964), 257–316.

12. Bowls, f-fibre-bundles and the alteration of critical values. *Anais Acad. Brasil. Cienc.* **36** (1964), 245–259. See Morse [16] for revision of §§1–4 of Morse [12].

13. "The Calculus of Variations in the Large," Vol. 18. 4th printing. *Am. Math. Soc. Colloq. Publ.*, Providence, Rhode Island, 1965 (First printing, 1934).

14. Focal sets of regular manifolds M_{n-1} in E_n. *J. Differential Geometry* **1** (1967), 1–19.

15. Non-degenerate functions on abstract differentiable manifolds M_n. *J. Analyse Math.* **19** (1967), 231–272.

16. Bowls, f-fibre-bundles and the alteration of critical values. *Proc. Natl. Acad. Sci. U.S.* **60** (1968), 1156–1159.

MORSE, M., and CANTWELL, J.

1. Diffeomorphism inducing automorphisms of $\pi_1(T_p)$. *Topology* **4** (1966), 323–341.

MORSE, M., and VAN SCHAACK, G. B.

1. The critical point theory under general boundary conditions. *Ann. of Math.* **35** (1934), 545–571.

MUNKRES, J. R.

1. "Elementary Differential Topology." Princeton Univ. Press, Princeton, New Jersey, 1963.

PALAIS, R. S.

1. Morse theory on Hilbert manifolds. *Topology* **2** (1963), 299–340.

PITCHER, E.

1. Inequalities of critical point theory. *Bull. Am. Math. Soc.* **64** (1958), 1–30.

POINCARÉ, H.

1. Sur les courbes définies par les équations différentielles (Pt. 3). *J. Math. Pures Appl.* [4], **1** (1885), 167–244.

PONTRJAGIN, L. S.

1. Characteristic cycles on differentiable manifolds. *Mat. Sb.* **21** (1947), 233–284.

REEB, G.

1. Sur certaines propriétés topologiques des variétés feuilletées. *Actualités Sci. Ind.* **1183** (1952), 91–154.

REMMERT, R.

1. Sur les espaces analytiques holomorphiquement convexes. *Compt. Rend.* **243** (1956), 118–121.

RICHARDSON, M.

1. The relative connectivities of symmetric products. *Bull. Am. Math. Soc.* **41** (1935), 528–534.

RICHARDSON, M., and SMITH, P. A.

1. Periodic transformations of complexes. *Ann. of Math.* **39** (1938), 611–633.

SARD, A.

1. The measure of the critical values of differentiable maps. *Bull. Am. Math. Soc.* **48** (1942), 883–890.

2. Hausdorff measure of critical images on Banach manifolds. *Am. J. Math.* **87** (1965), 158–174.

SEIFERT, H., and THRELFALL, W.

1. "Variationsrechnung im Grossen." Teubner, Leipzig and Berlin, 1938.

SÉMINAIRE H. CARTAN

1. E.N.S., 1951-1952, VII–XX.

SMALE, S.

1. The generalized Poincaré conjecture in higher dimensions. *Bull. Am. Math. Soc.* **66** (1960), 373–375.
2. The generalized Poincaré conjecture in dimensions greater than four. *Ann. of Math.* **74** (1961), 391–406.
3. A survey of some recent developments in differential topology. *Bull. Am. Math. Soc.* **69** (1963), 131–145.

TAKEUCHI, M.

1. Cell decompositions and Morse equalities on certain symmetric spaces. Journal of Faculty of Science. University of Tokyo, Sec. I, **12** (1965), Part 1, 81–192.

THOM, R.

1. Quelques propriétés globales des variétés différentiables. *Comment. Math. Helv.* **28** (1954), 17–86.
2. "La classification des immersions." Séminaire Bourbaki, 1957.

VALIRON, G.

1. "Cours d'Analyse Mathématique," Vol. II, Équations fonctionnelles. Applications. Masson, Paris, 1945.

VAN SCHAACK, G. B., AND MORSE, M.

1. See Morse, M., and Van Schaack, G. B. [1].

WALLACE, A.

1. "Differential Topology: First Steps." Benjamin, New York, 1968.

WHITE, M. B.

1. The dependence of focal points upon curvature for problems of the calculus of variations in space. *Trans. Am. Math. Soc.* **13** (1912), 175–198.

WHITEHEAD, J. H. C.

1. Combinatorial homotopy I. *Bull. Am. Math. Soc.* **55** (1949), 213–245.
2. Manifolds with transverse fields in euclidean space. *Ann. of Math.* **73** (1961), 154–212.

WHITNEY, H.

1. A function not constant on a connected set of critical points. *Duke Math. J.* **1** (1935), 514–517.
2. Differentiable manifolds. *Ann. of Math.* **37** (1936), 645–680.
3. "Geometric Integration Theory." Princeton Univ. Press, Princeton, New Jersey, 1957.

WHYBURN, W. M.

1. Non-isolated critical points of functions. *Bull. Am. Math. Soc.* **35** (1929), 701–708.

INDEX OF TERMS

Pure and Applied Mathematics

A Series of Monographs and Textbooks

Editors **Samuel Eilenberg and Hyman Bass**
Columbia University, New York

RECENT TITLES

ROBERT A. ADAMS. Sobolev Spaces

JOHN J. BENEDETTO. Spectral Synthesis

D. V. WIDDER. The Heat Equation

IRVING EZRA SEGAL. Mathematical Cosmology and Extragalactic Astronomy

J. DIEUDONNÉ. Treatise on Analysis: Volume II, enlarged and corrected printing; Volume IV; Volume V; Volume VI

WERNER GREUB, STEPHEN HALPERIN, AND RAY VANSTONE. Connections, Curvature, and Cohomology: Volume III, Cohomology of Principal Bundles and Homogeneous Spaces

I. MARTIN ISAACS. Character Theory of Finite Groups

JAMES R. BROWN. Ergodic Theory and Topological Dynamics

C. TRUESDELL. A First Course in Rational Continuum Mechanics: Volume 1, General Concepts

GEORGE GRATZER. General Lattice Theory

K. D. STROYAN AND W. A. J. LUXEMBURG. Introduction to the Theory of Infinitesimals

B. M. PUTTASWAMAIAH AND JOHN D. DIXON. Modular Representations of Finite Groups

MELVYN BERGER. Nonlinearity and Functional Analysis: Lectures on Nonlinear Problems in Mathematical Analysis

CHARALAMBOS D. ALIPRANTIS AND OWEN BURKINSHAW. Locally Solid Riesz Spaces

JAN MIKUSINSKI. The Bochner Integral

THOMAS JECH. Set Theory

CARL L. DEVITO. Functional Analysis

MICHIEL HAZEWINKEL. Formal Groups and Applications

SIGURDUR HELGASON. Differential Geometry, Lie Groups, and Symmetric Spaces

C. TRUESDELL AND R. G. MUNCASTER. Fundamentals of Maxwell's Kinetic Theory of a Simple Monatomic Gas: Treated as a Branch of Rational Mechanics

ROBERT B. BURCKEL. An Introduction To Classical Complex Analysis: Volume 1

In preparation

LOUIS HALLE ROWEN. Polynominal Identities in Ring Theory

JOSEPH J. ROTMAN. An Introduction to Homological Algebra

ROBERT B. BURCKEL. An Introduction To Classical Complex Analysis: Volume 2

B
C 8
D 9
E 0
F 1
G 2
H 3
I 4
J 5